EVOLUTION

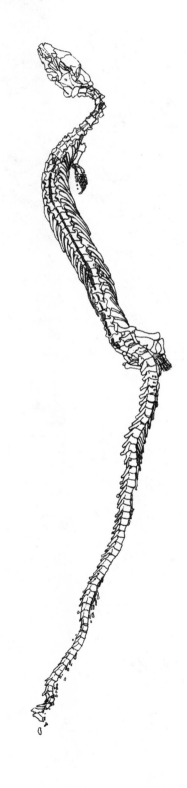

EVOLUTION

WHAT THE FOSSILS SAY AND WHY IT MATTERS

SECOND EDITION

DONALD R. PROTHERO

With Original Illustrations by Carl Buell

COLUMBIA UNIVERSITY PRESS   NEW YORK

CONTENTS

"creationist" views that "living things have existed in their present form since the beginning of time." Pew survey data at http://people-press.org/reports/display.php3?ReportID = 254.

[5] Kristol, quoted in Ron Bailey, Origin of the specious, *Reason*, July 1997.

[6] The three-hour briefing was held on May 10, 2000. Pearcey, quoted in David Wald, Intelligent design meets congressional designers, *Skeptic* 8, no. 2 (2000):16–17.

[7] Stephen Jay Gould, Kropotkin was no crackpot, *Natural History*, July 1988, 12–21.

[8] Peter Corning, Evolutionary ethics: An idea whose time has come? An overview and an affirmation, *Politics and the Life Sciences* 22, no.1 (2003):50–77.

[9] Petr Kropotkin, *Mutual Aid: A Factor in Evolution* (London: Heinemann, 1902).

[10] Daniel P. Todes, Darwin's Malthusian metaphor and Russian evolutionary thought, 1859–1917, *Isis* 78, no. 294 (1987):537–551.

Sit down before fact as a little child, be prepared to give up every preconceived notion, follow humbly wherever and to whatever abysses nature leads, or you shall learn nothing.
—Thomas Henry Huxley

The Bible tells you how to go to Heaven, not how the heavens go.
—Pope John Paul II

TO THE READER: Is Evolution a Threat to Your Religious Beliefs?

Speak to the earth and it shall teach thee.
—Job 12:8

Many people find the topic of evolution and religion troubling and confusing. Some were raised in very strict churches that preached that evolution is atheistic and that to even think about the evidence of evolution is sinful. Fundamentalists have long tried to drive a wedge between traditional Christians and science, arguing that their interpretation of the Bible is the only one and that anyone who accepts the evidence for evolution is an atheist.

But this is not true. The Catholic Church and most mainstream Protestant and Jewish denominations have long ago come to terms with evolution and accept it as the mechanism by which God created the universe. The Clergy Letter Project (http://theclergyletterproject .org) includes the signatures of more than 14,000 ministers, priests, and rabbis in the United States who accept evolution and do not view it as incompatible with religious belief. A number of studies have shown that about 50 percent of active scientists (Larson and Witham 1997) are also devoutly religious, including many of the prominent figures in evolutionary biology (Francisco Ayala, Kenneth Miller, Theodosius Dobzhansky, Francis Collins, and many, many others) and paleontologists (such as Peter Dodson, Richard Bambach, Anne Raymond, Mark Wilson, Patricia Kelley, Daryl Domning, Mary Schweitzer, and Simon Conway Morris), and they resent being called atheists by fundamentalists. As the late Stephen Jay Gould pointed out in his book *Rocks of Ages: Science and Religion in the Fullness of Life*, science and religion can be seen as nonoverlapping but equally valid means of understanding the world around us, and neither should encroach upon the domain of the other. Science helps us understand the natural world and the way it works, but it does not deal with the supernatural, and it does not make statements of what *ought* to be, as do morals and ethics. Religion, on the other hand, focuses on the supernatural and transcendent, with strong emphasis on the moral and ethical rules that humans should follow, but it is not a guide to understanding the natural world. When science tries to proscribe morals or ethics, it falters; when religion tries to interfere with our understanding of the natural world, it overreaches. For example, when Copernicus and Galileo showed that the earth is not the center of the universe, the Roman Catholic Church eventually had to recant its error and regret its persecutions.

If you find yourself puzzled by all this confusion and wondering who to believe, I welcome you to read these pages with an open mind. The fundamentalists have long been spreading myths and misconceptions and denying the self-evident facts about the fossil record. But they have no published research on fossils in peer-reviewed scientific journals, so they are no more qualified to write about fossils than they are qualified to write about auto mechanics or music theory. As a working paleontologist with firsthand familiarity with many of the fossils described here, I can testify and bear witness from personal experience that what I tell you about the fossil record in this book is based largely on my own observations and experience.

where elephants, horses, and rhinos came from; and how the first backboned animals evolved. We now have an amazing diversity of fossil humans, including specimens that show that we walked upright on two feet almost 7 million years ago, long before we acquired large brains. In addition to all this fossil evidence, we have new evidence from molecules that enables us to decipher the details of the family tree of life as never before.

Although scholars in 1859 may have considered Darwin's evidence from fossils weak, this is no longer true today. The fossil record is an amazing testimony to the power of evolution, with documentation of evolutionary transitions that Darwin could have only dreamed about. In addition, detailed studies of the fossils have even changed our notions about how evolution works and have fueled a lively debate in evolutionary biology about the mechanisms that drive evolution. The fossil record is now one of the strongest lines of evidence for evolution, completely reversing its subordinate status of only 150 years ago. Instead of the embarrassingly poor record that Darwin faced in 1859, we now have an embarrassment of riches.

PREFACE TO THE SECOND EDITION

The first edition of this book was written in 2005 and came out in 2007, and a lot has happened in the past decade. Most importantly, a large number of additional fossils that show the evolutionary transitions have been discovered and published. They make the case for evolution in the fossil record even stronger than before. I have incorporated most of these recent discoveries in the appropriate places in the book.

In addition, political events have changed the dialogue as well. The decisive defeat of the "intelligent design" creationism movement in the *Dover* decision in December 2005 had an even more profound effect than anyone anticipated. Even though the Discovery Institute in Seattle still relentlessly promotes it, ID creationism has completely vanished from the public discourse, and no school district has attempted to re-introduce it. Instead, the creationists have shifted tactics yet again, as described in this book. Even more surprisingly, political and demographic changes are beginning to turn the tide on creationism in this country, as described in the final chapter.

Most of the reviews and comments I have received about the first edition of this book were overwhelmingly positive (including hundreds of reviews on Amazon.com). Nevertheless, there was a persistent undercurrent of comments that I should ignore what the creationists say and not spend much time in the book dissecting their claims and debunking their falsehoods. But there are many good books out there that completely ignore the creationists and their claims and only discuss the positive evidence for evolution. My 2015 book *The Story of Life in 25 Fossils* (Columbia University Press) takes just such a positive upbeat approach to talking about important fossils, with almost nothing said about creationism. As I made it clear in the first edition of this book, I felt there was a need for someone to address the creationists' lies and distortions about the fossil record directly. No one else has undertaken this task, and it is important to set the record straight and clearly state the truth about the fossil record, so that anyone who is seriously curious about what the fossils tell us does not get a one-sided, unchallenged creationist version.

I hope that readers of this book will continue to look at the evidence with an open mind. That was the most encouraging outcome of the first edition of this book. Many readers and reviewers sit on the fence between their childhood religious beliefs and the evidence they never heard about (except the distorted false version from creationism). They were confused or puzzled about the fossil record of evolution. These were the people who told me that the book opened their eyes and changed their minds. They praised my book because it laid out the facts in a clear and straightforward fashion and showed them that there was more to the story than what they had learned from the creationists. It is for those people who are looking for a scientific answer without the filter of creationism and who are willing to openly explore the evidence that this book is written.

ACKNOWLEDGMENTS

Many people have guided my lifelong odyssey through science and religion. I am grateful to my brilliant pastor, the late Rev. Dr. Bruce Thielemann, who showed me that you could be religious and intellectual at the same time. I thank Dr. Don Polhemus for urging me to take Hebrew in high school and for keeping me on my toes with my "alephs" and "gimels," and Dr. Anastasius Bandy for teaching me to read the New Testament in its original Greek. I thank my former professors in philosophy, classics, anthropology, and religion at University of California–Riverside, who helped me understand not only the Bible but also world religions and the anthropology and sociology of religion. I thank my many mentors in college and in my profession who have taught me about the fossil record, including Drs. Michael Woodburne and Michael Murphy at UC Riverside and Drs. Earl Manning, Eugene Gaffney, Niles Eldredge, James Hays, and the late Drs. Malcolm McKenna, Richard Tedford, and Bobb Schaeffer at Columbia University and the American Museum of Natural History. I thank the late Dr. Stephen Jay Gould for his enthusiasm and encouragement when I was a struggling grad student and young professional and for being an inspiration to all of us in paleontology. I thank the late Stanley Weinberg, who first recruited me to battle the creationists in downstate Illinois in 1983 and inspired me to create my very successful Evolution, Creation, and the Cosmos course at Knox College. I thank my colleagues Dr. Dewey Moore and the late Dr. Larry DeMott for their support at the beginning of my professional career at Knox College when I tackled the creationist issue.

I thank my many professional colleagues, who generously provided images of their fossils for this book. They are all acknowledged in the figure captions. I would especially like to thank Carl Buell for his amazing illustrations. I also thank my father, the late Clifford R. Prothero, for his original art in chapter 14. I thank Patrick Fitzgerald, publisher for the life sciences; Ryan Groendyk, editorial assistant; Leslie Kriesel, assistant managing editor; Lisa Hamm, interior designer; and Julia Kushnirsky, who designed the cover. I want to thank the staff of Cenveo Publisher Services for all they did to make this book a reality. I thank Drs. Kevin Padian, Alan Gishlick, Bruce Lieberman, and Wilfred Elders for carefully reading the entire manuscript of the first edition and for reading the second edition, and Lisa Spoon and Dr. James W. Prothero for commenting on the religious aspects of the early chapters. Finally, I thank my wonderful sons, Erik, Zachary, and Gabriel, and my amazing wife, Teresa, for their love and support during this project. They have made it all worthwhile, and I hope this book will make a better world for them in the future.

EVOLUTION

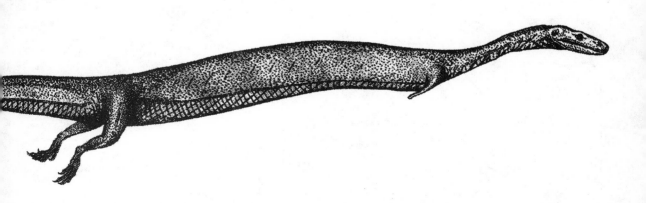

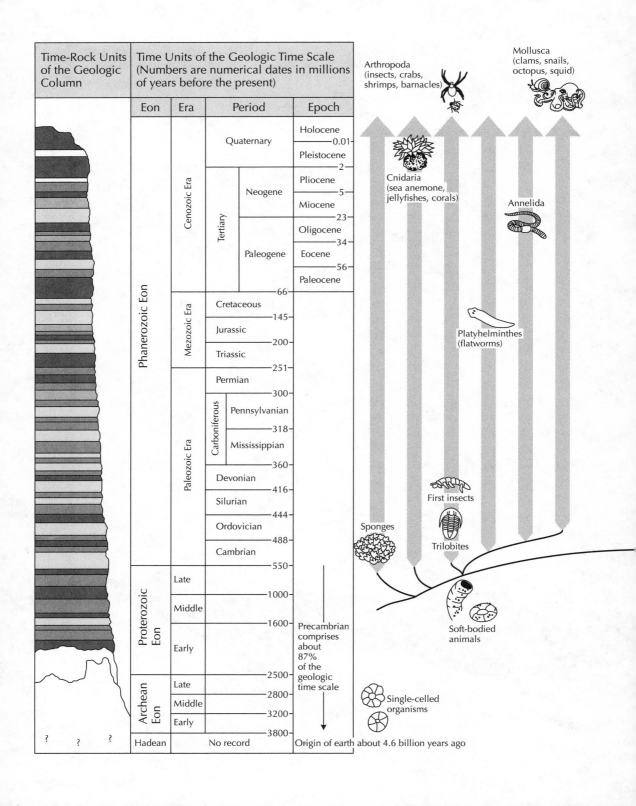

Time-Rock Units of the Geologic Column	Time Units of the Geologic Time Scale (Numbers are numerical dates in millions of years before the present)			
	Eon	Era	Period	Epoch
	Phanerozoic Eon	Cenozoic Era	Quaternary	Holocene 0.01 Pleistocene 2
			Tertiary — Neogene	Pliocene 5 Miocene 23
			Tertiary — Paleogene	Oligocene 34 Eocene 56 Paleocene 66
		Mezozoic Era	Cretaceous 145 Jurassic 200 Triassic 251	
		Paleozoic Era	Permian 300 Carboniferous — Pennsylvanian 318 Carboniferous — Mississippian 360 Devonian 416 Silurian 444 Ordovician 488 Cambrian 550	
	Proterozoic Eon	Late 1000 Middle 1600 Early 2500		
	Archean Eon	Late 2800 Middle 3200 Early 3800		
? ? ?	Hadean	No record		

Arthropoda (insects, crabs, shrimps, barnacles)

Mollusca (clams, snails, octopus, squid)

Cnidaria (sea anemone, jellyfishes, corals)

Annelida

Platyhelminthes (flatworms)

First insects

Trilobites

Sponges

Soft-bodied animals

Precambrian comprises about 87% of the geologic time scale

Single-celled organisms

Origin of earth about 4.6 billion years ago

Part I

EVOLUTION AND THE FOSSIL RECORD

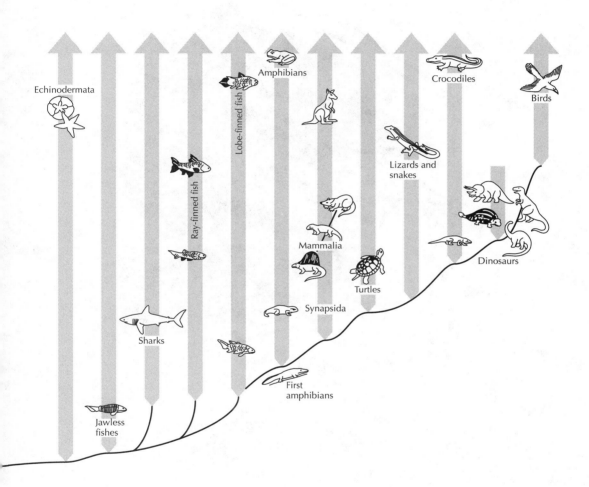

The modern geologic time scale (left) and a simplified "tree of life" showing the branching sequence of the major groups of modern animals. Contrary to creationist "flood geology" misconceptions, most lineages of primitive marine animals are not just found on the lower strata, but span all of geologic time, and have modern representatives as well.

FIGURE 1.1. Scientists cannot revert to supernaturalism and invoke miracles, or their explanations will lead nowhere. (Cartoon courtesy Sidney Harris)

THE NATURE OF SCIENCE

1

What Is Science?

The great tragedy of science—the slaying of a beautiful hypothesis by an ugly fact.
—Thomas H. Huxley

There are many hypotheses in science which are wrong. That's perfectly alright; they're the aperture to finding out what's right. Science is a self-correcting process. To be accepted, new ideas must survive the most rigorous standards of evidence and scrutiny.
—Carl Sagan

Before we discuss evolution and the fossil record in detail, we must clear up a number of misconceptions about what science is—and isn't. Many people get their image of science from Hollywood stereotypes of the "mad scientist," fiendishly plotting some diabolical creation with a room full of bubbling beakers and sparking electrical apparatuses. Invariably, the plot concludes with some sort of "Frankenstein" message that it's not nice for science to mess with Mother Nature. Even the positive stereotypes are not much better, with nerdy characters like Jimmy Neutron and Poindexter (always wearing glasses and the obligatory white lab coat) using the same bubbling beakers and sparking equipment but trying to invent something new or good.

In reality, scientists are just people like you and me. Most of us don't wear lab coats (I don't) or work with bubbling beakers or sparking Van de Graaff generators (unless they are chemists or physicists who actually work with that equipment). Most scientists are not geniuses either. It is true that, on average, scientists tend to be better educated than the typical person on the street, but that education is a necessity to learn all the information that allows a scientist to make discoveries. Still, there are geniuses, like Thomas Edison, who had minimal formal education (he only attended school for a few months) but a natural instinct for invention. So education is not always required if you have talent to compensate. Scientists are not inherently good or evil; nor are they trying to create Frankensteins, invent the next superweapon, or tamper with the operations of nature. Most are ordinary people who have the interest and curiosity to solve some problem in nature, and rarely do they discover anything that might threaten humanity.

Scientists are not characterized by *who* they are or *what* they wear, but *what* they do and *how* they do it. As Carl Sagan put it, "Science is a way of thinking much more than it is a body of knowledge." Scientists are defined not by their lab equipment but by the tools and assumptions they use to understand nature—the *scientific method*. The scientific method is mentioned even in elementary school science classes, yet most of the public still doesn't understand it (possibly because the mad scientist Hollywood stereotype is more powerful than the bland material from school). The scientific method involves making observations about the natural world, then coming up with ideas or insights (*hypotheses*) to explain them.

In that regard, the scientific method is similar to many other human endeavors, such as mythology and folk medicine, which observe something and try to come up with a story for it. But the big difference is that scientists must then *test* their hypotheses. They must try to find some additional observations or experiments that shoot their idea down (*falsify* it) or support it (*corroborate* it). If the observations falsify the hypothesis, then scientists must start over again with a new hypothesis or recheck their observations and make sure that the falsification is correct. If the observations are consistent with the hypothesis, then it is corroborated, but it is *not* proven true. Instead, the scientific community must continue to keep looking for more observations to test the hypothesis further (fig. 1.1).

This is where the public most misunderstands the scientific method. As many philosophers of science (such as Karl Popper) have shown, this cycle of setting up, testing, and falsifying hypotheses is unending. Scientific hypotheses must *always* be tentative and subject to further testing and can *never* be regarded as finally true or proven. Science is *not* about finding final truth, only about testing and refining better and better hypotheses so these hypotheses *approach* what we think is true about the world. Anytime scientists stop testing and trying to falsify their hypotheses, they also stop doing science.

One of the reasons for this is the nature of testing hypotheses. Lots of people think that science is purely *inductive*, making observation after observation until some general scientific law can be inferred. It is true that scientists must start with observations, but they do not arrive at scientific principles from induction. As Charles Darwin himself put it in 1861,

> About thirty years ago there was much talk that geologists ought only to observe and not theorize; and I well remember someone saying that at this rate a man might as well go into a gravel-pit and count the pebbles and describe the colours. How odd it is that anyone should not see that all observation must be for or against some view if it is to be of any service! (1903, 1:195)

Darwin correctly points out that all useful observations must be made in the framework of a hypothesis. What data are needed to test the hypothesis? How will they be useful in falsifying or corroborating it? Instead of inducing general principles of nature, most science is about *deductive* reasoning: we set up (*deduce*) a hypothesis, then try to test it. Philosophers use the cumbersome term "hypothetico-deductive method" to describe this process, but it is simple when you think about it.

The difference between inductive generalization and deductive hypothesis testing is easy to illustrate. Suppose we make the inductive statement that "all swans are white." We could observe thousands of swans for many years but never prove that statement true. All it takes is one nonwhite swan and we can easily falsify this hypothesis. Indeed, there are black swans (fig. 1.2) in Australia and elsewhere, so the statement has been falsified. As Karl Popper pointed out, there is an asymmetry between verification and falsification. It is easy to falsify something; all we need is one well-supported observation that proves the hypothesis wrong. But we can never prove something true (*verify* it). Additional corroborating observations may *support* the hypothesis but never finally prove it true. As Popper put it in the title of a book, science is about conjectures and refutations.

Most people think that science is about finding the final truth about the world and are surprised to find that science never *proves* something finally true. But that's the way the scientific method works, as philosophers of science have long ago demonstrated about the

FIGURE 1.2. Not all swans are white. This is the Australian black swan. (Photo by the author)

logic of the scientific method. Science is *not* about final truth or "facts"; it is only about continually testing and trying to falsify our hypotheses until they are extremely well supported.
At that point, the hypothesis becomes a *theory* (as scientists use the term), which is a well-
corroborated set of hypotheses that explain a larger part of our observations about the world.
Some well-known and widely accepted theories are the theory of gravitation, the theory of
relativity, the germ theory of disease, and of course, the theory of evolution.

As people, scientists must use common speech as well. An observation or explanation
that is extremely well supported is a *fact* in everyday language (even though we technically
cannot use the term within science). As we will discuss below, the evidence supporting the
hypothesis that life has evolved (and is still evolving now) is so overwhelming that it is a
fact in popular parlance. But the bigger problem is the different usages of the word "theory."
As we just explained, to a scientist, a theory is an extremely well-supported framework of
hypotheses that explains a large part of nature. But the public uses the word entirely differently to describe some sort of wild idea or harebrained guess or conjecture, such as theories of
how and why John F. Kennedy was assassinated, or how aliens landed in Area 51 in Nevada
or Roswell, New Mexico, and the entire episode was covered up by the U.S. government.

This confusion between the scientific and vernacular use of the same word has been a
common problem with misunderstanding what evolution is about. As Isaac Asimov put it,
"Creationists make it sound as though a 'theory' is something you dreamt up after being
drunk all night." For example, then-presidential candidate Ronald Reagan (speaking about

evolution during his 1980 campaign) said, "Well, it is a theory. It is a scientific theory only, and it has in recent years been challenged in the world of science—that is, not believed in the scientific community to be as infallible as it once was." Reagan (perhaps playing to his fundamentalist audience) was voicing the common confusion in the public mind about the two uses of the word "theory." To scientists, a theory is extremely well supported, has survived hundreds of tests and potential falsifications, and is accepted as a valid explanation of the world. But Reagan is confusing that meaning with the everyday meaning of theory as a "wild harebrained scheme." He is also showing his ignorance of another aspect of science. Science is *always* about challenging hypotheses and trying to test them and *never* reaches a point where a scientific idea is "believed" or is "infallible." These are words used in dogmatic belief systems, not in science. If scientists stop challenging theories and hypotheses, they stop doing science.

There is also another public confusion embedded in this quote: the confusion between the fact of evolution and the theory of evolution. The idea that life has evolved (and we can still see it evolving) is as much a descriptive fact about nature as the fact that the sky is blue. This was already established long before Darwin and represents an empirical observation about nature, and it is no longer disputed within the scientific community. What Darwin provided was a theory that included a mechanism for evolution he called *natural selection*. There has always been debate within the biological sciences about whether that mechanism is sufficient to explain the fact that life has evolved and is evolving. Argument and dispute is good in science; it's a sign that dogmas are being challenged and that no hypothesis is being accepted without question. But even if Darwin's mechanism, the theory of natural selection, were to be rejected by scientists, it would not change the fact that life has evolved. It is comparable to the theory of gravitation. We *still* do not have a full understanding of the mechanism by which gravity works, but that does not change the fact that objects fall to the ground.

Science and Belief Systems

A central lesson of science is that to understand complex issues (or even simple ones), we must try to free our minds of dogma and to guarantee the freedom to publish, to contradict, and to experiment. Arguments from authority are unacceptable.
—Carl Sagan

Humans have many systems of understanding and explaining the world besides science. In most cultures, religious beliefs provide the role of explaining how and why things work ("the gods did it"), and until the Enlightenment and Scientific Revolution, these beliefs tried to explain the physical and biological world. In some parts of the world, Marxism is the official "state religion," and every aspect of life is subjected to "dialectical materialism" and viewed through a Marxist filter. Likewise, there are many organizations (some would call them "cults") that explain the world through unusual perspectives, such as claiming that aliens are responsible for most of what we don't understand. These belief systems are not necessarily good or bad, but they are not science, because they are not testable and their main ideas cannot be falsified. Whenever a religious or Marxist dogma is challenged by some observation, that inconvenient fact is explained away or dismissed or ignored altogether, because maintaining the belief system is more important than allowing any inconvenient facts to undermine it. Many people find great comfort in such belief systems. That's fine, as long as they don't call these ideas "scientific." People around the world believe a wide variety of things, and they are entitled to do so. As long as they don't endanger themselves or others, that's OK.

The main exception, of course, is when a belief system is detrimental to the believers or to other people. There are cults of "snake handlers" in the Appalachians who caress poisonous rattlesnakes and copperheads during their religious ceremonies with the conviction that God will protect them from snakebite—and they are regularly bitten and killed (70 of these snake handlers have died from snakebites in the past 80 years, including the founder of the cult). In Darwinian terms, this belief system is so hazardous to the believers that they will eventually die out, and a harsh form of natural selection will weed out this self-destructive religion. There are cults that commit ritual suicide, such as Jim Jones's Peoples Temple, whose members drank cyanide-laced Kool-Aid in the Guyana jungle, killing 913 people in 1978, or the Heaven's Gate cult, with 39 people committing ritual suicide in 1997 at the urging of their leader, believing that aliens were about to take them to heaven. Likewise, there are ascetic monks who starve themselves to death or stare into the sun in search of enlightenment until they are blind; they, too, are harming themselves and endangering their own survival. Some would say that religious wars, such as the continual battles among Christians, Muslims, and Jews that have plagued the Middle East for over 1,000 years, or the Catholic-Protestant warfare in Ireland, or over much of Europe since the Reformation, or the horrors of the Inquisition, or the Muslim-Hindu wars in India since before Pakistan split away, are arguments that religious belief systems can be murderous and detrimental to the believers.

Many people who have strong belief systems that seem to conflict with science want it both ways. They accept their belief system in explaining most aspects of the world but still accept scientific explanations and advances where and when they need them. Some people in the Western world depend on modern scientific medicine for their improved health and chances of survival, yet they refuse to accept important aspects of science (fig. 1.3) that are a part of that great improvement in medicine (such as the rapid evolutionary change that makes viruses and bacteria dangerous to us each year). As Carl Sagan (1996:30) put it, "If you want to save your child from polio, you can pray or you can inoculate." Extreme fundamentalists push a strange model of the earth (discussed in chapter 3); they call it "flood geology"—yet, if they had any firsthand practical experience with real geology and accepted the results, they would see the absurdity of flood geology. More importantly, they would not benefit from the oil, coal, and natural gas that modern geology has provided all of us, and which flood geology would have no chance of discovering.

This is the strange situation in which we find ourselves. The modern world runs on science and technology, and our future economic and social well-being rely on continuing to make scientific and technological advances. Yet when extremists learn something from science that they don't want to hear (like evolution), they reject the very system that has made their lives better and that they willingly accept in most other circumstances. As science educator Bill Nye the Science Guy put it, "The natural world is a package deal; you don't get to select the facts which you like and which you don't." Or as astrophysicist Neil deGrasse Tyson said, "When different experiments give you the same result, it is no longer subject to your opinion. That's the good thing about science. It's true, whether or not you believe in it. That's why it works."

FIGURE 1.3. This *Doonesbury* cartoon eloquently expresses the inherent hypocrisy of the creationists, who try to have it both ways. They reject science and evolution, except when it benefits them. (Cartoon by Garry Trudeau, by permission of Universal Press Syndicate)

There are a whole range of ideas that might be considered what Al Gore very aptly called "inconvenient truths"—scientific evidence that conflicts with our belief systems, whether it be climate change or evolution. But scientists have nothing to gain by telling us what we don't want to hear, and they don't win grant funds or societal approval by bearing bad news. Scientists are not killjoys or spoilsports by nature. Instead, they are obligated by the scientific method to report what the data tell them, no matter whether we like it or not. Instead, the incentives for most people is to tell you good news you want to hear, so if a scientist tells you an inconvenient truth, it is almost certainly because the scientist *must* do so. A very amusing web cartoon shows this through a series of panels that depict scientific advances that society did not appreciate, from Archimedes being killed by a Roman soldier, to Giordano Bruno being burned at the stake for saying the earth revolves around the sun, to Darwinian evolution, to Einsteinian relativity. The final panel says it best: "Science: if you ain't pissin' people off, you ain't doing it right."

Science is not perfect, of course. Scientists are human, and as humans, we do make mistakes or may develop things that could harm us (like releasing the gases that have led to the ozone hole, among other things). But without science, we would be back in the Dark Ages. The next time you hear a fundamentalist preaching about the "evils of science and evolution," think hard: Would you rather go back to a world just a century or two ago (and still prevailing in many underdeveloped countries) when most children died before age two and many mothers died in childbirth; where the life expectancy was very short because many diseases were incurable; and where we had no conveniences like electricity, automobiles, airplanes, plastics, and telephones? For better or for worse, we now live in a scientific age, and most of us would not want to turn back time and renounce all the benefits that science has brought us.

Belly Buttons and Testability

What did Adam and Eve never have, yet they gave two of them to each of their children? Answer: parents.
—Old children's riddle

A classic example of an untestable theory to explain nature was the *Omphalos* hypothesis of Philip Henry Gosse. He was a well-respected naturalist in early nineteenth-century England who had written best-selling books about natural history. He was also a very devout member of a Puritanical sect called the Plymouth Brethren. As a good naturalist, Gosse was finding more and more evidence that life had evolved, but as a biblical literalist, he was obligated to follow creationism. Gosse resolved his problems by publishing *Omphalos: An Attempt to Untie the Geological Knot* in 1857, just two years before Darwin's *On the Origin of Species* was published. The curious term *Omphalos* means "belly button" or "navel" in Greek and refers to the common theological conundrum of the day: If Adam and Eve were specially created and did not have human parents (and therefore no umbilical cord), did they have navels or belly buttons? Many religious artists avoided this issue by painting Adam and Eve with a fig leaf not only over their genitalia but also over their midriffs. Gosse's answer was yes, of course, Adam and Eve had navels. According to Gosse, God created nature to *look* as if it had a history, to *look* as if it had evolved, but in reality, nature was created quite recently. In order for the world to be "functional," God would have created the earth with mountains and canyons, with trees that have growth rings, and with Adam and Eve with navels. No evidence that indicates the presumed age of the earth or events in the past can be taken at face value. In this manner, Gosse felt that he had solved his own dilemmas about the fact that nature appears to have evolved and that the earth was very old, and this solution allowed him to retain his creationist beliefs.

Naturally, an idea as bizarre as this didn't go over very well with most religious people of the time, because it implies that God created a fake world and makes God into a deceiver, not a benevolent deity. Gosse's son, Edmund Gosse, wrote in *Father and Son* (1907:92), "He offered it, with a glowing gesture, to atheists and Christians alike. . . . But alas! Atheists and Christians alike looked at it and laughed, and threw it away . . . even Charles Kingsley, for whom my father had expected the most instant appreciation, wrote that he could not . . . 'believe that God has written on the rocks one enormous and superfluous lie.' "

More importantly, it is a classic example of a completely unfalsifiable theory of the world. No observation could ever prove it wrong, because everything *looks* as if it evolved, but it was just created to look that way! As described by Martin Gardner (1952:126), "Not the least of its remarkable virtues is that while it won not a single convert, it presented a theory so logically perfect, and so in accordance with geologic facts that no amount of scientific evidence will ever be able to refute it." Some philosophers have argued that all of reality is an illusion, and it is perfectly logical to suggest that the world was created a few minutes ago, with everyone having memories of a past that doesn't exist. Any memories you might have of the past were created in your head when you were created, just as the fossils were placed in the rocks to look as if they were from the ancient past. This idea is nicknamed "Last Thursdayism" by the famous philosopher Bertrand Russell, as in "the world might have been created last Thursday—how would we know the difference?" Of course, this idea is just as untestable as Gosse's original hypothesis.

Gosse had high hopes that his ideas would resolve the growing divide between natural history and religion, but he was ignored or ridiculed. Just two years later, Darwin's book came out and made his ideas irrelevant. Gosse ended up an embittered old man whose natural history books were no longer important in a Darwinian world. His troubled later years were vividly described by his son, whose famous biography *Father and Son* (1907) is considered a classic of its genre.

One would think that such a bizarre and untestable idea, which was rejected and ridiculed even by the religious and devout in the days before Darwin, would never be revived. But modern creationists have brought in their own versions of the *Omphalos* hypothesis. When young-earth creationists are confronted with evidence that shows that galaxies are millions of light-years away and that their light is just reaching us after millions of years, they say that God created the universe as it is with the light from those galaxies already on the way! This seems like an extreme form of pretzel logic to explain away an inconvenient fact and salvage their cherished hypothesis.

The *Omphalos* story, however, raises an important point about our models of the world. If we want them to make sense and not violate what we have learned about nature, we have to be true to the conclusions to which nature leads us. We cannot twist and bend our explanations into pretzels like the *Omphalos* hypothesis just to save some cherished belief. We will discuss how some creationists have done just this by giving strange and contorted explanations for things that are most simply explained by evolution.

Natural and Supernatural

For all the controversies over these issues, however, there is a basic philosophical point on which the evolutionary biologists all agree. Some say new mechanisms have to be introduced and others say the old mechanisms are adequate, but nobody with a reputation to lose proposes to invoke a supernatural creator or a mystical "life force" to help out with the difficulties. The theory in question is a theory of naturalistic evolution, which means that it absolutely rules out any miraculous or supernatural intervention at any point. Everything is conclusively presumed to have happened through purely material mechanisms that are in principle accessible to scientific investigation, whether they have yet been discovered or not.
—Phillip Johnson

Another issue in the philosophy of science has come up as a result of the creationism controversy. Berkeley lawyer (*not* a scientist *or* a theologian *or* a philosopher of science) Phillip Johnson criticizes science because it makes the "naturalistic assumption." In other words, science assumes only natural, not supernatural, processes in the attempt to understand nature. To Johnson, this is unfair; if science excludes the supernatural even before the debate can begin, then it excludes the possibility of any conclusion but evolution. In his book *Darwin on Trial*, Johnson writes, "Naturalism assumes the entire realm of nature to be a closed system of material causes and effects, which cannot be influenced by anything from 'outside.' Naturalism does not explicitly deny the mere existence of God, but it does deny that a supernatural being could in any way influence natural events, such as evolution, or communicate with natural creatures like ourselves" (Johnson 1991:114–115). Johnson's attack concludes by arguing that if supernatural causes were allowed, other conclusions besides evolution (such as creationism) might be reached.

This argument is so confused and disingenuous that it almost doesn't require rebuttal. As Pennock (1999:190) points out, Johnson has conflated two entirely different naturalistic concepts. *Ontological* or *metaphysical naturalism* (the kind mentioned in the Johnson quotes above) makes the bold claim that the natural is all that exists and that there is no supernatural. That is an interesting philosophical issue, but that does not reflect what scientists are doing. Instead, scientists practice *methodological naturalism*, where they use naturalistic assumptions to understand the world but make no philosophical commitment as to whether the supernatural exists or not. Scientists don't exclude God from their hypotheses because they are inherently atheistic or unwilling to consider the existence of God; they simply cannot consider supernatural events in their hypotheses. Why not? Because, as we saw with Gosse's *Omphalos* hypothesis, *once you introduce the supernatural to a scientific hypothesis, there is no way to falsify or test it*. Once you introduce the untestable supernatural explanation, you are no longer doing science—it's a "science stopper." We might want to say, "It is this way because God willed it so" or "And then a miracle occurs" (fig. 1.1), and for some religious people, that is all that need be said. But scientists are not allowed to do this, because it is completely untestable and therefore outside the realm of science. If scientists do offer evidence that falsifies the statement that "God did it this way," do you think that fundamentalists would accept the evidence? As we will demonstrate throughout this book, the evidence for evolution supplies just such a falsification, but creationists must deny it to salvage their untestable hypotheses. Ironically, Johnson spends a whole chapter (chapter 5) in *Darwin on Trial* talking about Karl Popper and the falsifiability criterion, but he completely misses the point as to why it requires methodological naturalism for science to work in the first place.

In fact, there have been many scientific tests of supernatural and paranormal explanations of things, including parapsychology, ESP, divination, prophecy, and astrology. All of these nonscientific ideas have been falsified when subjected to the scrutiny of scientific investigation (see Isaak 2006; also 2002 for a review). Johnson loudly complains that the supernatural has been unfairly excluded from the debate, but this is clearly not true. Every time the supernatural has been investigated by scientific methods, it has failed the test.

As Isaak (2002) put it,

Indeed, many supernatural explanations are rejected not because they are supernatural but because they cannot or do not lead anywhere. It is possible to come up with any number of possible explanations for anything—lost socks could be caused by extradimensional vortices which our observations prevent from forming; hiccups could be caused by evil spirits inside us trying to escape; stock market fluctuations could be caused by the secret manipulations of powerful extraterrestrials. Scientists reject such claims on the grounds of parsimony. All of those claims are possible, but they require adding complicated entities which there is no adequate evidence for. To make matters worse, the nature of those entities effectively prevents investigation of them, and the impossibility of investigation prevents us from learning anything new about them. We cannot conclude that any of those explanations are wrong. But from a scientific standpoint, they are worse than wrong; they are useless.

Johnson also shows that he has not read much about the history or philosophy of science (which is odd, considering that the whole debate is about a central point in the philosophy of science). Methodological naturalism arose out of necessity more than 400 years ago when early scientists tried to make sense of their universe and realized that as long as the prevailing attitude was "God did it this way—end of story," our scientific understanding of nature would go nowhere. Yet all of these early scientists were religious men, not atheists trying to dispose of God. Indeed, Isaac Newton is probably more responsible for scientific naturalism than anyone. Newtonian physics showed how the universe could function completely without supernatural intervention. Yet Newton spent far more of his time and energy exploring religious questions than doing physics! In the centuries that followed, methodological naturalism became ingrained in science. When the great mathematician and astronomer Pierre Laplace presented a copy of his 1799 book on celestial mechanics to Napoleon, who asked where God fit in, Laplace replied, "I have no need of that hypothesis." He wasn't saying that he was an atheist—only that supernatural intervention did nothing to help us understand the motion of the heavenly bodies and that the entire enterprise would become unscientific if supernaturalism were introduced.

A similar transformation from supernaturalism to naturalism took place in many other fields of science over this period. For example, until about 1780, geologists tried to explain the record of the earth's history by stories such as Noah's flood. But in 1788 the great Scottish geologist James Hutton introduced a naturalistic view of the earth (often called *uniformitarianism* or *actualism*) that used present-day understanding of natural earth processes to decipher the past. Even though he was devout, Hutton did not resort to Bible stories to explain the rock record, because he could see that they had led nowhere after centuries of theological debate, whereas naturalistic explanations provided a whole new view of the earth. For about 40 years, there was continual strife between the uniformitarians and the old-line "catastrophists" such as German mineralogist Abraham Gottlob Werner, who still used untestable supernatural explanations for the earth. But by the time Charles Lyell (who was devoutly religious, as were most British scientists of his time) published *Principles of Geology* in 1830–1833, the case for a naturalistic explanation of the earth was overwhelming, and supernaturalism in geology died soon thereafter. Today, the label "catastrophism" is so tainted with supernaturalism and untestability that when there is evidence that natural catastrophes happened on the earth (such as the impact of asteroids or gigantic glacial floods), many geologists were reluctant to accept the evidence.

Finally, Johnson makes another assumption that reveals his bigotry and lack of understanding of religion. He writes (1991:115) that "scientific naturalism makes the same point by starting with the assumption that science, which studies only the natural, is our only reliable path to knowledge. God who can never do anything that makes a difference, and of whom we can have no reliable knowledge, is of no importance to us." As Pennock (1999: 192) points out, this may describe the fundamentalist deity who is constantly intervening in nature and performing miracles, but not the deity of theistic evolutionists who are willing to say God used evolution as his tool to modify nature. Nor does it jibe with the Deistic view (the religious attitude of Washington, Jefferson, Madison, and the Founding Fathers that the fundamentalists like to quote), which claims that God created the universe long ago but no longer interferes with his creation. In addition, many religious believers (including many Christians) view God as a universal life force or mystical unity, not an omniscient, omnipotent old man with a beard who constantly meddles with the universe. To Johnson, all these people (including the Founding Fathers) are virtual atheists. Apparently, if they don't believe in an activist deity, they don't qualify as being religious!

Science, Pseudoscience, and Baloney Detection

Skeptical scrutiny is the means, in both science and religion, by which deep thoughts can be winnowed from deep nonsense.
—Carl Sagan

There's a sucker born every minute.
—Phineas T. Barnum

Science may provide some of the most powerful explanations of the universe we have and may have provided humans with the benefits of modern civilization, but people still have an ambivalent attitude toward science. They readily accept most of its benefits, but they are easily suckered into believing "weird things," or what is known as *pseudoscience*, as well. Pseudoscience tries to masquerade as science (knowing the prestige that we now attach to scientific things), but when you examine the claims closely, they do not hold up to scientific scrutiny. Humans appreciate many of the advantages of living in the scientific age, but apparently they also have a deep-felt need for answers to questions that science cannot answer, but for which pseudoscience will willingly sell them an answer. Sometimes these beliefs are harmless, but often they involve the pseudoscience practitioner swindling the gullible victims out of valuable things, such as their time, energy, and money. Some pseudoscientists are clearly con artists, while others truly believe in their lies, but they will take your money just the same. Unfortunately, there are aspects of pseudoscience that are not just expensive but deadly to the follower. These parasites prey on people in all cultures and all walks of life, feeding the need for the mystical and miraculous, yet causing more harm than the little bit of psychological good feeling and reassurance that they may temporarily provide.

Ironically, most humans are already equipped with a skeptical filter for such con artists in many parts of life. When we bargain for items, or negotiate a price or a contract, we expect the bargaining to be somewhat adversarial and tricky. We are constantly on the lookout for someone who might cheat or shortchange us. We are bombarded with commercials everywhere we go, yet our skeptical filters tend to screen out most commercial appeals, just like a good spam filter on our computer keeps our email from being overwhelmed by junk. *Caveat emptor*—"let the buyer beware"—is a slogan we normally live by in such negotiations. Yet when it comes to claims that appeal to our sense of mystery, or to our need to connect with the unknown or with dead loved ones, humans readily suspend these skeptical filters and will believe (and pay for) almost anything, as long as it makes them feel better. That's when we are marks to be swindled. The world is full of con artists who will take your money and violate your trust by appealing to your gullibility—if you let them.

Thus, even though American citizens benefit from one of the highest standards of living and one of the best educational systems in the world, poll after poll shows that a high percentage of Americans still believe in UFOs, in ESP, in astrology, in Bigfoot and Nessie and the Yeti, in psychic phenomena, in palm reading and tarot cards, and so on. It doesn't seem to matter that the evidence for UFOs or astrology or psychic powers has been debunked and discredited over and over again. As humans, we apparently have a *need* to believe such things. It is understandable how a "psychic" who claims to be able to talk to your dead relatives or an astrology column that predicts your future has a deep-seated appeal to people who would otherwise not believe such drivel. But it is harder to understand why people are sucked into belief systems, such as the anti-Semitic Holocaust deniers, who claim that the Holocaust never happened or did not actually kill 6 million Jews and many more Poles, Gypsies, and other groups; or the beliefs in UFOs and alien abductions, which sound bizarre when we hear them, yet many people still accept that these phenomena are real; or the widespread acceptance of cryptozoology and its weird and nonbiological catalogue of beasts that have never been adequately documented, from the Loch Ness monster to Bigfoot to the Yeti. It may be that we have a need for things that are mystical and unexplained by the mundane, naturalistic process of science, but how anti-Semitism, UFOs, and the Loch Ness monster fill that need is beyond me.

If we want to avoid deception and try to determine what is likely to be true and what is clearly baloney (the magicians and entertainers Penn & Teller used the more direct term "bullshit" in their TV series of the same name), we need a set of "baloney filters" to enhance our skeptical screening of all the ideas we hear about, good, bad, and indifferent. Carl Sagan (1996:10) gives a list of tools for his "baloney detection kit," and Michael Shermer (1997:48) provides an interesting list of many of the common fallacies of reasoning employed by pseudoscience, so it is not necessary to repeat those lists here. However, there are several more important principles that we all need to remember to avoid being duped by pseudoscience.

1. Extraordinary Claims Require Extraordinary Evidence

This simple statement by Carl Sagan (a paraphrase of earlier versions of the same statement) makes an important point. Everyday science produces hundreds of small hypotheses that only require a small extension of what is already known to test their validity. But crackpots, fringe scientists, and pseudoscientists are well known for making extraordinary claims about the world and arguing that they are true. These include the many believers in UFOs and aliens, whose evidence is flimsy at best, but they are firmly convinced (as are a majority of Americans, according to polls) that such UFOs have landed here repeatedly and that aliens have interacted with humans. Never mind the fact that such "aliens" seem only to make themselves known to gullible individuals with no other witnesses present or that the "physical evidence" for aliens landing in Area 51 in Nevada or in Roswell, New Mexico, has long ago been explained as caused by secret military experiments. Just think for a moment: If you were part of a superior alien culture and able to travel between galaxies, would you only pick up a few isolated individuals out in the boonies, or would you contact the heads of the governments on this planet and let your existence be known? Think about our extraordinary network of satellites and radar; we can detect virtually anything moving in the skies anywhere in the world now—yet we have never received a reliable indication of a UFO, only these unverifiable claims made by random plane or ground observers and photos that have been documented as fakes. Nearly everyone carries a cell phone camera with them at all times nowadays, yet there are no better photos than before—in fact, the quality of the evidence is getting worse now that everyone has a camera. Certainly, it is *possible* that aliens have visited us, but such an extraordinary claim requires higher levels of proof than ordinary science, and the evidence provided so far is pretty flimsy.

Likewise, the claim that odd "monsters" live in remote places and have escaped scientific attention (*cryptozoology*) is an extraordinary claim that requires more than the usual level of documentation required to describe a new species of insect. Take, for example, the Loch Ness monster. We give great credence to a few blurry photos (most of which have now turned out to be frauds) or to eyewitness accounts (which only show how easily the human mind is deceived), but every thorough study of Loch Ness has failed to produce anything conclusive. No one seems willing to answer the harder questions: If there is a plesiosaur or similar beast living in Loch Ness, how does it survive alone? Large beasts like the "monster" that have supposedly lived for centuries would require a whole population of such beasts, yet there is no conclusive evidence that even one exists. The believers also ignore another inconvenient geologic fact: Loch Ness is a glacial valley that was filled with ice 20,000 years ago and only now has water in it during our current interglacial period. So where did the monster live before it became trapped in Loch Ness? And if there were enough of them to form a population, why have they never been found in any other body of water anywhere in the world?

Similar flaws apply to the claims of the Bigfoot or Sasquatch in the Pacific Northwest, or the Abominable Snowman or Yeti in the Himalayas, or the supposed living sauropod dinosaur Mokele-Mbembe in the Congo jungle. All the "evidence" for these beasts is inconclusive or admitted to be fraudulent, and there would have to be a big population of them (never detected) for us to still witness them. We don't know for sure that such beasts *don't* exist, but they are so remarkable that they require much better proof of their existence (especially in this overpopulated world where the truly unexplored, dark regions are nearly gone) than has been presented so far.

2. Burden of Proof

Related to the first principle is the idea of *burden of proof*. In a court of law, one side (usually the prosecution or plaintiff) is assigned the task of proving its case "beyond a reasonable doubt," (in a criminal case) or "based on the preponderance of the evidence" (in a civil case), and the defense often needs to do nothing if the other side has not met this burden of proof. Similarly, for extraordinary claims that appear to overthrow a large body of knowledge, the burden of proof is also correspondingly greater. In 1859 the idea of evolution was controversial, and the burden of proof was to show that evolution had occurred. By now, the evidence for evolution is overwhelming, so the burden of proof on the anti-evolutionists is much larger: they must show creationism is right by overwhelming evidence, not just simply point out a few inconsistencies or problems with evolutionary theory. Likewise, the evidence that the Holocaust occurred is overwhelming (with many eyewitnesses and victims still alive and many Nazi documents that describe what they did), so the Holocaust denier has to provide overwhelming evidence to prove that it did not occur.

3. Anecdotes Do Not Make Science

As storytelling animals, humans are prone to believe accounts told as stories by "witnesses." The telemarketers know that if they get a handful of celebrities or sincere-sounding customers (or actors) to praise their products, we will believe these people and go out and buy their merchandise, even if there have been no careful scientific studies or FDA approvals to back up their claims. One or two anecdotes may sound convincing, and the experience of your back-fence neighbor may be interesting, but to truly evaluate claims made in science (and elsewhere), you need a detailed study with dozens or hundreds of cases. In addition, there must be a "control" group that does not receive the treatment but a placebo instead, yet think that they did get the real medicine (so the power of suggestion is not responsible for the alleged benefit). Anything approved by the FDA has met this standard; most stuff sold in the "new age" or "health food" stores has not been so carefully studied. When it has been analyzed, there usually turn out to be either marginal benefits or none at all. (The con artists and snake oil salesmen will take your money all the same.) If you listen closely to the language of the ads for some of these "medicines," it carefully avoids the terminology of medicine and pharmacology, and instead uses phrases likes "supports thyroid health" or "promotes healthy bladder function." These phrases are not true medicinal claims, and so they are not subject to FDA regulations. Nonetheless, the great majority of these products that have been scientifically analyzed turn out to be worthless and a waste of money, and every once in a while, they turn out to be harmful or even deadly.

Similarly, the evidence for UFOs or alien abductions or Sasquatch sightings is largely anecdotal. One person, usually alone, is a witness to these extraordinary events and is convinced they are real. However, studies have shown again and again how easily people can hallucinate or be deceived by common natural phenomena into "seeing" something that really isn't there. A handful of "eyewitnesses" means nothing in science when the claims are unusual; much more concrete evidence is needed.

4. Arguments from Authority and Credential Mongering

Many people try to win arguments by quoting some "authority" on the subject in an attempt to intimidate and silence their opponents. Sometimes they are accurately quoting people who really are experts in a subject, but more often than not the quotation is out of context and does not support their point at all, or the authority is really not that authoritative. As we shall see in the chapters that follow, this is the usual problem with creationist "quote mining": when you go back and look at the source, the quote is out of context, and means just the opposite of what they claim, or the source itself is outdated or not very credible. As Carl Sagan puts it, there are no true authorities; there are people with expertise in certain areas, but nobody is an authority in more than a narrow range of human knowledge.

One of the principal symbols of authority in scholarship and science is the Ph.D. But you don't need a Ph.D. to do good science, and not all people who have Ph.D.s are good scientists either. As those of us who have gone through the ordeal know, a Ph.D. only proves that you can survive a grueling test of endurance in doing research and writing a dissertation on a very narrow topic. It doesn't prove that you are smarter than anyone else or more qualified to render an opinion than anyone else. Because earning a Ph.D. requires enormous focus on a specific area, many people with that degree have actually *lost* a lot of their scholarly breadth and knowledge of other fields in the process of focusing on their theses.

In particular, it is common for people making extraordinary claims (like creationism or alien abductions or psychic powers) to wear a Ph.D. (if they have one) like a badge, advertise it prominently on their book covers, and feature it in their biographies. They know that it will impress and awe the listener or reader into thinking they are smarter than anyone else or more qualified to pronounce on a topic. Nonsense! Unless the claimant has earned a Ph.D. *in the subject being discussed*, the degree is entirely irrelevant to the controversy. For example, leading creationists include the late Duane Gish, who had a doctorate in biochemistry, and the late Henry Morris, who had a doctorate in hydraulic engineering. However, they both earned their degrees almost 50 years ago, so they were not up-to-date in these rapidly changing fields that they have not practiced in decades. If they stuck to discussing just those topics, they might be halfway believable, but all of their criticisms focus on the fossil record, geology, thermodynamics, and so on—topics in which they have absolutely no firsthand experience, published research, or training. Their entire knowledge of these fields (vividly demonstrated by reading their books) consists of skimming and misquoting popular books by real experts in those fields who did the actual work, not going out and doing the research themselves or publishing in peer-reviewed journals. They are no more qualified to comment on paleontology or geology, based on their irrelevant degrees, than they are qualified to fix a car or critique music theory! Yet they always flaunt their Ph.D.s to awe the masses and try to intimidate their opponents. The same goes for creationists like Jonathan Sarfati (physical chemistry), Michael Behe (biochemistry), and Jonathan Wells (cell biology)—none of those subjects gives them *any* background in fossils or paleontology, and none of these scientists has published in any peer-reviewed paleontological journals, so they are complete amateurs when it comes to fossils.

Similarly, there are many fringe and crackpot ideas in anthropology and paleontology, and the more "way-out" they are, the more likely the author has put "Ph.D." on the cover. There is even a maverick paleontologist who does this on all his book covers and flaunts it on the lecture circuit (where he is highly successful), even though he has been repeatedly dumped by one academic institution after another, has not had an article published in a peer-reviewed journal in many years, and has no credibility in professional organizations like the Society of Vertebrate Paleontology. By contrast, legitimate scientists never put their degrees on their book covers and seldom list their credentials on a scientific article. If you doubt this, just look at the science shelves in your local bookstore. The quality of the research must stand by itself, not be propped up by an appeal to authority based on your level of education. To most scientists, credential mongering is a red-flag warning. If the author puts his or her Ph.D. on the cover, beware of the stuff between the covers!

5. Bold Statements and Scientific-Sounding Language Do Not Make It Science

People who want to promote their radical ideas are prone to exaggeration and famous for making amazing pronouncements, such as "a milestone in human history" or "the greatest discovery since Copernicus" or "a revolution in human thinking." Our baloney detection alarms should go off automatically when we hear politicians or actors try to hype policies or movies that turn out to be much less than claimed. They should also scream in alarm when we hear people making claims about human knowledge or science that seem overblown.

Another strategy to make a wild idea acceptable to the mainstream is to cloak it in the language of science. This cashes in on the goodwill and credibility that science has in our culture and attempts to make outrageous ideas more believable. For example, when the creationists realized that they could not pass off their religious beliefs in public school science classrooms as science, they began calling themselves "creation-scientists" and eliminating the overt references to God in their public school textbooks (but the religious motivation and source of the ideas is still transparently obvious). Several churches (including Christian Science and Scientology) appropriate the aura of scientific authority by using the word "science" in their names, even though they are not falsifiable and do not fit the criteria of science as discussed here. Similarly, the snake oils and nostrums peddled by the telemarketers and by the "New Age" alternative medicine fanatics are often described in what appears to be scientific lingo, but when you examine it closely, they are not actually following scientific protocols or the scientific method. The are the famous examples of television commercials that show an actor in a white lab coat, often with a stethoscope around his neck, saying, "I'm not a doctor, but I play one on TV" and then promoting a product that he has no medical training to analyze. However, just the appearance of scientific and medical authority is sufficient to sway people to buy his product.

6. Correlation Is Not Causation

Human beings are programmed by our genes to see patterns in nature and to recognize connections between things. Yet sometimes these instincts lead us astray. We wear a particular item of clothing one day, and our team wins; we forget to do it once, and they lose. Then we are convinced that wearing that item is "lucky" for the team, and we wear it every time, no matter whether the team wins or loses. We cannot shake this superstition, and no amount of falsification from future failed predictions will change it. Many people believe in "earthquake weather," because they recall one or two very strong earthquakes that happened to occur on hot mornings. They are not dissuaded when you point out that the daily temperature fluctuation due to weather is not felt more than a few feet underground, while earthquakes come from faults that are many miles underground. One or two coincidences are enough to reinforce this "urban myth." Seismologists have done rigorous statistical analysis again and again and have shown conclusively that earthquakes of all sizes occur in any weather and at any time of day or night. The most common form of this superstition is known as the *post hoc, ergo propter hoc* ("after this, therefore because of this") fallacy.

Scientists are also prone to believe that there is some connection when they see one or two positive results in a row. But as scientists, we are trained early in our careers to study the mathematics of probability and statistics, so that we can analyze in a rigorous way whether an apparent connection between events is truly significant or still could be due to chance. Although scientists still use hunches and intuition to guess that phenomena might be related, once they try to publish their ideas in a peer-reviewed journal, they had better do the appropriate statistics, or their article will be quickly rejected!

A good example of this was the furor caused in the 1980s, when paleontologists David Raup and Jack Sepkoski (1984, 1986) made the claim that there were mass extinction events every 26 million years and that some of them were caused by impacts of asteroids from space. Astronomers quickly jumped on the bandwagon even before the data were published, with "explanations" for this extinction "periodicity" that ranged from a mysterious Planet X to an undetected companion star to the sun called "Nemesis" to the motions of the solar system within the galactic plane to periodic pulses of mantle overturn triggering global volcanism. But, as the original data were scrutinized more closely, the correlation began to fall apart. Several statistical analyses showed that there was no correlation at all; many of the "extinction peaks" turned out not to be real or were millions of years too early or too late to have been part of a regular astronomical cycle; and most were found not to have any evidence of an extraterrestrial impact (see Prothero 1994a). Sepkoski (1989) made one last valiant attempt to answer the critics and salvage the hypothesis, but Stanley (1990) gave a much simpler explanation that better fits the data. In a truly gigantic mass extinction, there are so few survivors living in the aftermath that the world is populated with opportunistic "weed-like" species, ecological generalists that thrive in disturbed habitats with little or no competition. Eventually, however, more complicated, specialized species and ecosystems re-evolve and replace those that the mass extinction wiped out. It apparently takes about 20 million years or more for the planet to recover from a mass extinction and for all these extinction-prone specialized species to evolve again. If some great disturbance happens only a million years after a mass extinction, you would never see it, because there are few species that are vulnerable to extinction. Only after enough time has passed do they evolve; this is why the mass extinctions described by Raup and Sepkoski were spaced *roughly* 20–30 million years apart, but no shorter.

7. The World Is Not Black and White, but Shades of Gray

Also known as the "either-or" fallacy or the "false dilemma," this is a common strategy in which the arguer tries to present his or her case as a choice between one extreme and another. This thought is reflected in the famous slogans "If you are not with us, you're agin' us" or "If you are not part of the solution, you are part of the problem." By dichotomizing their world into only two positions, they create the false dilemma that evidence *against* one point of view is evidence *for* another. This is the principal tactic of the creationists, who try to create the false dilemma that they are the only true Christians, and anyone who does not agree with them is an atheist.

But as mature adults know, most matters in life are not black and white but shades of gray. Arguments against one position do not necessarily support the opposite position. Creationism is not the only form of Christianity, and there are many Christian evolution- ists. Indeed, there is a wide spectrum of beliefs, from literalistic young-earth creationists to "day-age creationists," who allow the "days" in Genesis to be geologic ages, to theistic evolutionists and so on. As in any other aspect of life, there are many possible answers or possible viewpoints, and we should not be suckered into the dilemma of believing that there are only two alternatives.

A corollary of the "false dilemma" principle is what Shermer (1997:52) calls "the unex- plained is not inexplicable." Many people (such as the "intelligent design" creationists dis- cussed in the next chapter) argue that if *they* cannot explain something, then *nobody* can. This is not only arrogant, but it is built on the false "either-or" premise that if no explanation is currently available, then the phenomenon will never be explained. But just because we don't have an explanation now doesn't mean that we won't find one someday. In the meantime, we do science and knowledge a disservice by defaulting to supernatural explanations sim- ply because we still have an unsolved mystery in front of us. Scientists are used to dealing with uncertainty and realizing that their answers are tentative and subject to change, but the general public seems to prefer the comfort of *any* kind of answer (even if it is wrong) to the insecurity of living with the uncertain and unknown. As H. L. Mencken put it, "For every problem, there is a solution that is simple, neat, and wrong."

8. Special Pleading and Ad Hoc Hypotheses

In science, when an observation comes up that appears to falsify your hypothesis, it is a good idea to examine the observation closely or run the experiment again, to be sure that it is real. If the contradictory data are sound, then the original hypothesis is falsified, dead, kaput. It is time to throw it out and come up with a new, possibly better hypothesis.

In the case of many nonscientific belief systems, from religions to mysticism to Marxism, it does not work this way. Belief systems often have a profound emotional and mystical significance for people. They exist in spite of contradictory observations, and people refuse to let rationality or the facts shake them. As Tertullian put it, "I believe *because* it is incredible." Saint Ignatius Loyola, the founder of the Jesuits, wrote, "To be right in everything, we ought always to hold that the white which I see is black, if the Church so decides it." That's fine, if you are willing to accept that system and suspend disbelief of some of their claims for a more important benefit of emotional and mystical connections.

If you pass off your belief system as science, however, you must play by the rules of science. When con artists try to sell you snake oil, and someone points out an inconvenient fact about it, the con artists will try to attack this fact or explain it away with an "after the fact" or ad hoc (Latin for "for this purpose") explanation. If the snake oil fails to work, they might say "you didn't use it right" or "it doesn't work on days when the moon is full." If the séance fails to contact the dead, the medium might scold the skeptic by saying, "you didn't believe in it sufficiently" or "the room wasn't dark enough" or "the spirits just don't feel like talking today." If we point out that there are millions of species on earth that could not have fit into the biblical Noah's ark, the creationist tries to salvage their hypothesis by saying "only the created kinds were on board" or "insects and fish don't count" or "God miraculously crammed all these animals into this tiny space, where they lived in harmony for 40 days and 40 nights" or some similar garbage.

As we shall see in the chapters that follow, ad hoc hypotheses are common when the conclusion is already accepted and the believer must find any explanation to wiggle out of inconvenient contradictory facts. But they are not acceptable in science. If the conclusion is a given and cannot be rejected or falsified, then it is no longer scientific.

9. Not All "Persecuted Geniuses" Are Right

People trying to promote wild ideas that seem crazy to us will often point to the persecution of Galileo (arrested and tried for advocating Copernican astronomy) or Alfred Wegener (ridiculed for his ideas about continental drift) and take solace in how these geniuses were eventually proven right. But as Carl Sagan (1996:64) put it, "The fact that some geniuses were laughed at does not imply that all who are laughed at are geniuses. They laughed at Columbus, they laughed at Fulton, they laughed at the Wright Brothers. But they also laughed at Bozo the Clown." The annals of science are full of wild and crackpot notions that didn't survive testing and were eventually abandoned, and they far outnumber those of the handful of "misunderstood geniuses" who were vindicated in the end.

These misunderstood geniuses often turn to Schopenhauer, who wrote, "All truths pass through three stages. First, it is ridiculed. Second, it is violently opposed. Third, it is accepted as self-evident." But Schopenhauer is wrong. Many revolutionary and radical ideas (such as Einstein's theory of relativity) were never ridiculed or violently opposed. In the case of Einstein, his theories were mostly ignored as interesting but untested until scientific observations made in 1919 corroborated them.

Science is open to all sorts of ideas, from the conventional to the wacky. It doesn't matter where they came from, but they all have to pass muster. If your ideology has failed the test of science, you can't just claim you're a misunderstood genius—it is more likely that your cherished hypothesis is just plain wrong. Scientists are too busy, and there are too many worthwhile and important scientific goals for them to pursue for them to waste their time testing and evaluating every wild scheme that comes along. The fringe element might wail that they are persecuted and misunderstood geniuses. But if they want to be taken seriously, they must play by the rules of science: get to know other scientists, exchange ideas, be willing to change your own ideas, present your results at scientific conferences, and submit them to the scrutiny of peer-reviewed journals and books. If your ideas can survive this rigorous gantlet, then they will get the attention of scientists that they deserve.

The list of logical and scientific fallacies goes on and on (see Sagan 1996:210–217; Shermer 1997:44–61), so I will not try to cover all of them here. As we review the scientific evidence discussed in the latter part of this book, we should always have in the back of our minds questions such as: How do we test this hypothesis? Is it falsifiable or unfalsifiable? Is the evidence strong enough to support an extraordinary claim? Is the claim supported by multiple cases and statistical testing, or is it just anecdotal? Does the arguer use quotations out of context or flaunt their credentials? Does the arguer present a false dilemma? Does the arguer attempt to rescue their failing belief system with special pleading and ad hoc hypotheses, or are they willing to accept that their conclusions might be wrong?

Follow the Evidence Wherever It May Lead

Sit down before a fact as a little child, be prepared to give up every preconceived notion, follow humbly wherever and to whatever abysses nature leads, or you shall learn nothing.
—Thomas Henry Huxley

Belief systems are so powerful that many people have difficulty when the evidence begins to accumulate against them. As we just discussed, it's human nature to cling to a cherished idea and not reject a belief system in the face of new evidence but to explain it away with ad hoc rationalizations. To many people, the comfort of the belief system is more important than the self-deception they are willing to employ to salvage it. Most of the time, we don't worry about people who have their own beliefs, as long as they don't try to impose them on us, or as long as their belief systems don't lead to dangerous actions, such as flying airplanes into buildings.

It's another thing when people reject scientific evidence because of their belief systems, yet keep claiming to be scientists. The American creationist movement is a good example of this. Proponents try to find as many members of their group as possible who have advanced degrees and then advertise these people prominently, as if the fact that a few of them passed through to the Ph.D. level of education makes them experts on everything. As we already discussed, the Ph.D. is irrelevant unless it is in the field of study that is being argued. Likewise, these creationist "authorities" almost always freely admit that they are driven by their fundamentalist religious views and reject evolution because their predetermined belief system forced them to. For example, in Ashton's 2000 book, *In Six Days: Why 50 Scientists Choose to Believe in Creation*, the "scientists" interviewed confess that they started out as fundamentalists, then wrestled with the evidence of evolution (usually not to any great depth), and then came back to their original belief system. I know of absolutely no scientists who rejected evolution on purely scientific evidence without the powerful force of religious fundamentalism operating behind the scenes. Instead, these creationist "scientists" all came to their conclusions because their religious beliefs demanded it, and afterward began to take seriously the phony "evidence" against evolution that we'll discuss in the rest of the book.

By contrast, true scientists *must* reject a cherished belief if enough evidence goes against it. A classic example was the revolutionary idea of continental drift and plate tectonics, which swept through geology in the 1950s and 1960s, and by 1970 was as well established as any idea in science. The reactions of the geologic community say a lot about the sociology of science. The "Old Guard" who had a lot of time and research invested in fixed continents tended to be skeptical the longest, and many held out until the evidence became overwhelming. Eventually, they all had to concede their cherished beliefs were wrong. In contrast, the first to accept the new ideas were the "Young Turks," mostly younger scientists (especially graduate students) who did not have emotional connections to the old ways of thinking and were more willing to try out new concepts.

One of the bravest examples of this process was the famous geologist Marshall Kay, who passed away just before I arrived at Columbia University to take classes from him. He had spent his entire life explaining the complexities of geology based on the assumption that continents did not move; he even published a major book in 1951 that detailed the nature of the thick sedimentary basins, assuming fixed continents. Yet when the evidence for plate tectonics and continental drift became overwhelming in the 1960s, he wholeheartedly embraced plate tectonics. Even though he was near retirement age, he began redoing his life's work using the new concepts. Such intellectual honesty and bravery is an admirable and rare trait in humans. How many people do you know nearing retirement age who are willing to redo their entire life's work because they've realized that the assumptions they followed for 50 years were wrong?

Richard Dawkins (2006) points to another admirable example. In his words,

> It does happen. I have previously told the story of a respected elder statesman of the Zoology Department at Oxford when I was an undergraduate. For years he had passionately believed, and taught, that the Golgi apparatus (a microscopic feature of the interior of cells) was not real: an artifact, an illusion. Every Monday afternoon it was the custom for the whole department to listen to a research talk by a visiting lecturer. One Monday, the visitor was an American cell biologist who presented completely convincing evidence that the Golgi apparatus was real. At the end of the lecture, the old man strode to the front of the hall, shook the American by the hand and said—with passion—"My dear fellow, I wish to thank you. I have been wrong these fifteen years." We clapped our hands red. No fundamentalist would ever say that. In practice, not all scientists would. But all scientists pay lip service to it as an ideal—unlike, say, politicians who would probably condemn it as flip-flopping. The memory of the incident I have described still brings a lump to my throat.

Contrast this with the way in which creationists operate. Those who have actually looked closely at the overwhelming abundance of evidence for evolution have several choices:

they can close their eyes and stop looking; they can wrestle with it and eventually deny what is self-evident to salvage their beliefs; they can distort it to fit their preconceptions; or they can face reality. A good example is Kurt Wise, who is famous as one of the few young-earth creationists with a legitimate background in paleontology; he actually got his Ph.D. at Harvard. But his advanced training did not lead him to creationism. In fact, he was raised with a fundamentalist background and describes in his autobiography (in the Ashton book cited above) how he wrestled with the inherent contradictions between pale- ontology and his fundamentalist beliefs in high school. He still had those doubts through his undergraduate days at the University of Chicago. He entered Harvard as a student of Stephen Jay Gould but apparently did not reveal his creationism to Gould or Harvard when he was admitted. I am not aware of what Gould thought when he found out that he had a creationist among his students, but several former Gould students have told me that their advisor was very fair minded and open to challenges. They speculate that Gould must either have thought that Wise would eventually see the problems with his creation- ist viewpoint or accepted him as a grand experiment in intellectual freedom. According to some of Gould's students who knew Wise in graduate school, he was polite and par- ticipated in discussions (although he was considered very arrogant and standoffish), but he was clearly only going through the motions and was not really learning anything new or opening his mind to new ideas. Instead, as he told one of his fellow graduate students, he was treating his Harvard experience as a sort of "Monopoly" game, playing the part of paleontologist to get his degree but not taking any of it very seriously or really absorbing the implications of what he was studying—any more than "Monopoly" players are really bankers or landowners or go to jail.

What's the point of going through the long ordeal of obtaining a Ph.D. if you're never going to learn something new or be challenged and think hard about your beliefs? More importantly, if you're doing all this work to obtain the degree but don't believe in any of the stuff you said or wrote, isn't that dishonest and fraudulent? Science is a social network based on trust and reputation, and nobody has time to constantly check the work of most other scientists to see if it was done honestly or if it is corrupted by biases and fraud. If someone like Kurt Wise is just going through the motions, how can other scientists trust whether he biased his data collection or analysis or whether he just made it up in order to fit his precon- ceptions? We will see elsewhere in this book how this kind of dishonest science is common among other creationists whose backgrounds are irrelevant to the stuff they are promoting, but the question is now relevant even with Harvard-trained scientists like Wise.

Armed with his Harvard Ph.D., Wise has become the most prominent creationist who actually has seen and studied real fossils and geology. But is he really a scientist by the standards we have just discussed? Apparently, his Harvard experience never caused him to truly examine his beliefs. In his autobiography, he freely admits that his entire creationist viewpoint comes from the literalistic interpretation of the Bible, not from actual scientific evidence, and that his belief system forces him to reject whatever he doesn't want to accept. As he wrote in his autobiography (in Ashton 2000):

> I am a young-age creationist because that is my understanding of the Scripture. As I shared with my professors years ago when I was in college, if all the evidence in the universe turned against creationism, I would be the first to admit it, but I would still be a creationist because that is what the Word of God seems to indicate.

No amount of evidence could ever turn him away from creationism? What kind of real scientist talks like this? If Marshall Kay and the Oxford professor mentioned above could learn from new evidence and reject their old beliefs (as good scientists are supposed to do), why can't Kurt Wise?

I have no problem with his belief system. He's entitled to believe whatever he wants. But when he completely rejects the data and methods of science in order to follow his rigid belief system, he's not acting as a scientist anymore—he's just another preacher. If he labeled his ideas as religiously inspired, that would be fine. But he continues to pretend that he is following the rules of science; he wears the label of scientist and promotes his particular brand of "science" to unsuspecting people who are impressed with his Harvard Ph.D. but don't realize that he admits that he stopped doing science a long time ago.

As Dawkins (2006:323) puts it,

> I find that terribly sad; but whereas the Golgi apparatus moved me to tears of admiration and exultation, the Kurt Wise story is just plain pathetic—pathetic and contemptible. The wound, to his career and his life's happiness, was self-inflicted, so unnecessary, so easy to escape. . . . Poor Kurt Wise reminds me more of Winston Smith in 1984—struggling desperately to believe that two plus two equals five if Big Brother says it does. Winston, however, was being tortured.

For Further Reading

Darwin, C., F. Darwin, and A. C. Seward. 1903. *More Letters of Charles Darwin*. London: John Murray.

Gardner, M. 1952. *Fads and Fallacies in the Name of Science*. New York: Dover.

Gardner, M. 1981. *Science: Good, Bad, and Bogus*. Buffalo, N.Y.: Prometheus.

Popper, K. 1935. *The Logic of Scientific Discovery*. London: Routledge Classics.

Popper, K. 1963. *Conjectures and Refutations: The Growth of Scientific Knowledge*. London: Routledge Classics.

Sagan, C. 1996. *The Demon-Haunted World: Science as a Candle in the Dark*. New York: Ballantine.

Shermer, M. 1997. *Why People Believe Weird Things: Pseudoscience, Superstition, and Other Confusions of Our Time*. New York: Freeman.

Shermer, M. 2005. *Science Friction: Where the Known Meets the Unknown*. New York: Times Books.

FIGURE 2.1. The "intelligent design" movement is simply a disguise for creationism to sneak into public schools and avoid the constitutional separation of church and state. (Cartoon by Steve Benson, copyright © Creators Syndicate)

SCIENCE AND CREATIONISM 2

Mything Links

These bits of information from ancient times, which have to do with themes that have supported human life, built civilizations, and informed religions over the millennia, have to do with deep inner problems, inner mysteries, inner thresholds of passage, and if you don't know what the guide-signs are along the way, you have to work it out for yourself.

—Joseph Campbell, *The Power of Myth*

Nearly every culture on earth has some form of a creation story or myth that it uses to explain its place in the universe and its relationships to its god or gods. As Joseph Campbell wrote in *The Power of Myth* (1988), these stories are essential for a culture to understand itself and its role in the cosmos and for individuals to know what their gods and their culture expect of them. At one time, myths served the role of explaining how the world came to be, usually with the subtext that it explained how that culture fit within the universe. In our modern technological scientific age, we tend to scoff at the stories that were believed by the Sumerians, Norse, and Greeks, but in their time, those stories served both as a metaphor and allegory for humanity's place in the universe and a rational explanation for how things came to be.

Many creation stories have common elements or themes that are universal across cultures and time. They often have elements of birth or eggs in them, because these are very powerful symbols of the creation of life in our world. In some versions of the Japanese creation myths, a jumbled mass of elements appeared in the shape of an egg, and later in the story, Izanami gives birth to the gods. In the beginning of one of the Greek myths, the bird Nyx lays an egg that hatches into Eros, the god of love. The shell pieces become Gaia and Uranus. In Iroquois legend, Sky Woman fell from a floating island in the sky because she was pregnant and her husband pushed her out. After she landed, she gave birth to the physical world. There are many Hindu creation stories. In one of them, the god Brahma created the primal waters as the womb for a small seed, which grew into a golden egg. Brahma split it apart and made the heavens from one half and the earth and all its creatures from the other. The Chinook Indians of the Pacific Northwest were created out of a great egg laid by the Thunderbird. Similar stories of a cosmic egg are known from Chinese, Finnish, Persian, and Samoan mythology.

Many stories often have mother and father figures who are responsible for the creation. The mother figure is often some form of "Mother Earth," and her fertility is symbolic of the earth's fertility. The Greek creation myths, for example, have the world arising from the mating of the earth goddess Gaia with the sky god Uranus, and their union created the pantheon of Greek gods, who in turn created the physical universe. In Japan, Izanagi and Izanami mated, and the mother goddess Izanami gave birth to three children, Amaterasu, the sun; Tsukiyumi, the moon; and Susano-o, their unruly son. The Australian aborigines believed in the Sun-Mother who created all the animals, plants, and bodies of water at the suggestion of the Father of All Spirits. Primordial parents are found in the myths of many other cultures, including the Egyptians, Cook Islanders, Tahitians, and the Luiseño and Zuni Indians.

Most creation myths, such as the Hebrew, Greek, and Japanese myths mentioned earlier, as well as the Sumerian-Babylonian myths discussed later, have some form of chaos or nothingness at the beginning that is organized or separated into sky and earth by their gods. Other myths, however, imagine a world that existed before our present world, and one of their gods from the earlier world brings our universe into existence. The Bushmen of Africa, for example, imagined a world where people and animals lived together in peace and harmony. Then the Great Master and Lord of All Life, Kaang, planned a wondrous land above theirs and planted a great tree that spread over it. At the base of the tree, he dug a hole and brought the people and animals up from below. In the Hopi myth, there were past worlds beneath ours. When life became unbearable in those worlds, the people and animals climbed up the pine trees to reach new, unspoiled worlds where they could live. This ladder is endless, so some creatures may still be climbing out of this world and into the next. The Navajo creation myth is similar, but instead of climbing pine trees from one world to the next, they climb through a great hollow reed.

The theme of humans breaking some sort of divine edict from the gods and causing pain and suffering by their disobedience is also common. In addition to Adam and Eve in the Garden of Eden story, there is the Greek story of Pandora, who was given to Epimetheus as a gift from Zeus, along with a box she was not allowed to open. But Zeus also gave her curiosity, so when she did open the box, she released all the sins and troubles of the world. The African Bushmen were told by their gods not to build fire, and when they disobeyed, their peaceful relationships with animals were destroyed forever. According to the Australian Aborigines, the Sun-Mother created the animals, which she demanded must live peacefully together. But envy overcame them, and they began to quarrel. She came back to earth and gave them a chance to change into any shape they wanted, resulting in the strange combination of animals in Australia. But because the animals had disobeyed the Sun-Mother's instructions, she created two humans who would rule over the animals and dominate them.

The theme of a great flood that destroys nearly all of life is common to nearly all mythologies. In addition to the Sumerian story of Ziusudra and Babylonian story of Utnapishtim (described later) and the Hebrew legend of Noah (probably derived from the Sumerian or Babylonian account), the Greeks talked of Deucalion, who survived the great flood and seeded the land with the humans after the floodwaters receded. There are similar flood legends in Norse, Celtic, Indian, Aztec, Chinese, Mayan, Assyrian, Hopi, Romanian, African, Japanese, and Egyptian mythology. Scholars suggest that this may be because most cultures that live near large bodies of water (which nearly all do, except those in the mountains) have experienced some catastrophic flood in their distant past. It also wiped out much of their culture and tradition, so that flood achieves legendary status when its story is told generation after generation. Only a few cultures or religions, including the Jainists of India and the Confucianists of China, have no creation myth whatsoever.

This brief thematic summary does not do justice to the details and the imagery of the original myths or to the power of the language in which they were written. If you have never done so, I strongly recommend that you pick up a book of comparative mythology or examine some of the many texts that are now available on the Internet. Through all this discussion, we have seen how mythologies often reflect universal themes about human existence and about how humans fit into the universe. None of these stories is necessarily "true" or "false"—they are products of their own cultures and were essential to those people for giving their world a context. All humans hunger for an understanding of their origins, so they generate some sort of story to explain those origins. Once that story has been passed from generation to generation, it acquires its own sort of reality or "truth," and it is important to the members of that culture so that they understand their own role in the world and their relationship to their gods.

As Michael Shermer (1997:30) sums it up, "Does all this mean that the biblical creation and re-creation stories are false? To even ask the question is to miss the point of the myths, as Joseph Campbell (1949, 1982) spent a lifetime making clear. These flood myths have deeper meanings tied to re-creation and renewal. Myths are not about truth. Myths are about the human struggle to deal with the great passages of time and life—birth, death, marriage, the transitions from childhood to adulthood to old age. They meet a need in the psychological or spiritual nature of humans that has absolutely nothing to do with science. To turn a myth into science, or science into a myth, is an insult to myths, an insult to religion, and an insult to science. In attempting to do this, the creationists have missed the significance, meaning and sublime nature of myths. They took a beautiful story of creation and re-creation and ruined it."

The Genesis of Genesis

When on high heaven was not named,
And the earth beneath did not yet bear a name,
And the primeval Apsu, who begat them,
And chaos, Tiamat, the mother of them both
Their waters were mingled together,
And no field was formed, no marsh was to be seen;
When of the gods none had been called into being,
And none bore a name, and no destinies were ordained;
Then were created the gods in the midst of heaven,
Lahmu and Lahamu were called into being . . .
Ages increased . . .
—*Enuma Elish*, about 3000 B.C.E.

The origin of the Hebrew creation stories in the Bible has been studied for nearly 200 years and is well known and accepted by most Bible scholars. In the 1860s and 1870s, archeologists excavated several ancient Sumerian cities in Mesopotamia (what is now Iraq) and found clay tablets written in cuneiform. This is the oldest written language on earth, created with marks in soft clay made by a wedge-shaped stylus. Some of the stories date back at least to 4000 B.C.E., and most were recycled by the mythology of the Akkadian, Babylonian, and Assyrian cultures that replaced the Sumerians in Mesopotamia. The longest and best known of these stories is the *Enuma Elish* (in Babylonian, the first two words of the story, translated "When on high . . ."), which describes a creation epic that bears remarkable similarity to Genesis 1, including a formless void and chaos, with gods dividing the waters from the land and naming the creatures. Since the story predates any of the Hebrew creation stories by centuries, there is little doubt that the early Hebrews were influenced by this powerful epic accepted by all Mesopotamian civilizations for over two millennia. Psalm 74 also borrows heavily from the *Enuma Elish*, where Yahweh destroys the Leviathan and splits its head open in an almost word-for-word copy of the way in which Marduk, the chief god of Babylon, splits open the head of Tiamat, the goddess of the ocean.

Another source is *The Epic of Gilgamesh*, which dates to about 2750 B.C.E. The Sumerians had a hero called Ziusudra (called Atrahasis by the Akkadians and Utnapishtim by the Babylonians), who is warned by the earth goddess Ea to build a boat because the god Ellil was tired of the noise and trouble of humanity and planned to wipe them out with a flood. When the floodwaters receded, the boat was grounded on the mountain of Nisir. After Ziusudra's boat was stuck for seven days, he released a dove, which found no resting place and returned. He then released a swallow that also returned, but the raven that was released the next day did not return. Ziusudra then sacrificed to Ea on the top of Mount Nisir. The story is nearly identical to that of Noah's flood, not only in its plot and structure, but also in the details of its phrasing. Only the names of the characters and gods and a few details have been changed to suit the differences between the monotheistic Hebrew culture and the polytheistic cultures of the Sumerians, Akkadians, and Babylonians.

Two centuries of detailed study by scholars has also revealed the way in which the Bible was put together. In its original Hebrew, the Old Testament (especially the first five books, or Pentateuch) show unmistakable signs of different authors writing different parts, and then someone later patching the whole thing together. Someone reading a later translation (especially the outdated King James translation) cannot pick up these differences easily, but they are obvious to those who read Hebrew. In high school, I was troubled by the contradictions between what I learned in my Presbyterian Sunday School and what I had learned from science; I decided to find out about the Bible myself. Not only did I read many books about biblical scholarship, but I also learned to read Hebrew so I could decipher Genesis on my own, making my own judgment about translations. In college, I also learned ancient Greek, and I can still read the New Testament in the original text and recognize when someone is mistranslating or misinterpreting the original.

To Hebrew scholars, the most obvious signs of different authorship are their choices of certain phrases and words, especially the word they use for God. One source is known as the "J" source, after *Jahveh*, a common name for God. This name is also spelled and pronounced "Yahweh" or "YHWH" for those who dare not speak God's name (since early written Hebrew had no vowels or even the modern system of vowel points, only the consonants are used). This name was mispronounced and misspelled as "Jehovah" by later authors. The authors of the J document were priests of the southern kingdom of Judah, who wrote sometime between 848 B.C.E. and the Assyrian destruction of Israel in 722 B.C.E. They use terms such as "Sinai," "Canaanites," and phrases such as "find favor in the sight of," "call on the name of," and "bring *out* from the land of Egypt." The J authors were probably religious leaders associated with Solomon's temple, very concerned with delineating the guiding hand of Jahveh in their history but not so concerned with the miraculous.

The second main source is known as the "E" document, after their name for God, *Elohim*, "powerful ones" in Hebrew. The priests who composed the E document were interested in different issues, used a different set of phrases, and can be traced to the northern kingdom of Israel, sometime between 922 B.C.E. and the Assyrian conquest in 722 B.C.E. The E authors use such terms as "Horeb" instead of Sinai, "Amorites" instead of Canaanites, and the phrase "bring *up* from the land of Egypt." Most scholars think that the E authors were Ephraimite priests, who were more interested in the righteousness that God requires of his people. When people sinned they must repent. Moses is the central focus of these accounts, along with miraculous aspects of their history.

A third source, the "P" source, or Priestly Code, was apparently written by Aaronid priests around the time of the Babylonian captivity in 587 B.C.E. It is the youngest of the sources of the Old Testament. The P source emphasizes the role of Aaron and diminishes the role of Moses in the early books of the Bible. This source frequently uses long lists and is characterized by long boring interruptions to the narrative and cold unemotional descriptions. To Hebrew scholars, the P source is also distinctive in its low, clumsy, inelegant literary style. The P source views God as distant and transcendental, acting and communicating only through the priesthood. According to P, God is just but also unmerciful, using brutal, abrupt punishment when laws are broken.

Sometime during the reign of King Josiah around 622 B.C.E., the Hebrews began combining these different traditions along with other sources (such as the "D" source of the Deuteronomic code). All of these documents date from the period before Judah was captured, Jerusalem burned, the Temple destroyed, and the Hebrews dragged off to captivity by Nebuchadnezzar of Babylon in 587 B.C.E.

Verse by verse, scholars can tease apart the way in which each book of the Old Testament was woven together (see Friedman 1987 or Pelikan 2005). As a result, the Bible is full of internal contradictions that make it impossible for anyone who reads it closely to take it literally, but only makes sense in the context of different sources being blended together. For example, Genesis 1 (largely from the P source) gives the order of creation as plants, animals, man, and woman, but Genesis 2 (from the J source) gives it as man, plants, animals, and woman. According to Genesis 1:3–5, on the first day, God created light, then separated light and darkness, but according to Genesis 1:14–19, the sun (which separates night and day) wasn't created until the fourth day. Genesis 6–7 gives the story of Noah twice, once from the J source and once from the P source, with verses from the two sources intermingled so that they sometimes contradict each other. Genesis 6:5–8 are from the J source, but Genesis 6:9–22 are from the P source. Then Genesis 7:1–5 are from the J source, and Genesis 7:6–24 are alternately from the J and P sources every other line or so (Friedman 1987:54). This leads to many contradictions, such as Genesis 7:2 (from the J source), saying that Noah took seven pairs of each clean beast in the ark, but Genesis 7:8–15 (from the P source), saying he took only one pair of each beast in the ark. In Genesis 7:7, Noah and his family finally enter the ark, and then in Genesis 7:13 they enter it all over again (the first verse from the J source, the second from the P source). According to Genesis 6:4, there were Nephilim (giants) on the earth before the Flood, then Genesis 7:21 says that all creatures other than Noah's family and those on the ark were annihilated—but Numbers 13:33 says there were Nephilim after the Flood.

Many more examples could be cited, but the basic point is clear: the Bible is a composite of multiple sources that did not always agree on details. This was no problem to the ancient Hebrew culture, which used the Bible for inspiration but was not concerned with literal consistency. It is a big problem for modern fundamentalists (most of whom have never read the Bible in the original Hebrew or Greek, so they are in no position to argue) who believe that every word of the Bible is literally true. Most nonfundamentalist Christians, Catholics, Jews, and Muslims have accepted what scholarship has shown us about the origin of the Bible and use it as a book for understanding their relationship to their God but not as a science textbook or literal account of history. As Joseph Campbell and many other later authors have pointed out, these religious stories are important to believers for their meaning and symbolism and connection to the inner mysteries of life, not as detailed literal accounts of events. Only in our modern scientific age, with its obsession with literalism and detail, have fundamentalists made such a gross error about the spirit and meaning of the Scriptures (see papers in Frye 1983).

What Is Creationism?

It not infrequently happens that something about the earth, about the sky, about other elements of this world, about the motion and rotation or even the magnitude and distances of the stars, about definite eclipses of the sun and moon, about the passage of years and seasons, about the nature of animals, of fruits, of stones, and of other such things, may be known with the greatest certainty by reasoning or by experience, even by one who is not a Christian. It is too disgraceful and ruinous, though, and greatly to be avoided, that he [the non-Christian] should hear a Christian speaking so idiotically on these matters, and as if in accord with Christian writings, that he might say that he could scarcely keep from laughing when he saw how totally in error they are. In view of this and in keeping it in mind constantly while dealing with the book of Genesis, I have, insofar as I was able, explained in detail and set forth for consideration the meanings of obscure passages, taking care not to affirm rashly some one meaning to the prejudice of another and perhaps better explanation.

—Saint Augustine, *The Literal Interpretation of Genesis* 1:19–20

The United States is home to a unique and peculiar form of religious extremism known as creationism. As a movement, it has almost no following in Canada, Europe, Asia, or most of the rest of the world, but in America it has had a long influence on science education and public understanding of evolution (fig. 2.1). As a result, most Americans still don't understand or accept the evidence of evolution.

Ironically, the creationist movement is not only a uniquely American phenomenon, but it is also the latest form of backlash against the inevitable forces of change and modernity. For most of the past 2,000 years, people did not question the literal accounts of creation in the first books of Genesis. Even as early as 426 C.E., however, the great Christian philosopher Saint Augustine wrote that the Genesis account of creation was an allegory and should not be interpreted literally, as adherence to a literal reading of Genesis might discredit the faith.

As more and more scientific discoveries were made, some of the literalistic readings of the Bible had to be rethought. Once people accepted that the spherical earth revolved around the sun, it was no longer plausible to think that Joshua had made the sun stand still, that the earth was flat, or that it was the center of the universe, as described in the Bible. By the mid-1700s, enough facts about nature had accumulated that many educated people doubted the literal accounts of the Bible. During the French Enlightenment of the mid-1700s, writers such as Denis Diderot, Voltaire, and Jean-Jacques Rousseau rejected the Roman Catholic Church's dogma, and in 1749, the great naturalist Georges-Louis Leclerc, the count of Buffon, even speculated that the earth was 75,000 years old, that life evolved, and that humans and apes were closely related.

By the early 1800s, the idea that Genesis was a literal account of earth's history was widely questioned by educated people, especially in England, France, and Germany. As a backlash to this widespread skepticism, a number of ministers and naturalists tried to write accounts that reconciled nature with the Bible (the *Bridgewater Treatises*) or tried to use the apparent design and perfection in nature as evidence of a divine designer (natural theology). But literal belief in Genesis was already widely discredited long before Darwin published *On the Origin of Species* in 1859.

Darwin, of course, changed the terms of the debate entirely, polarizing the Western world into those accepting evolution and those rejecting it. At first, the argument was intense, as the shock of Darwin's ideas began to sink in, but by the time Darwin died in 1882, the fact that life had evolved was no longer controversial in any European scientific or intellectual community. Darwin's ideas had become so respectable when he died that he was buried with honors in the Scientists' Corner of Westminster Abbey, right next to Isaac Newton and other famous British scientists.

Most American scholars and scientists also came to terms with Darwin by the 1880s or created their own form of compromise between their own religious beliefs and the idea that life had evolved. For example, in 1880, the editor of one American religious weekly estimated that "perhaps a quarter, perhaps a half of the educated ministers in our leading Evangelical denominations" believed "that the story of the creation and the fall of man, told in Genesis, is no more the record of actual occurrences than is the parable of the Prodigal Son" (Numbers 1992:3). At the same time, a skeptical analytical approach known as "higher criticism" was being applied to the Bible itself, and scholars (especially in Germany) were able to show by careful analysis of the original texts and their language that the Old Testament is a composite of several schools of thought in Hebrew history, not the words of Moses and the Prophets.

Higher criticism alarmed the devout biblical literalists even more than Darwinism and evolution, so in 1878, ministers met in the First Niagara Bible Conference. Beginning in 1895 and concluding by 1910, they had published 90 pamphlets that were known as *The Fundamentals* of their faith (hence the term "fundamentalist"). Most of *The Fundamentals* concerned the miracles of Jesus, his virgin birth, his bodily resurrection, his death on the cross to atone for our sins, and finally, that the Bible is the directly inspired word of God. Fundamentalism was largely a reaction to the higher criticism of the Bible, and its early proponents were not quite as strongly against evolution, because evolution was already widely accepted not only by scientists but also by most ministers. A. C. Dixon, the first editor of *The Fundamentals*, wrote that he felt "a repugnance to the idea that an ape or an orangoutang was my ancestor" but was willing "to accept the humiliating fact, if proved" (Numbers 1992:39). Reuben A. Torrey, who edited the last two volumes of *The Fundamentals*, acknowledged "for purely scientific reasons" that a man could "believe thoroughly in the absolute infallibility of the Bible and still be an evolutionist of a certain type" (Numbers 1991:39). Although the early fundamentalists were not happy with evolution, they were willing to live with it; they were not as stridently opposed to the idea as they would be a generation later. More importantly, evolution was accepted by most of the science textbooks of the time, so even if the parents were fundamentalists who rejected evolution, their children accepted it. Even in the conservative Baptist South, evolution was taught without much resistance in many educational institutions (Numbers 1992:40).

Twentieth-Century Creationism

Congress shall make no law respecting an establishment of religion.
—First Amendment to the U.S. Constitution

"Creation science" . . . is simply not science.
—Judge William Overton, *McLean vs. Arkansas*

The first two decades of the twentieth century were a time of global turmoil, with the progressive politics of Presidents Theodore Roosevelt and Woodrow Wilson, the bloodshed of World War I, and the great influenza epidemic of 1918. Then came the Roaring Twenties and a national conservative backlash. It was also a "return to normalcy" as Warren Harding promised when he won the presidency in 1920. With the conservative backlash came Prohibition. This did nothing to stop alcohol consumption in the United States, but it did make profitable careers for gangsters and moonshiners and the owners of illegal speakeasies. Another conservative movement, however, was the backlash against evolution by the resurgent fundamentalist movement. The movement was led by William Jennings Bryan, one of the most popular and powerful political figures in the United States, who had run for president on the Democratic ticket three times and lost. By the 1920s, however, Bryan was in his sixties, in failing health, and beginning to promote conservative causes that were becoming popular in the 1920s. Bryan campaigned vigorously for laws to outlaw the teaching of evolution. By the end of the 1920s, more than 20 states had debated such laws, and five (Tennessee, Mississippi, Arkansas, Oklahoma, and Florida) had banned or curtailed the teaching of evolution in their public schools. It went so far that the U.S. Senate debated, but never passed, a resolution that banned radio broadcasts favorable to evolution.

Ironically, Bryan himself was not a biblical literalist. He confided to a friend shortly before his death that he had no objection to evolution, as long as it didn't include man (Numbers 1992:43). He was also less than literal about the meaning of Genesis 1, subscribing to the common "day-age" theory that each "day" in Genesis 1 was actually a long period of geologic time, or "age." Nevertheless, he became the national spokesman for a witch hunt that hounded many biologists out of their jobs in Southern universities and destroyed the careers of many other scientists.

The climax of the creationist movement in the 1920s was the infamous Scopes Monkey Trial of 1925, long called the "trial of the century" until the O. J. Simpson trial eclipsed it in notoriety. Not only was it a titanic struggle between two of the giants of the time, Bryan and the legendary defense attorney Clarence Darrow, but it was also one of the first trials to be covered live on radio and in newsreels, beginning the modern trend toward celebrity trial journalism. Among the press covering the trial was none other than the famous satirist and essayist H. L. Mencken, who wrote many savage columns and editorials for the *Baltimore Sun*, ridiculing the biblical literalism and backward habits and racism of the South.

The trial itself was originally planned as a publicity stunt by the town fathers of Dayton, Tennessee. Anxious to rake in tourist dollars, garner attention, and provide a test case to challenge the recently passed Tennessee Butler Act, or "monkey laws," that banned the teaching of evolution, the civic leaders recruited a local high school teacher, John T. Scopes, to be their guinea pig. Scopes volunteered to take time off from teaching gym to teach biology for one day so that he could test the law, although later he admitted that he wasn't sure he had actually taught anything about evolution. He did, however, use the classic textbook, *Hunter's Civic Biology*, which mentioned evolution prominently. Once the trial was underway, Darrow's defense plans collapsed because Judge John T. Raulston would not allow the testimony of any of the expert scientific witnesses whom Darrow had brought. The judge ruled that the case only concerned whether Scopes had broken the law, and witnesses challenging the law itself were irrelevant. In desperation, Darrow turned this defeat into one of the greatest legal tours de force in history. He baited Bryan into taking the stand as an expert witness on the Bible. Under a blistering cross-examination (vividly portrayed in the famous play and movie *Inherit the Wind*), Darrow got Bryan to admit to many of the logical absurdities of a literalistic interpretation of the Bible. Bryan could not explain how Joshua had gotten the sun (and therefore the earth) to stand still or where Cain had gotten his wife (when there were only supposed to be four people on earth, Adam and Eve and Cain and Abel), or the many other problems with a literal interpretation of the Bible. Even more devastating, Bryan admitted under oath that the "days" of Genesis were not 24-hour days but could be long geologic "ages," a revelation that shocked most of his fundamentalist followers. Soon, fundamentalism and the monkey law itself were subject to ridicule. Bryan died a week after the trial, which occurred during a torrid heat wave and aggravated his already failing health. More importantly, the fundamentalist monkey laws had taken a bad beating in the press and in the public eye, and most Americans were embarrassed that our nation had been portrayed as so scientifically backward.

The trial itself, however, was inconclusive. The judge mistakenly levied a $100 fine on Scopes that was supposed to be levied by the jury, so his verdict was thrown out on this technicality. As a result, it was not possible to take the case to higher courts and have a verdict examined on appeal. Scopes never had to pay the judge's fine. Eventually, Scopes went to college and became a successful oil geologist. Meanwhile, the Tennessee monkey law stayed on the books for decades and was not declared unconstitutional until 1968. In that year, Susan Epperson, a young biology teacher in Arkansas, got her case heard before the Supreme Court, who then struck down all laws forbidding the teaching of evolution—43 years after the Scopes trial!

By 1929 the Great Depression had changed the mood of the country, and creationism was no longer in the forefront. Fundamentalists were more concerned about issues like sex education, and no further legal cases challenging the monkey laws were filed, although the old laws remained on the books. Instead, the fundamentalists focused their attention on making sure evolution vanished from the biology textbooks, which it did shortly after the Scopes trial (due to pressure by a few determined creationists on textbook publishers and on local school boards). Creationism and evolution existed in an uneasy truce until the Soviets shook America with the launch of Sputnik in 1957. Then Americans were shocked to discover how far behind our science and technology had fallen, and by 1958, the Republican Congress and Eisenhower administration had begun pouring big money into scientific research and science education. Science also became more and more respected by the American public, especially after the technological advances of World War II, the atomic bomb, and eventually the space race. Federal funding from scientific research went from 0.02 percent of the gross national product during the Hoover administration (1929–1933) to 1.5 percent of the GNP by 1960. With this new emphasis on science came biology textbooks that reflected the new ideas in evolution represented by the neo-Darwinian synthesis of the 1940s and 1950s (see chapter 4). The new generation of science textbook authors was not as cowed by creationist pressure to dilute or eliminate coverage of evolution when the nation was facing a crisis in science education, in large part caused by the lackadaisical coverage of science in public schools.

The newly Darwinized biology textbooks roused the creationists from their inactivity. In 1961, John C. Whitcomb and Henry M. Morris published *The Genesis Flood*, which represented a whole new approach of creationists to not only discredit evolution but also geology (see chapter 3). By 1963, they had founded the Creation Research Society near San Diego, followed by the Institute for Creation Research (ICR), which was the main base of operations for fundamentalist creationists until its founders and leaders all died, and it was eclipsed by other groups and faded into irrelevance. Through their books, debate appearances, and public speeches, they raised awareness of their literalist views to a new level, although they had no impact on the community of science yet.

Creationists, however, still faced one major hurdle: the Constitution and the legal system. By 1968, the Supreme Court had struck down all the old anti-evolutionary "monkey laws," and the creationists no longer had the backing of conservative legislatures as they had in the 1920s. Because they could no longer legally exclude evolution from the classroom, they tried a tactic of demanding equal time for their ideas. However, in court case after court case, they were turned back, because their ideas were clearly religious in origin, with no scientific content, and the Constitution prohibits the government from establishing a state religion or favoring one religion over another. Led by fundamentalist lawyer Wendell Bird, the creationists changed tactics yet again. They began calling their ideas "scientific creationism" and claimed that their ideas were as scientific as evolution and deserved equal time in science classes. Of course, this is simply "bait and switch," because the creationist literature is full of references to God and the Bible. They even published two editions of the same textbook, one of which was labeled "Public School Edition" and deleted the overt references to God and Bible, but otherwise the text was the same.

Their main spokesmen seemed to be talking out of both sides of their mouths. In public, they argued that "creation science" is good science, but when speaking to a religious audience, they let their fundamentalist beliefs show. For example, Henry Morris (1972:preface) writes, "Creation, on the other hand, is a scientific theory which does fit all the facts of true science, as well as God's revelation in the Holy Scriptures." On page 58 of the same book, he writes "we conclude that special creation theory is the best theory, strictly on the scientific merits of the case." Yet the ICR's principal debater and spokesman, Duane Gish, wrote (1973:40) "we cannot discover by scientific investigations anything about the creative processes used by the Creator," and "creation is, of course, unproven and unprovable by the methods of experimental science. Neither can it qualify as a scientific theory" (8).

The climax came when Arkansas and Louisiana passed bills that mandated "equal time" for creationism in science classes, and these laws were promptly challenged in federal court. The American Civil Liberties Union (ACLU), challenging the Arkansas law, put not only distinguished scientists and philosophers of science on the witness stand, but also a group of ministers and theologians, and parents of children in the school district. In fact, the lead plaintiff challenging the law was a minister, the Reverend Bill McLean of Little Rock. The witnesses showed example after example of how there was no difference between "creation science" and religion, and how the nature of science forbids any belief system that twists the facts to fit its preexisting conclusions. The creationist case was further hampered by the fact that they had no credible scientific witnesses to bring to the stand. One of their star witnesses, the maverick British astrophysicist N. Chandra Wickramasinghe, openly scoffed at the idea of creation science. On January 5, 1982, Judge William R. Overton gave his ruling on *McLean vs. Arkansas Board of Education*. Judge Overton saw through the thin disguise of creation science and ruled that the Arkansas law "was simply and purely an effort to introduce the Biblical version of creation into the public school curricula." According to Overton, the law "left no doubt that the major effect of the Act is the advancement of particular religious beliefs." The law that required balanced treatment "lacks legitimate educational value because 'creation science' as defined in that section is simply not science." In 1985, Federal Judge Adrian Duplantier ruled in a summary judgment (thus not requiring even a trial or witnesses) that Louisiana's equal time law was also unconstitutional. In the 1987 *Edwards vs. Aguillard* case, the U.S. Supreme Court, in a 7–2 vote, upheld the decisions of the lower courts, and the creationists lost their last legal battle in this round.

For about ten years, the creationists stayed away from the courts and stopped trying to force their way into education by legal means. Instead, they focused their energies on school boards and textbook publishers. Every week, those of us in the front lines of the creationism battle heard news of another school district that was under pressure to teach creationism or put anti-evolutionary stickers in biology textbooks. Most of these battles were eventually decided against the creationists, but they are a determined and well-funded minority that has nothing but time, energy, and money to push their cause, while most scientists are too busy doing legitimate research to pay attention to the problem.

"Intelligent Design"—or "Breathtaking Inanity"?

> The evidence, so far at least and laws of Nature aside, does not require a Designer. Maybe there is one hiding, maddeningly unwilling to be revealed. But amid much elegance and precision, the details of life and the Universe also exhibit haphazard, jury-rigged arrangements and much poor planning. What shall we make of this: an edifice abandoned early in construction by the architect?
>
> —Carl Sagan, *Pale Blue Dot*

The label "scientific creationist" was seen as a fraud by Judges Overton and Duplantier and by the Supreme Court. So the creationists resorted to a new strategy: "intelligent design" (commonly abbreviated ID). In order to find a way to make their ideas constitutional and legal, they had to try to eliminate any signs of religion from their dogmas, not simply dress up biblical ideas as "scientific creationism." In the 1990s, a new generation of creationists came up with a different strategy that focused on the apparent "design" in nature, arguing that it requires some sort of "intelligent designer." Led by Berkeley lawyer Phillip Johnson, Lehigh biochemist Michael Behe, and former Baylor professor William Dembski, they published a number of books that promoted their views. They argued that nature was full of things that not only showed intelligent design but also were "irreducibly complex" and could not have evolved by chance. They pointed to a number of examples, such as the flagellum and the eye, which they believed could not be explained by chance events or by gradual evolution.

In most ways, their arguments are recycled from over two centuries ago, when many devout naturalists ascribed to the school of thought known as "natural theology." (Ministers were often naturalists back then because they had lots of time for studying nature as evidence of God's handiwork, and there were no professional scientists.) The most famous advocate of natural theology was the Reverend William Paley, who in 1802 wrote *Natural Theology*, the classic treatment of the subject. His most famous metaphor is the "watchmaker" analogy. If you were to find a watch on a beach, you would immediately recognize that it was "intricately contrived" and infer that it had a maker. To Paley, the "intricate contrivances" of nature were evidence that there was a Divine Watchmaker, namely God.

In its day, the natural theology school of thought was very influential, and Darwin himself knew Paley's book almost by heart. Yet the basic arguments had been discredited even before the time of Paley. In 1779, the Scottish philosopher David Hume published *Dialogues Concerning Natural Religion*, which demolished the whole argument from design. Using dialogues between characters to voice different points of view, Hume puts the standard natural theology arguments in the mouth of a character called Cleanthes, then he tears them down in the words of a skeptic named Philo. Philo notes that pointing to the design in nature is a faulty analogy, because we have no standard to compare our world to, and it is possible to imagine a world much better designed than the one in which we live. Even if we concede that the world looks designed, it does not follow that the designer is the Judaeo-Christian God. It could have been the god of another religion or culture, or the work of a committee of gods, or a juvenile god who makes mistakes. Jews and Christians simply assumed that if there was a Designer, it must be their God, but there is no compelling evidence to show that it wasn't some other god.

More importantly, evidence was already in existence in Hume's and Paley's times that did not reflect well on the Divine Designer. For all the examples of beauty or symmetry in nature, one could also point to many examples where nature is poorly designed or jury-rigged so that it just barely works, or where nature shows astonishing cruelty that does not reflect a caring, compassionate God. Stephen Jay Gould pointed to examples such as *Lampsilis* (fig. 2.2A), the freshwater clam that sticks a brood sac full of eggs

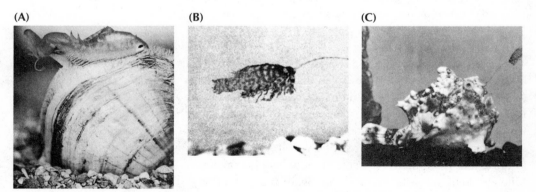

(A)   **(B)**   **(C)**

FIGURE 2.2. Nature is full of examples of jury-rigged adaptations that work just well enough to serve a purpose but are not perfectly designed. (A) The freshwater clam *Lampsilis* has a brood pouch that looks somewhat like a fish and lures fish to bite it. When they do, the clam's larvae then hook onto the fish's gills and complete their life cycle. (Photograph by J. H. Welsh, from the cover of *Science* magazine, v. 134, no. 3472, 1969; copyright ©1969 American Association for the Advancement of Science. Reprinted with permission.) (B and C) The anglerfish has a spine above its mouth with a fringed tip that looks vaguely fishlike. When prey comes near to bite the lure, the anglerfish sucks its victim into its mouth. (Photos from Pietsch and Grobecker, *Science* 201:369–370, 1978; copyright ©1978 American Association for the Advancement of Science. Reprinted with permission.)

out of its shell that looks vaguely like a fish. It's not a very good fishing lure, but it's good enough to get fish to bite it and transfer the eggs to their gills, where they are passed on to another generation. Similarly, the anglerfish (fig. 2.2B and C) has a crude fringe on the tip of a spine above its eyes that looks vaguely fishlike, and when flicked around, is just good enough to lure prey close enough to be gulped down. Again, it's not a very good facsimile of a fish, but it's good enough to lure prey within reach. Gould's favorite example is the panda's "thumb" (fig. 2.3). Pandas, like most cats, dogs, bears, and members of the order Carnivora, have all five fingers united in a paw, yet pandas

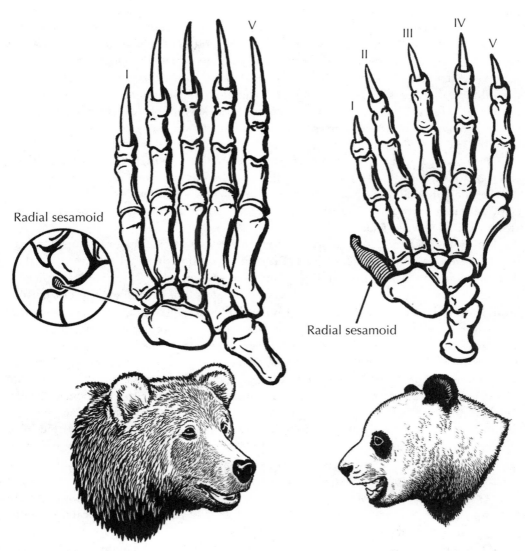

FIGURE 2.3. The panda, like all Carnivora, has all five fingers forming a paw, but unlike other Carnivora, it eats bamboo. Consequently, it has modified a wristbone, the radial sesamoid, into a crude "thumb" that enables it to strip the leaves off bamboo. It works just well enough to feed a panda; it is not beautifully designed, but crude, clumsy, and jury-rigged. (Drawing by Carl Buell)

are almost the only carnivore that eats plants (bamboo). Consequently, pandas have modified a wristbone, the radial sesamoid, into a crude thumb-like device, which is not jointed and not very flexible or strong, but just strong enough to allow pandas to strip off the leaves from the bamboo as they eat. Once again, a clumsy, poorly designed jury-rigged device—good enough to allow the survival of pandas (although we're now driving them to extinction due to habitat destruction in China), but evidence of a very clumsy designer at best.

Examples of poor or at least very puzzling design can be accumulated endlessly. Many cave-dwelling fish and salamanders have the rudiments of eyes but are completely blind. If God specially created these creatures to live in totally dark caves, why bother to give them nonfunctional eyes in the first place? Even more peculiar is the course of the recurrent laryngeal nerve, which connects the brain to the larynx and allows us to speak. In mammals, this nerve avoids the direct route between brain and throat and instead descends into the chest, loops around the aorta near the heart, then returns to the larynx (fig. 2.4). That makes it seven times longer than it needs to be! For an animal like the giraffe, it traverses the entire neck twice, so it is 15 feet long (15 feet of which are unnecessary!). Not only is this design wasteful, but it also makes an animal more susceptible to injury. Of course, the bizarre pathway of this nerve makes perfect sense in evolutionary terms. In fish and early mammal embryos, the precursor of the recurrent laryngeal nerve attached to the sixth gill arch, deep in the neck and body region. Fish still retain this pattern, but during later human embryology, the gill arches are modified into the tissues of our throat region and pharynx. Parts of the old fishlike circulatory system were rearranged, so the aorta (also part of the sixth gill arch) moved back into the chest, taking the recurrent laryngeal nerve (looped around it) backward as well.

In fact, the more one looks at nature, the more one finds examples of clumsy or jury-rigged design because, unlike a Divine Designer, evolution does not require perfection. Any solution that ensures the survival of an organism long enough to breed is sufficient. We humans are classic examples of an organism not optimally designed to our current lifestyles. Our backs and our feet are not well adapted to walking upright, as those of us who suffer with back and foot pain know. Our knees are poorly constructed and easily damaged, as those who have had knee surgery can attest. Our eyes are designed backward, with several layers of cells and tissues blocking and distorting the light hitting the retina in the back of our eye before the light finally reaches the photoreceptor cells on the very bottom layer. We have vestigial organs, such as our tiny tailbones, tonsils, and appendix, the latter two of which no longer perform an important function but can become infected and be deadly to us. These only make sense if they were inherited from ancestors who had functioning versions of these organs. Our genome is full of nonfunctional DNA, including inactive pseudogenes that were active in our ancestors. Humans, like most primates, cannot make vitamin C and must get it from their diet. We still carry all the genes for making vitamin C but no longer use them, probably because our primate ancestors got it from their fruit-rich diets instead. Finally, ask any ID advocate: Why did God give men nonfunctional nipples?

ID creationists may want to think twice before pointing to God's handiwork as evidence of a benevolent God, because it is full of examples of not only poor or incompetent design but also outright cruelty. The most famous example is the family of wasps known as the Ichneumonidae, which consist of about 3,300 species who all reproduce in a distinctive way.

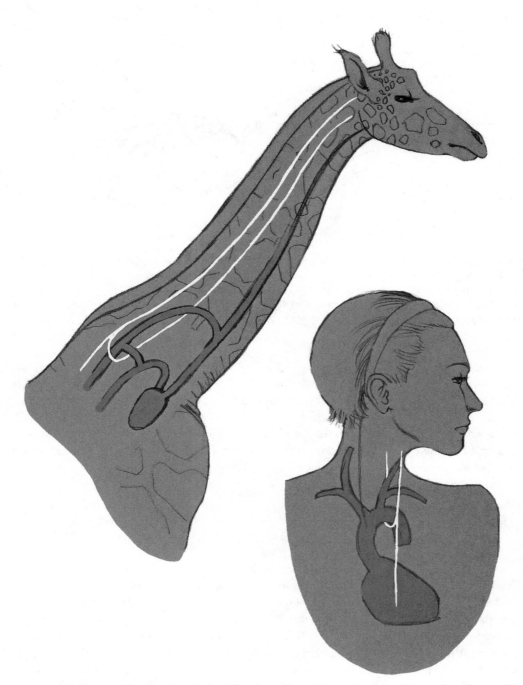

FIGURE 2.4. The recurrent laryngeal nerve branches from the spinal cord and sends nerve impulses to the vocal cords, as well as to other parts of the esophagus (digestive tube) and trachea (breathing tube). During embryology, it is associated with the front part of the gills in a fish, so it loops over the aorta, the artery that supplies most of the blood to be pumped by your heart. But fish don't have necks, so this pathway is short. In fish, it travels from the brain past the heart to the gills. For humans, this same nerve must loop from the spinal column at the base of the head, down through the neck and to the heart, where it still loops around the aorta, then back up through the neck to reach the throat, where it controls your voice box. In a giraffe, the nerve branches out from the spinal column just below the head. It then must travel more than 7 feet down the neck, where it can loop around the aorta, and then 7 more feet back up the neck to reach the throat region and control the vocal cords and other parts of the esophagus and trachea. That's a total of 15 feet of unnecessary length!

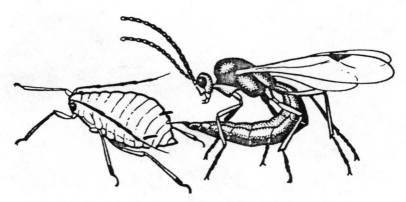

FIGURE 2.5. The cruelty of the reproductive habits of the ichneumonid wasps horrified the Victorians and mocked the idea of a benevolent God. The female wasp stings its prey and paralyzes it; then she inserts her eggs into the body of the living prey. The eggs then hatch into larvae, which eat the prey from the inside, consuming less essential organs first and only killing the victim at the very end—at which point, the prey's body becomes a cocoon containing the baby wasps about to hatch.

The female wasp (fig. 2.5) stings a prey animal with her ovipositor and lays her eggs inside the paralyzed prey. After the eggs hatch, the larvae slowly eat the living prey animal from the inside, destroying the less essential parts first and only eating the essential parts (and killing the host) at the very end, when they are ready to hatch out of its dead shell. (Shades of the creepy extraterrestrials in the movie *Alien*.) The Victorians were horrified when this example became well known and were at a loss as to how to square this fact of nature with their idea of a benevolent God who looks after the tiny sparrow and cares about all of his creation. As Charles Darwin wrote,

> I cannot persuade myself that a beneficent and omnipotent God would have designedly created parasitic wasps with the express intention of their feeding within the living bodies of Caterpillars.

But that has long been a problem for those who would believe in a God who is all-knowing and all-powerful. If so, why does he allow innocents to suffer and die? Why can't he stop great natural disasters? What about the 2004 Indian Ocean tsunami, which killed about a quarter of a million innocent people? This is the classic "problem of pain" (*theodicy*) that has always tortured Christian apologetics, but many skeptics consider it good evidence against a Divine Designer who watches his handiwork closely. As Darwin himself put it (in an 1856 letter to Thomas Henry Huxley): "What a book a devil's chaplain might write on the clumsy, wasteful, blundering low and horridly cruel works of nature!"

Reading the ID creationists closely, you find that they don't offer any new scientific ideas or a true alternative theory of life competing with evolution. All they argue is that some parts of nature seem too complex for them to imagine an evolutionary explanation. This is the classic "god of the gaps" approach: concede to science that which it has already explained but reserve to supernatural forces that which hasn't been explained—yet. Back in the Middle Ages, people thought that God made the heavens run and the stars and planets move until Copernicus, Galileo, Newton, and Johannes Kepler showed that it could all be explained by natural laws and processes without divine intervention. So theology retreated from explaining that part of nature, and it has been retreating ever since. But nature is always full of things that we have not explained. Explaining the unexplained is the goal of science—to continue solving those unsolved mysteries, not to stop and throw up our hands and say, "Oh, well, I can't think of an explanation now, so God must have done it." As Michael Shermer (2005:182) points out, the ID approach is actually quite arrogant: if the ID creationists can't think of a natural explanation, then they are asserting that *no* scientist can either, and the problem cannot be solved. Needless to say, giving up on hypotheses and testable explanations, shrugging our shoulders, and going home while saying "God works in mysterious ways" is not how science operates.

ID creationists actually concede that they don't have a real alternative theory to evolution. Leading ID creationist Paul Nelson said at a meeting at Biola College in Los Angeles in 2004: "Easily the biggest challenge facing the ID community is to develop a full-fledged theory of biological design. We don't have such a theory right now, and that's a problem. Without a theory, it's very hard to know where to direct your research focus. Right now, we've got a bag of powerful intuitions, and a handful of notions such as 'irreducible complexity' and 'specified complexity'—but, as yet, no general theory of biological design." Nor is their "research program" legitimate. During cross-examination in an ID creationism trial in Dover, Pennsylvania, Behe was forced to confess, "There are no peer-reviewed articles by anyone advocating for intelligent design supported by pertinent experiments or calculations which provide detailed rigorous accounts of how intelligent design of any biological systems occurred." Behe also conceded that there were no peer-reviewed articles supporting some of the other claims that systems (such as the blood-clotting cascade, the immune system, and the bacterial flagellum) were irreducibly complex or intelligently designed. Their entire literature consists of books and articles published by their own supporters or for the general trade book market, where there are no scientific standards of peer review. (The one exception I'm aware of is discussed later in this chapter.)

But all this talk about intelligent design is actually a smokescreen for what is still fundamentally a religious dogma. For public consumption, the ID advocates may say that the designer need not be the Judaeo-Christian God but could also be an alien or some other supernatural entity. Dembski claims that "scientific creationism has prior religious commitments whereas intelligent design does not." But in reality, the ID creationists are all evangelical Christians, who clearly have used intelligent design as a smokescreen (fig. 2.1) for their real agenda: get religion into science classrooms and evolution out—or weaken it, at least. In public, they try to hide these religious convictions, but when speaking to their fellow fundamentalists, they let their true colors show. In an article in the Christian magazine *Touchstone*, Dembski wrote, "Intelligent design is just the Logos theology of John's Gospel restated in the idiom of information theory." In 1999, Dembski wrote, "Any view of the sciences that leaves Christ out of the picture must be seen as fundamentally deficient. . . . The conceptual soundness of a scientific theory cannot be maintained apart from Christ." On February 6, 2000, Dembski told the National Religious Broadcasters: "Intelligent Design opens the whole possibility of us being created in the image of a benevolent God. . . . The job of apologetics is to clear the ground, to clear obstacles that prevent people from coming to the knowledge of Christ. . . . And if there's anything that I think has blocked the growth of Christ as the free reign of the Spirit and people accepting the Scripture and Jesus Christ, it is the Darwinian naturalistic view." At the same conference, Phillip Johnson said, "Christians in the twentieth century have been playing defense. They've been fighting a defensive war to defend what they have, to defend as much of it as they can. It never turns the tide. What we're trying to do is something entirely different. We're trying to go into enemy territory, their very center, and blow up the ammunition dump. What is their ammunition dump in this metaphor? It is their version of creation." In 1996, Johnson said, "This isn't really, and never has been, a debate about science. . . . It's about religion and philosophy." One of the ID creationist authors, Jonathan Wells, is a follower of the Reverend Sun-Myung Moon and his Unification Church (which is vehemently anti-evolutionary). As Wells wrote, "When Father chose me (along with about a dozen other seminary graduates) to enter a Ph.D. program in 1978, I welcomed the opportunity to prepare myself for battle."

Ironically, most ID creationists accept some microevolutionary change and conventional geology and the great age of the earth, and regard the "young-earth" literalist creationists of the ICR or Answers in Genesis as irrelevant dinosaurs, relics of the past. In 2005, Dembski actually debated the dean of the old-guard creationists, Henry Morris, where Dembski said, "Thus, in its relation to Christianity, intelligent design should be viewed as a ground-clearing operation that gets rid of the intellectual rubbish that for generations has kept Christianity from receiving serious consideration."

Even though the ID creationists pretend to be dispassionately following the truth, when you look closely at their internal documents, it is clear that they are waging outright warfare on science by whatever dirty tactics and PR techniques that are necessary. Brown and Alston (2007) in their book *Flock of Dodos: Behind Modern Creationism, Intelligent Design, and the Easter Bunny* detail some of the more dishonest activities of the Discovery Institute and print in full the infamous "Wedge Document" of the ID creationists, which details their devious political and PR strategy to force their viewpoints on the American scientific community and educational system. (Although ID creationists tried to hide it, the Wedge Document is easy to find online with a simple search. Nothing ever vanishes in cyberspace). As Brown and Alston summarize, the Discovery Institute "is willing to mischaracterize the results achieved by real scientists in order to achieve short-lived propaganda victories, and it is willing to continue to do so even after these real scientists object and even after it has apologized and promised to stop doing so. Above all, it is willing to cloak its true socio-political goals behind a consciously-crafted veil of dispassionate scientific inquiry, even while denouncing science itself. If the Discovery Institute tells a lie, it does so in order to advance the Truth. Because the Discovery Institute fights for morality, it is above morality. Indeed, the intent of the Discovery Institute is simple enough. Con men are rarely complicated" (136–137).

If the words of the ID creationists were not evidence enough, we can always heed the warning of "Deep Throat" (in *All the President's Men*): "Follow the money." The ID movement is largely based at the Center for the Renewal of Science and Culture (CRC), a part of the Discovery Institute in Seattle. The CRC receives most of its funding from right-wing evangelical and religious organizations and from rich individuals and foundations whose expressed goals are to promote evangelical Christianity. These include $750,000 from the Ahmanson Foundation, whose executor, Howard Ahmanson Jr., said that his goal was "the total integration of biblical law into our lives." The MacClellan Foundation gave $450,000 to promote "the infallibility of the Scripture"; they give grants to organizations "committed to furthering the Kingdom of Christ." The Stewardship Foundation gives $200,000 a year, and their goal is "to contribute to the propagation of the Christian Gospel by evangelical and missionary work." Most of the 22 organizations funding the CRC were politically and religiously conservative, according to the *New York Times*. The *Times* also reported that the CRC received $4.1 million in 2003 and spends about $50,000 to $60,000 a year on about 50 researchers. That money buys a lot of airtime on radio and TV and provides funding for their advocates to speak and debate around the country and publish their books, get hearings at various conservative school boards—and file lawsuits promoting their ideas.

By 2005, ID creationism had reached a peak in publicity, when they made the cover of *Time* magazine and received the endorsement of President George W. Bush. They also got their ideas heard by the conservative Kansas State Board of Education (which endorsed them) and tried to push their ideas onto the Dover, Pennsylvania, school board (which also tried to follow them, until sued by the parents in the district). In December 2005, however, they suffered a great defeat in the case of *Kitzmiller et alia vs. the Dover Area School District*. Federal Judge John E. Jones III, a traditional Christian appointed by President Bush in 2002 (and *not* a liberal activist judge), saw through their smokescreen and ruled that ID creationism was clearly an unconstitutional establishment of a particular religion in public schools. His 139-page ruling was very thorough and detailed. Judge Jones castigated the evangelical Christian school board that rammed through the creationist curriculum and drew the lawsuit by the parents of the students. In his words, "The breathtaking inanity of the board's decision is evident when considered against the factual backdrop which has now been fully revealed through the trial. The students, parents, and teachers of the Dover Area School District deserved better than to be dragged into this legal maelstrom, with its resulting utter waste of monetary and personal resources." Judge Jones was particularly irritated by the hypocrisy of the ID creationists, who attempted to sound secular when the Constitution was involved, but crowed about their religious motives when not in court. "The citizens of the Dover area were poorly served by the members of the board who voted for the intelligent design policy. It is ironic that several of these individuals who so staunchly and proudly touted their religious convictions in public would time and again lie to cover their tracks and disguise the real purpose behind the intelligent design policy." And in another passage, "We find that the secular purposes claimed by the board amount to a pretext for the board's real purpose, which was to promote religion in the public school classroom." Still later he wrote, "Any asserted secular purposes by the board are a sham and are merely secondary to a religious objective."

The judge also pointed out that the ID advocates had tried to paint evolution as atheism, which was preposterous. "Both defendants and many of the leading proponents of intelligent design make a bedrock assumption which is utterly false. Their presupposition is that evolutionary theory is antithetical to a belief in the existence of a supreme being and to religion in general." Finally, the judge was puzzled by the fact that there was no real theory behind intelligent design creationism, just criticisms of evolutionary biology and a vague god of the gaps idea. In Judge Jones's words, "Defendants' asserted secular purpose of improving science education is belied by the fact that most if not all of the board members who voted in favor of the biology curriculum change concealed that they still do not know, or nor have they ever known, precisely what intelligent design is."

Brown and Alston (2007) dissect the absurdities of the Dover trial. Their first chapter gives a detailed account of the trial and uses the court transcripts and the creationists' own words to expose their lying and dishonesty. Brown and Alston quote extensively from the confused and convoluted testimony of William Buckingham, the creationist school board chairman who openly promoted his religious motivations before trial, then lied under oath repeatedly to cover his tracks, apparently at the instructions of the lawyers from the Discovery Institute. As Brown and Alston sum it up, "To know William Buckingham is to know the millions of our fellow Americans who are ignorant not only of the theory they'd like to discredit, but also of the pseudo-theory with which they'd like to replace it; who, knowing full well that they lack the basic data to make a decision between the two, do so anyway and loudly at that; and who lie through their teeth when asked exactly what it is that motivates them to do these sorts of things in the first place. William Buckingham lied because he believed it was necessary to do so in order to preserve the truth as he saw it—that literalized Christianity is the one true religion, and that Darwinism is its greatest threat" (26).

During the course of the trial, the witnesses for ID creationism were repeatedly exposed for their bad science, and the documents that were introduced show the clear imprint of having been recycled from older creationist documents. The most striking evidence of this was the discovery of different editions of the textbook *Of Pandas and People* (Davis and Kenyon 2004). Early drafts were full of conventional creationism, but when a federal case struck down young-earth creationism for the final time, the authors just cut and pasted a few phrases here and there to remove the references to "God," "creation," and "creationism." In one place, the plaintiffs' legal team exposed a tell-tale palimpsest: the phrase "cdesign proponentsists," where the phrase "design proponents" has been clumsily and incompletely pasted over the word "creationists."

At the time of Judge Jones's decision, analysts thought that the Dover verdict would be the death knell for future attempts by ID creationists to win victories by legal means, because in most cases courts follow precedents established by other courts (especially if they are thorough and well-reasoned). But ID creationism was extinguished even more completely than anyone could have imagined. Even though the Discovery Institute keeps pushing it, no school district has tried to adopt any ID school materials in the 11 years since the original *Dover* decision. Even more surprising, the entire notion of ID creationism has abruptly died from American discourse. A simple Google Trends search of terms like "intelligent design" done by Nick Matzke shows that it virtually vanished from the Internet after 2006, with almost no real discussion since then. "Intelligent design" creationism is really and truly dead.

Since the *Dover* decision, the Discovery Institute refused to accept this, and keeps on cranking out books and literature and propaganda. As Nick Matzke (2015) wrote:

> Of course, the Discovery Institute is still around, still desperately trying to re-write history, claiming that they never supported teaching ID in public schools (when they clearly did, as even the Thomas More Law Center noted), that they never supported what the school board in Dover was doing (never mind that it was the DI's care package of ID materials, particularly Icons of Evolution stuff, that ginned up the school board in the first place, which was exactly the intent of all of the emotional language about "fraud" etc. in Icons), that the Dover Area School Board was a bad place for a test case because of obvious religious motivations (never mind that ID is and always has been mostly a wing of apologetics for conservative evangelicals, and in fact that audience is still the only one where ID events, books, etc. have much of an audience today), and that ID isn't creationism relabeled.

Now the Discovery Institute is trying to put on a brave face and claim that "intelligent design" is alive and well. On the tenth anniversary of the *Dover* decision, their site had a long post bragging about their "accomplishments" in the past ten years that is a monument to special pleading and selective misuse of facts. They mostly brag about how their lawyers have won nuisance suits against real scientists and real museums who crossed them, and about their books (mostly by Stephen Meyer, one of their leading authors), which have been roundly criticized and mostly ignored by the real scientific community for their scientific incompetence, dishonesty, and outright deception.

And what about their actual scientific research program? Back in the late 1990s when the Discovery Institute was founded, their Wedge Document proposed to get 100 scientific articles published in the next ten years. Now, almost 20 years later, their latest post touts their "80 peer-reviewed publications" as if it's some great accomplishment. Most productive individual scientists have at least that many papers, and the Discovery Institute is a giant propaganda mill with many contributors. In fact, I have more than 300 peer-reviewed publications, almost four times their total, all by myself. Moreover, if you look through the list in the Discovery Institute post, it is almost entirely papers for their own house journal *BIO-Complexity*, or unreviewed online fringe sites like the *Journal of Cosmology*, or some other predatory online journals that will publish anything for a fee. Only one or two articles are found in reputable journals, and the titles of those articles indicate that their content isn't really about ID at all.

Ironically, the point is largely moot in Dover, Pennsylvania, because in November 2005, the citizens of Dover (embarrassed by all the negative publicity) voted the conservatives off the school board and voted in a new school board that was opposed to teaching intelligent design in its schools. Naturally, this new school board did not wish to appeal the judge's ruling, but applauded it—but they were still stuck with the legal bills that the folly of the old school board had generated.

The Monkey Business of Creationism

Creation is, of course, unproven and unprovable by the methods of experimental science. Neither can it qualify as a scientific theory.
—Duane Gish, "Creation, Evolution, and the Historical Evidence"

Creation isn't a theory. The fact that God created the universe is not a theory—it's true. However, some of the details are not specifically nailed down in Scripture. Some issues—such as creation, a global Flood, and a young age for the earth—are determined by Scripture, so they are not theories. My understanding from Scripture is that the universe is in the order of 6,000 years old. Once that has been determined by Scripture, it is a starting point that we build theories upon.
—Kurt Wise, 1995

Ultimately, creationism has nothing to do with science, except that the creationists want to replace a valid scientific idea with their own religious dogmas. It is all about politics and power and promoting their cherished ideas whatever the cost. Creationists don't do normal science, don't publish their anti-evolutionary ideas in peer-reviewed scientific journals, don't present their results at legitimate scientific meetings, and more importantly, don't even begin to follow the basic precept of science: *there is no final truth, and all ideas must be subject to testing and falsification.* Creationists have their conclusions already determined. The ICR even makes their members swear a loyalty oath that predetermines their conclusions. No real scientist would ever do this, since in real science, the conclusions must remain tentative and subject to change. Creationists will do whatever it takes to twist and jumble and distort the evidence to support their case. Indeed, the term "creation science" is an oxymoron, a contradiction in terms, like "jumbo shrimp." Creationists are not really doing science as long as their conclusions are predetermined and they are unwilling to test and falsify their conclusions. The quotes from Duane Gish above and cited earlier in the chapter clearly confess this.

As Shermer (1997:131) pointed out, the creationists have much in common with the neo-Nazi Jew-hating Holocaust deniers, who refuse to acknowledge that millions of Jews were killed by the Nazis. Like the creationists, Holocaust deniers pretend to be legitimate objective scholars and deny their underlying motives in public, but in private they reveal their true anti-Semitic hatred that drives them to distort and deny the truth. The principal strategy of Holocaust deniers is to find small errors in the scholarship of historians (or scientists, in the case of creationists) and imply the entire field is wrong, as if scholars never disagreed or made mistakes. Holocaust deniers often quote other people out of context (Nazis, Jews, other Holocaust scholars) to make them seem as if they are supporting the deniers' position; creationists do the same to evolutionists' publications. The existence of a debate over details is used by the Holocaust deniers to suggest that the Holocaust didn't happen, or the scholars can't get their stories straight, and creationists do the same to the legitimate scientific debate among evolutionary biologists. However, as Shermer says, the Holocaust deniers can at least be partially right in that the number of Jews killed may be revised, but the creationists cannot. Once you introduce supernaturalism to the debate, it is no longer scientific.

Because they have lost every battle in the courts, creationists resort to other tactics: pressuring school boards, intimidating textbook publishers, harassing people who oppose them, and disguising their religious motives by such flimsy ruses as intelligent design. Because their unscientific ideas could never pass peer review in scientific journals or make it into a university curriculum, they publish their own books and journals, and create their own educational institutions to reflect their dogmas. Because their ideas would not withstand the scrutiny of peer review in scientific meetings, they seldom attend real scientific meetings but preach to the choir instead.

The exceptions to these statements prove the rule. In August 2005, Stephen Meyer's ID creationist article on the "Cambrian explosion" appeared in the obscure *Journal of the Biological Society of Washington.* According to reports, the peer reviews were scathing and recommended rejection of the article, but the editor, Richard Sternberg, had creationist sympathies and let it be published anyway. Once the rest of the editorial board and the Smithsonian scientists became aware of what had been slipped past them, they repudiated the article, and the editor resigned. To my knowledge, this is the only openly creationist paper that has ever appeared in a legitimate scientific peer-reviewed journal—and only because the editor was sympathetic to their cause and violated journal policy by overruling his reviewers. The other papers published by Fritz, Baumgardner, Austin, and creationist "flood geologists" don't appear in peer-reviewed scientific journals, only in the creationists' own publications. Any of their writings that do appear in a legitimate peer-reviewed journal concern some minor issue (such as the polystrate trees of Yellowstone or the fossil concentrations in some places), and nowhere do the authors reveal their creationist agenda in the research.

Likewise, creationists do not present their arguments at legitimate professional meetings of respected scientific societies; they use stealth tactics instead. Forrest and Gross (2004) describe the sneaky efforts of an ID creationist, Paul Chien, to organize a conference in China, ostensibly about the amazing Precambrian and Cambrian fossils that have been found there. But respected scientists, such as Dr. David Bottjer of the University of Southern California and Dr. Nigel Hughes of University of California–Riverside, arrived only to find that the meeting was funded by the Discovery Institute and full of ID creationist speakers. The whole conference was a deliberate ruse to get the papers of legitimate scientists published alongside those of creationists and to lend creationists some respectability.

Whenever they do try to engage the scientific community, creationists do so through a debate format. At first, this seems like a fair strategy, because for many fields we have a long tradition of using debate to explore evidence and clarify ideas. But in fact, the debate format does nothing to sort out the evolution/creation dispute, except to resolve who has better rhetorical and debating skills. Creationists are very skilled at this, since they do it all the time and have a lot of practice. By contrast, scientists never actually engage in a true formal debate (complete with pro and con positions, moderator, rebuttals, etc.) at scientific meetings. In addition, creationists dictate the terms of the debate by constantly attacking their evolutionist opponents with one charge after another, jumping from astronomy to thermodynamics to paleontology to biology to anthropology. This is known as the "Gish Gallop," after Duane Gish, who was a master of this tactic. The scientists opposing the creationist debaters cannot possibly answer all of the misconceptions and distortions of complex concepts that they have introduced in the short debate format, because they can't teach an audience the actual science as fast as the creationists can distort it. When the evolutionist debater tries to go on the offensive, the creationist quickly dodges the question and continually tries to make his or her (mostly religious) audience believe that they must believe the creationist or become an atheist. When a scientist with good debating skills (especially one with religious convictions who can't be called an atheist) pins them down, creationists crumble, because their knowledge of scientific subjects is superficial and learned by rote, so they really don't understand what they are talking about. But their skill in debating is such that they seldom get pinned down or rattled for very long. Most scientists won't even bother to debate them, because it's a no-win situation; everyone who attends the debate has already made up their minds. In addition, most scientists are poorly trained at debating, and we don't want to treat creationists as scientific peers (they aren't) and dignify their arguments with the pretense of a debate. Plus, we all have much better things to do, such as real scientific research. Consequently, creationists taunt scientists and claim they are afraid to defend evolution.

Stephen Jay Gould said it best,

> Debate is an art form. It is about the winning of arguments. It is not about the discovery of truth. There are certain rules and procedures to debate that really have nothing to do with establishing fact—which they are very good at. Some of those rules are: never say anything positive about your own position because it can be attacked, but chip away at what appear to be the weaknesses in your opponent's position. They are good at that. I don't think I could beat the creationists at debate. I can tie them. But in courtrooms they are terrible, because in courtrooms you cannot give speeches. In a courtroom you have to answer direct questions about the positive status of your belief. We destroyed them in Arkansas. On the second day of the two-week trial, we had our victory party. (from a 1985 Caltech lecture, quoted in Shermer 1997:153)

My own experience with debating creationists was truly eye-opening. In October 1983, I was asked by the late Stanley Weinberg, head of the Committees of Correspondence (predecessor of the National Center for Science Education, which battles creationism) to represent down-state Illinois for the committee. I had responsibility for fighting the creationists over the entire state except Chicago, all while I was doing research and teaching a full load of geology classes at Knox College in Galesburg, Illinois. Luckily, for most of my term it was "all quiet on the downstate front," except when ICR's top debater Duane Gish made a tour through the area. He was invited to a debate at Purdue University, just across the Illinois border in West Lafayette, Indiana. When no faculty at Purdue would take him on, I agreed to their invitation, just for the experience and to say I'd done it once.

First, though, I talked with people who had debated (and beaten him) before. A week before the debate, I went to see Gish give an unopposed lecture at the University of Illinois in Champaign–Urbana. At an open lecture with no opponent, he gives his standard canned speech and slides. Since attendance was free, it drew a crowd of hundreds of bright, skeptical college students who hissed and booed and heckled him regularly at his outrageous misstate-ments and distortions. For me, it was valuable, because I saw his slides and talk in advance, and I'd been told that he was a robot. He never changes a line of his memorized script or the slide sequence, never acknowledges his opponent, and never realizes that his arguments have just been demolished. (In 1995, I saw Gish debate again, and his patter had not changed in 12 years, except that the slides looked noticeably faded.) I knew that I was going to get the first and third half-hours of our first two hours (in a four-hour debate!), so I prepared to attack and discredit his positions before he even got to them. Sure enough, he never noticed that I had done so and did not change a line of his standard litany, even though its credibility was already demolished. I was even tempted to steal his inevitable lame jokes and use them first to see if he noticed and was forced to change. I realized, however, that putting a picture of a chimpanzee on the screen and saying (as he does in every lecture), "How did that picture of my grandson get in there?" would make no sense coming from a 29-year-old unmarried scientist.

More startling than this, however, was the behavior of the creationists who organized the debate. They bused in hundreds of people from all the nearby churches, yet discouraged others from attending by charging heavily for Purdue students, who have a lot better (and cheaper) things to do on campus on a Saturday night. As a result, the audience was already 95 percent creationists when I arrived after driving for hours with my five students from Illinois. During introductions, the organizers had the gall to say, "We couldn't get Stephen Jay Gould or Niles Eldredge to debate Dr. Gish, so we got . . ." implying that I was some unknown sucker from the boonies of Illinois. Of course, this is their way of taunting evolu-tionists by suggesting that Gould and Eldredge are afraid of Gish, but they never considered the possibility that it was an insult to me as a scientist. Then, after I had demolished Gish during the first two hours (many people came up to me and said so, and said they had been converted from creationism), they went to a question-and-answer format. After the break, they handed me a stack of 3 × 5 inch index cards with questions from the audience, and they put questions like "What are your religious beliefs?" and "Are you a sexual pervert?" and "Are you going to hell?" at the top. Clearly, the debate was never about science from the beginning, and the mostly fundamentalist audience didn't come to hear about science, but only to cheer their champion and to pity the soul of the poor damned evolutionist. Gish, of course, shuffled the cards until he found sympathetic questions, which he used to add points that he had not discussed in the first two hours.

Even though I beat Gish badly, I decided not to debate him again because it was a waste of time preaching to the converted, and I didn't want to continually dignify their position. But in 2002, I was invited to take part in a panel debating evolution on public television in Los Angeles. Two creationists from ICR were the opponents, and I spent the entire debate canceling out their outrageous lies and distortions. Luckily, they made the mistake of mentioning the fossil record, so I had a huge advantage over them. I'd learned from past experience never to let their misstatements go unchallenged, so I interrupted and cut them off as soon as they said their lies. My debating partner was a lawyer from the ACLU, and he ended up winning the debate by calmly pointing out again and again the simple fact that creationism is religion and forbidden from public school science classes by the Constitution.

What startles you most about the creationist debaters is that they never learn anything new or come up with different arguments. They repeat the same old, tired lines like a mantra, seemingly unaware or unwilling to admit that their argument has long been discredited. For example, nearly every creationist debater will mention the Second Law of Thermodynamics and argue that complex systems like the earth and life cannot evolve, because the Second Law seems to say that everything in nature is running down and losing energy, not getting more complex. But that's *not* what the Second Law says; every creationist has heard this but refuses to acknowledge it. The Second Law *only applies to closed systems*, like a sealed jar of heated gases that gradually cools down and loses energy. But the earth is *not* a closed system—it continually gets new energy from the sun, and this (through photosynthesis) is what powers life and makes it possible for life to become more complex and evolve. It seems odd that the creationists continue to misuse the Second Law of Thermodynamics when they have been corrected over and over again, but the reason is simple: it sounds impressive to their audience with its limited science education, and if a snow job works, you stay with it.

Gish was particularly dishonest in this regard. If he is beaten in a debate in one city and forced to admit that an argument is not true, he will still use the same invalid argument the next night in front of a different audience, since they didn't see him recant the argument the previous night (Arthur 1996; Petto 2005). Gish has been repeatedly caught in lies and deliberate deceptions (Arthur 1996; see www.holysmoke.org/gish.html), yet he refuses to change even one line of his deceptive and discredited ideas. How honest or truthful can a debater be if he cynically uses an argument he knows has been proven wrong on the next unsuspecting audience?

Likewise, creationists will use spurious arguments from probability to claim that evolution cannot occur, again counting on their followers to be impressed with math and statistics. Creationists will cite the improbability that a monkey could type the works of Shakespeare as an analogy for evolution building complex systems by random chance. Gish's favorite analogy (stolen from maverick astronomer Fred Hoyle) concerns the improbability that a hurricane blowing through a junkyard could assemble a 707. But these analogies are completely off base. Evolution is not "random chance" but a process whereby natural selection weeds out unfavorable variations and greatly improves the likelihood of events. A better analogy is a monkey with a word processor, whose program (like your spell-checker) automatically deletes or fixes mistakes, so that even by typing random keys, the monkey will eventually assemble a recognizable string of words. Richard Dawkins (1986, 1996) has provided many interesting examples and computer models that show just how easily this can be done.

Besides, as anyone who really understands the mathematics of probability knows, you can't make this kind of argument after the fact. If you do so, then *any complex sequence of events is extremely improbable, even though they actually occur.* A good analogy is the one I used in the Gish debate. I asked the audience of several hundred to estimate the probability *after the fact* that all of the events that had happened in their lives would actually happen, and the probability that among all those unlikely events, they would all end up in this room at this particular moment. Naturally, the improbability of this event is enormous. I pointed out to the audience that by Gish's probability arguments, they could not exist!

Most of the standard shopworn creationist arguments are debunked elsewhere in this book, so I will not discuss them here. Suffice it to say that if real scientists never learned anything new in 50 years of research and never changed their position once they'd been proven wrong, they wouldn't last long in the scientific community.

For Further Reading

Alters, B., and S. Alters. 2001. *Defending Evolution*. Sudbury, Mass.: Jones and Bartlett.

Berra, T. 1990. *Evolution and the Myth of Creationism*. Stanford, Calif.: Stanford University Press.

Brockman, J., ed. 2006. *Intelligent Thought: Science Versus the Intelligent Design Movement*. New York: Vintage.

Brown, B., and J. P. Alston. 2007. *Flock of Dodos: Behind Modern Creationism, Intelligent Design, and the Easter Bunny*. Cambridge, U.K.: Cambridge House.

Eldredge, N. 1982. *The Monkey Business: A Scientist Looks at Creationism*. New York: Pocket Books.

Eldredge, N. 2000. *The Triumph of Evolution and the Failure of Creationism*. New York: Freeman.

Forrest, B., and P. R. Gross. 2004. *Creationism's Trojan Horse: The Wedge of Intelligent Design*. Oxford: Oxford University Press.

Franz, M.-L. von. 1972. *Creation Myths*. Zurich: Spring.

Friedman, R. 1987. *Who Wrote the Bible?* New York: Harper & Row.

Frye, R. M., ed. 1983. *Is God a Creationist? The Religious Case Against Creation-Science*. New York: Scribner.

Futuyma, D. 1983. *Science on Trial: The Case for Evolution*. New York: Pantheon. Godfrey, Laurie, ed. 1983. *Scientists Confront Creationism*. New York: Norton.

Graves, R., and R. Patai. 1963. *Hebrew Myths: The Book of Genesis*. New York: McGraw-Hill.

Heidel, A. 1942. *The Babylonian Genesis*. Chicago: University of Chicago Press.

Heidel, A. 1946. *The Gilgamesh Epic and Old Testament Parallels*. Chicago: University of Chicago Press.

Humes, E. 2007. *Monkey Girl: Evolution, Education, Religion, and the Battle for America's Soul*. New York: Ecco.

Isaak, M. 2006. *The Counter-Creationism Handbook*. Berkeley: University of California Press.

Kitcher, P. 1982. *Abusing Science: The Case Against Creationism*. Cambridge: MIT Press.

Larson, E. 1985. *Trial and Error: The American Controversy Over Creation and Evolution*. New York: Oxford University Press.

Matzke, N. 2015. "Kitzmas Is Coming!" Panda's Thumb (blog), https://pandasthumb.org/archives/2015/12/kitzmas-is-comi.html.

McGowan, C. 1984. *In the Beginning: A Scientist Shows Why the Creationists Are Wrong*. Buffalo, N.Y.: Prometheus.

Miller, K. 1999. *Finding Darwin's God: A Scientist's Search for Common Ground Between God and Evolution*. New York: HarperCollins.

Numbers, R. 1992. *The Creationists: The Evolution of Scientific Creationism*. New York: Knopf.

Olasky, M., and J. Perry. 2005. *Monkey Business: The True Story of the Scopes Trial*. New York: B&H.

Pelikan, J. 2005. *Whose Bible Is It? A History of the Scriptures Through the Ages*. New York: Viking.

Pennock, R. 1999. *Tower of Babel: The Evidence Against the New Creationism*. Cambridge: MIT Press.

Perakh, M. 2004. *Unintelligent Design*. Buffalo, N.Y.: Prometheus.

Pigliucci, M. 2002. *Denying Evolution: Creationism, Scientism, and the Nature of Science*. Sunderland, Mass.: Sinauer.

Ruse, M. 1982. *Darwinism Defended*. New York: Addison-Wesley.

Ruse, M. 1988. *But Is It Science? The Philosophical Questions in the Creation/Evolution Controversy*. Buffalo, N.Y.: Prometheus.

Ruse, M. 2003. *Darwin and Design: Does Evolution Have a Purpose?* Cambridge: Harvard University Press.

Ruse, M. 2005. *The Evolution-Creation Struggle*. Cambridge: Harvard University Press.

Sarna, N. 1966. *Understanding Genesis: The Heritage of Biblical Israel*. New York: Schocken.

Scott, E. C. 2005. *Evolution vs. Creationism: An Introduction*. Berkeley: University of California Press.

Shanks, N. 2004. *God, the Devil, and Darwin: A Critique of Intelligent Design Theory*. Oxford: Oxford University Press.

Shermer, M. 2006. *Why Darwin Matters: Evolution and the Case Against Intelligent Design*. New York: Henry Holt/Times Books.

Shulman, S. 2007. *Undermining Science: Suppression and Distortion in the Bush Administration*. Berkeley: University of California Press.

Smith, C. M., and C. Sullivan. 2007. *The Top Ten Myths About Evolution*. Buffalo, N.Y.: Prometheus.

Smith, H. 1952. *Man and His Gods*. New York: Little, Brown.

Young, M., and T. Edis, eds. 2005. *Why Intelligent Design Fails: A Scientific Critique of the New Creationism*. Piscataway, N.J.: Rutgers University Press.

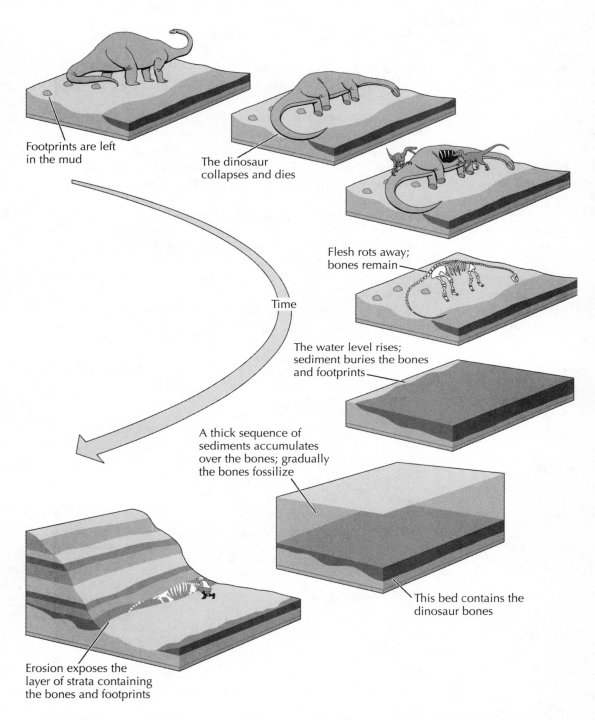

Footprints are left
in the mud

The dinosaur
collapses and dies

Flesh rots away;
bones remain

Time

The water level rises;
sediment buries the bones
and footprints

A thick sequence of
sediments accumulates
over the bones; gradually
the bones fossilize

This bed contains the
dinosaur bones

Erosion exposes the
layer of strata containing
the bones and footprints

FIGURE 3.1. The process of fossilization destroys 99 percent of the bones and shells of most organisms, so less than 1 percent of all the species that have ever lived are preserved as fossils—and then have the great luck to have been spotted in the last 200 years when a paleontologist happens to be out collecting.

THE FOSSIL RECORD

3

Lucky Accidents of Fossilization

Now let us turn to our richest geological museums, and what a paltry display we behold! That our collections are imperfect is admitted by everyone. Many fossil species are known from single and often broken specimens. Only a small portion of the earth has been geologically explored, and no part with sufficient care. Shells and bones decay and disappear when left on the bottom of the sea where sediment is not accumulating. We err when we assume that sediment is being deposited over the whole bed of the sea sufficiently quickly to embed fossil remains. The remains which do become embedded, if in sand or gravel, will, when the beds are upraised generally be dissolved by rainwater charged with carbonic acid.

—Charles Darwin, *On the Origin of Species*

To debunk creationist distortions about fossils, we must start with a clear understanding of the fossil record and the process of fossilization. As discussed in the prologue, the fossil record was embarrassingly incomplete when Darwin published *On the Origin of Species* in 1859, but it soon became one of his strongest lines of evidence. During the twentieth century, our fossil collections were vastly improved and hundreds of evolutionary sequences and transitional forms were documented.

This transformation from an embarrassing fossil record in 1860 to an embarrassment of riches by 1960 represented the hard work of thousands of dedicated paleontologists and geologists. Yet they battle against enormous odds. Creationists often assert that the fossil record is nearly complete and should show the innumerable insensibly graded transitions that Darwin expected in 1859. Yet, even with nearly 200 years of collecting behind us, the fossil record is relatively complete only in certain areas as indicated by the quote from Darwin. Fossilization is still a highly improbable event, and most creatures that have ever lived do not become fossils.

How do we know this? A whole subfield of paleontology, known as *taphonomy* (Greek for "laws of burial"), is dedicated to understanding how and why organisms become fossils (see Prothero 2013a: chap. 1). Consider the chain of events that happen to an organism after it dies (fig. 3.1). First, there are the biological agents (bacteria, fungi, insects, and other decomposers, scavengers, etc.) that break down or destroy an organism after death. The soft parts of animals decay or are eaten quickly, so they almost never fossilize. Only the hard parts, the shell or skeleton, have a reasonable chance of preservation. After an animal dies, its bones are typically scavenged and broken, so few or no remnants of the skeleton may actually survive. Taphonomists have done lots of research in places such as East Africa, where they have observed and documented the details of how hyenas and other scavengers tear up a carcass and break up nearly every bone. When the taphonomists mark and photograph these sites and return a year later to document the changes, even the bones that survive scavenging may have been broken or scattered by trampling or by other agents of destruction.

In the marine realm, there are many agents of destruction as well. Soft-bodied organisms, like sea jellies and marine worms, almost never leave a fossil record. Even organisms with hard parts, like mollusks and corals, are prone to destruction. Waves and currents wash the shells back and forth and pound them into pieces, so only the most durable shells survive. Many shells are broken by predators, such as crabs and lobsters, which use their claws or pincers to crack the shells or peel them open to get at the prey inside. Abandoned shells are degraded by organisms that use them as anchors or places to attach. A whole group of boring organisms, including sponges and algae, drill or dissolve holes in shells and reuse their minerals, thus weakening the shells even further.

Once a bone or shell survives this grueling gantlet, there are still further hazards. After burial, the shell or bone might be dissolved by water percolating through the sediments. Many fossils are actually made of new minerals that have replaced the original minerals, showing how little of the original material remains in the fossil record. As the potential fossil gets buried deeper and deeper, the huge pressure on the pile of sediments above it may distort the fossil or crush it entirely. Many deeply buried sedimentary rocks are actually transformed by high heat and pressure into metamorphic rocks, and then all original fossil traces vanish entirely.

If the shell or bone avoids or survives all these ordeals—dissolution, replacement, distortion, pressure, metamorphism—there are still more hazards ahead. Once the fossil-bearing sediment is uplifted and exposed again, the fossil is prone to erosion. If it weathers out at any time except when a paleontologist happens to wander by (which has only been happening at rare intervals in the past 200 years), then the fossil will be destroyed and lost forever. There are only a few thousand paleontologists in the entire world who can only devote at most a few weeks or months a year to collecting fossils, so most fossiliferous exposures go unexamined and their fossils are lost. If you think hard about it, the odds that any given organism will be fossilized and actually end up in our collections is minuscule. It is a miracle that we have any fossil record at all.

There are other ways to estimate the quality of our fossil record (see Prothero 2013a:22). At this moment, biologists know and have described and named about 1.5 million species on earth (mostly insects), and some estimates say that the earth harbors at least 4 or 5 million species in total. Yet there are at best only about 250,000 known species of fossil animals and plants, or about 5 percent of the species living today. But today is only one time slice among millions in the past 600 million years during which multicellular life has existed. If we total up all those time slices, then the total number of species that are represented in the fossil record is a tiny fraction of 1 percent.

Consequently, the fossil record of some groups that are entirely soft-bodied without hard skeletons or shells (especially insects, worms, sea jellies, and the like) is so poor that most paleontologists do not study them much and do not attempt to say much about their evolution. In certain groups with hard skeletons, however, the potential for preservation is much higher. If we focus just on groups with excellent skeletons and a good chance for preservation (including microfossils, sponges, corals, mollusks, sea stars and sea urchins and their relatives, trilobites, the "lamp shells" or brachiopods, and "moss animals" or bryozoans), the fossil record is not nearly so incomplete. These groups have about 150,000 living species but more than 180,000 fossil species. Depending on how you do the calculation, between 2 and 13 percent of all the species that have ever lived in these groups may be fossilized. That's still not great but much better than the fraction of 1 percent estimate we just discussed.

In some places, the record of fossil shells is very dense and continuous (fig. 3.2), and these are places where paleontologists focus their attention in studying things like evolution. They know that not every species is preserved, of course, but they have enough data to see how evolution occurs in the groups that do fossilize.

(A)

(B)

(C)

FIGURE 3.2. In some places the fossil record is very dense and continuous, with extraordinary numbers of specimens. (A and B) The cliffs of Chesapeake Bay in Maryland are legendary for their incredibly dense shell beds. (Photos courtesy S. Kidwell) (C) The famous Pliocene Leisey shell beds in central Florida, with thousands of exquisitely preserved fossil shells. (Photo courtesy Warren D. Allmon, Paleontological Research Institution)

Most single-celled organisms, like amoebas and paramecia, are soft-bodied and never fossilize. But a few groups, such as the amoeba-like foraminiferans and radiolarians, have beautiful shells made of calcium carbonate or silica that fossilize very well (fig. 8.1). These single-celled protistans live by the millions in the oceans, and their shells are so abundant that on many parts of the seafloor the entire sediment is made of nothing but the shells of foraminiferans. The coccolithophorid algae secrete tiny button-shaped plates of calcite only a few microns in diameter. In shallow marine waters, however, coccolithophorids can live in enormous densities, and they accumulate thick piles of the limy sediment we know as chalk.

For organisms as abundant as these, the fossil record is extremely good. All the micro-paleontologists need do is collect a few grams of sediment from the outcrop or from a core drilled in the deep-sea bottom and put them on microscope slides and they have thousands of specimens spanning millions of years of time. With a record as good as this, micropaleon-tologists can document evolution in great detail and tell how old the sediment is and show how the microfossils respond to climate change and whether the ocean waters in a given area grew deeper or shallower. Indeed, micropaleontology is the single largest subfield in paleontology because the work is indispensable to oil companies who need to know the age of the rocks that produce oil. In addition, micropaleontology is critical to marine geology in studying how climates and oceans have changed over geologic time. Without microfossils, we would have no oil and would still not understand the causes of the ice ages or earth's past climatic changes. We will look at some of the amazing examples of evolutionary change in microfossils in chapter 8, but microfossils are the ultimate answer to the usual complaint that the fossil record is too incomplete to document evolution.

Faunal Succession or "Flood Geology"?

> Let us now see whether the several facts and laws relating to the geological succes-
> sion of organic beings accord best with the common view of the immutability of spe-
> cies, or with that of their slow and gradual modification through natural selection. . . .
> Yet if we compare any but the most closely related formations, all the species will
> have been found to have undergone some change. When a species has once disap-
> peared from the face of the earth, we have no reason to believe that the same identical
> form ever reappears.
>
> —Charles Darwin, *On the Origin of Species*

One of the common myths that creationists spread about the fossil record is that geologists shuffle the layers of strata and their sequence of fossils to *prove* evolution, and then the evo-lutionists point to the sequence as *proof* of evolution (Gish 1972; Morris 1974:95–96). Accord-ing to creationists, this is a circular argument. But it is manifestly untrue and shows how little creationists actually know about the history of geology—and creationism.

The geologists who first discovered the fact that assemblages of fossils change through time, or *faunal succession*, were actually devoutly religious men who were not trying to prove evolution (an idea that would not be published for 50 to 70 years after they discovered fau-nal succession). One of them was William Smith, who was not an independently wealthy gentleman-scientist (like most of the early geologists and paleontologists) but a humble working man, an engineer for a local canal-digging company in the south of England. He had a keen eye for rocks and fossils and a talent for recognizing what he was digging up. Smith got one of the first good looks at a cross section of fresh rock through the normally heavily vegetated landscape of England as his canals were being dug. About 1795, he noticed that every formation excavated had a completely different assemblage of fossils, and he could tell what formation any given fossil came from because they were all distinct. He got so good at it that he would amaze the gentleman fossil collectors by telling them from exactly which stratum every fossil in their collection had been derived. He soon realized that the sequence of fossils through the rock layers of England was a powerful tool because the fossils representing each age were consistent, whereas the rock layers changed across distance. This allowed Smith to map the distinctive formations and their fossils, and by 1815 he had published the first geologic map of England, the first truly modern geologic map ever constructed (see Winchester 2002).

Because he was a common working man, Smith faced enormous difficulties getting his ideas published and recognized by the gentleman-geologists who dominated the field at that time. A few of them realized the importance of his discovery and tried to steal the credit for themselves. But after losing his impressive fossil collection and all his possessions trying to get his maps published, spending time in debtor's prison, and having difficulties with failing health, Smith finally got credit for his crucial discovery. By 1831, shortly before his death, Smith was hailed as the "father of British geology." Smith never attempted grand theological explanations for the change of fossils through time; he simply documented the pattern of change as an empirical fact about the rock record and a powerful tool for mapping and correlating strata all around the world.

Across the English Channel, similar ideas were being developed in France. The great anatomist and paleontologist Baron Georges Cuvier was studying the fossils found around Paris and describing the distinctive rock layers beneath the city. Together with Alexandre Brongniart, who specialized in fossil mollusks, he too began to realize that each formation had a distinctively different assemblage of fossils. Some scholars point to this apparently independent discovery of faunal succession as a classic example of how, when the time is ripe for a new idea, it will emerge in several places at once. Others suggest that Brongniart may have heard about it when he made a trip to England in 1802 (one of the few times that France and England were not hostile during the period from the French Revolution through the battles with Napoleon that did not end until 1815). Either way, faunal succession was a powerful tool that was adopted by geologists all over Europe and eventually the world, so that by 1850, they had worked out the succession of rocks and fossils in many places, and named the periods of the geologic time scale (fig. 3.3). Evolution was still a radical notion floating around among French and British biologists (but not geologists), and it was still a decade before Darwin would publish his ideas.

Indeed, Cuvier himself was staunchly against the evolutionary ideas of his colleague Lamarck and tried to use the fossil record against him. Cuvier pointed to mummified animals recovered from the Egyptian tombs (recently robbed by Napoleon's soldiers). These mummified cats and ibises had not changed since the time of the ancient Egyptians. To Cuvier, this was proof that life was not constantly changing and evolving, as Lamarck had suggested. As the most prominent man in French science, Cuvier also had to avoid the speculative approaches of Geoffroy and Lamarck. Instead, he proposed his own solution

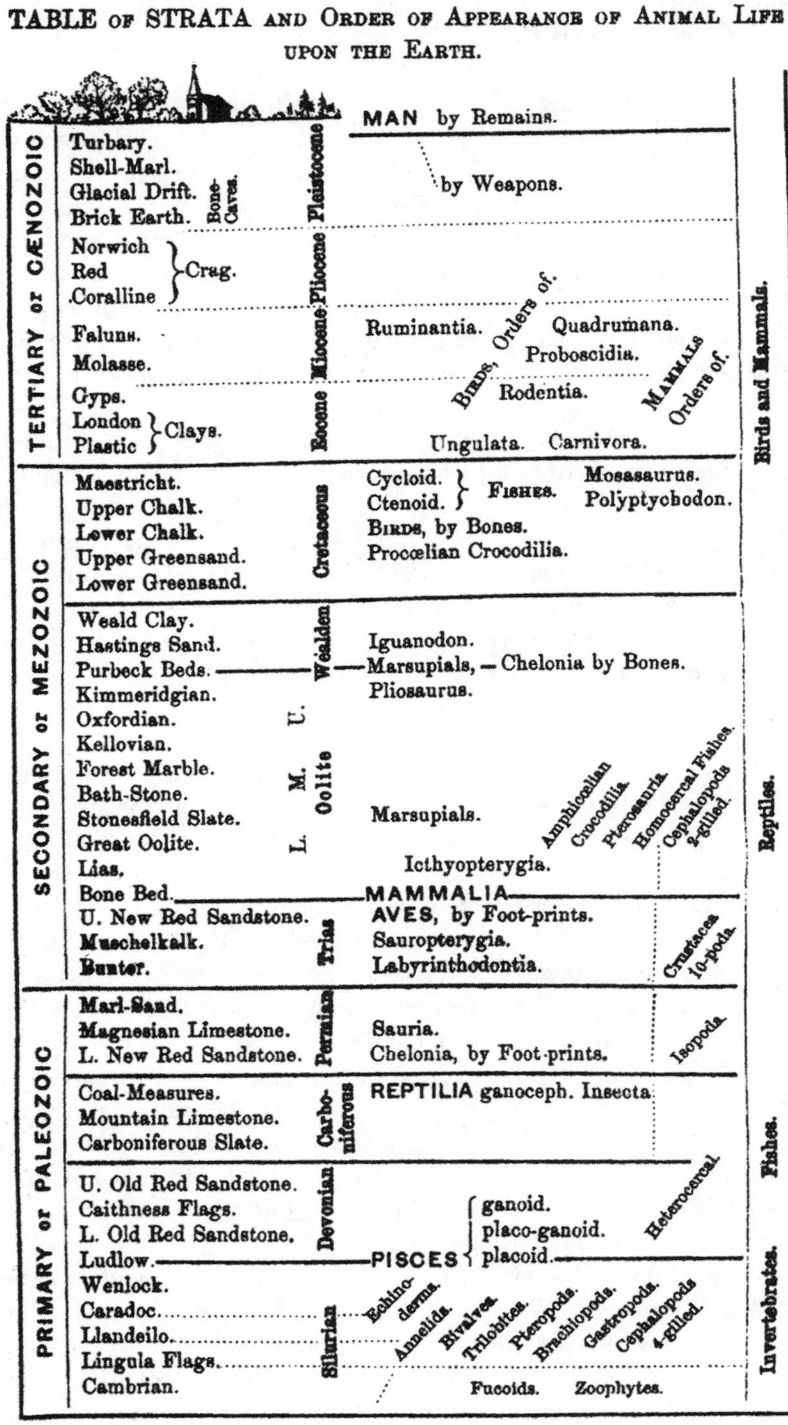

TABLE OF STRATA AND ORDER OF APPEARANCE OF ANIMAL LIFE UPON THE EARTH.

MAN by Remains.

Era	Strata	Period	Life
TERTIARY or CÆNOZOIC	Turbary. Shell-Marl. Glacial Drift. Brick Earth. (Bone-Caves.)	Plaistocene	MAN, by Weapons.
	Norwich, Red, Coralline } Crag.	Pliocene	
	Faluns. Molasse.	Miocene	Ruminantia. Quadrumana. Proboscidia. Birds, Orders of. Rodentia. MAMMALS, Orders of.
	Gyps. London, Plastic } Clays.	Eocene	Ungulata. Carnivora.
SECONDARY or MEZOZOIC	Maestricht. Upper Chalk. Lower Chalk. Upper Greensand. Lower Greensand.	Cretaceous	Cycloid. Ctenoid. } FISHES. Mosasaurus. Polyptychodon. BIRDS, by Bones. Procœlian Crocodilia.
	Weald Clay. Hastings Sand. Purbeck Beds.	Wealden	Iguanodon. Marsupials, — Chelonia by Bones. Pliosaurus.
	Kimmeridgian. Oxfordian. Kellovian. Forest Marble. Bath-Stone. Stonesfield Slate. Great Oolite. Lias.	U. M. L. Oolite	Marsupials. Amphicœlian Crocodilia. Pterosauria. Homocercal Fishes. Cephalopoda 2-gilled. Icthyopterygia.
	Bone Bed. U. New Red Sandstone. Muschelkalk. Bunter.	Trias	MAMMALIA. AVES, by Foot-prints. Sauropterygia. Labyrinthodontia. Crustacea 10-poda.
PALEOZOIC	Marl-Sand. Magnesian Limestone. L. New Red Sandstone.	Permian	Sauria. Chelonia, by Foot-prints. Isopoda.
	Coal-Measures. Mountain Limestone. Carboniferous Slate.	Carboniferous	REPTILIA ganoceph. Insecta.
PRIMARY or PALEOZOIC	U. Old Red Sandstone. Caithness Flags. L. Old Red Sandstone.	Devonian	ganoid. placo-ganoid. placoid. Heterocercal.
	Ludlow. Wenlock. Caradoc. Llandeilo. Lingula Flags. Cambrian.	Silurian	PISCES. Echinoderma. Annelida. Bivalves. Trilobites. Pteropoda. Brachiopoda. Gastropoda. Cephalopoda 4-gilled. Fucoids. Zoophytes.

(Right margin: Birds and Mammals. Reptiles. Fishes. Invertebrata.)

FIGURE 3.3. The geologic time scale was originally reconstructed by devout creationist geologists who realized that the record of fossils was too complex to be explained by a single Genesis flood. This version of the time scale was published by Richard Owen (1861), one of the last legitimate creationist biologists.

to the dilemma. The layers of rock with fossils of extinct animals represented a dark, dangerous period before the Creation and flood of Genesis (the *antediluvian period*, Latin for "before the flood") not described in the Bible. God had created and destroyed these earlier antediluvian worlds before the Genesis record begins. This solution was not too heretical for the time; it allowed Cuvier to recognize that the rock record was full of fossils of extinct organisms that could not have made it to Noah's ark and that were certainly not alive in his day.

Other geologists and paleontologists followed Cuvier's lead and tried to describe each layer with its distinctive fossils as evidence of yet another Creation and flood event not mentioned in the Bible. In 1842, Alcide d'Orbigny began describing the Jurassic fossils from the southwestern French Alps and soon recognized ten different stages, each of which he interpreted as a separate nonbiblical creation and flood. As the work continued, it became more and more complicated until 27 separate creations and floods were recognized, which distorted the biblical account out of shape. By this time, European geologists finally began to admit that the sequence of fossils was too long and complex to fit with Genesis at all. They abandoned the attempt to reconcile the fossil sequence with the Bible. Remember, these were devout men who did not doubt the Bible and were certainly not interested in shuffling the sequence of fossils to prove Darwinian evolution (an idea still not published at this point). They simply did not see how the Bible could explain the rock record as it was then understood.

Instead of worrying about theology, geologists and paleontologists realized that faunal succession was an extremely powerful tool that allowed them to date and correlate rocks all over the world. The principle of faunal succession grew into the discipline of *biostratigraphy*, where the distribution of fossils in the different strata helps us determine their age (see Prothero 2013a: chap. 10). Biostratigraphy, in turn, helps us map the distribution of rocks on earth. It is the principal tool used by oil and coal geologists to date and correlate the rocks they are drilling and exploring for valuable resources. Without biostratigraphy, we would have no oil or gas, and our modern industrial age, dependent as it is on cheap petroleum, would never have occurred.

As we saw with Smith and Cuvier, biostratigraphy does not require evolution or any other theoretical explanation for why the fossils change through time. It simply deals with the empirical fact that they *do* change. Biostratigraphic theory considers how to decipher these patterns of fossil distribution and get the most reliable results and has little or no biological component at all. Indeed, many fossils that are valuable biostratigraphic time indicators are actually treated as unusual objects of curious shapes, not like remains of extinct organisms. If they were nonbiological objects such as nuts, bolts, and screws and changed predictably through time, they would work just as well. The proof of this is that many of the best fossils for biostratigraphy are poorly understood biologically (this includes most microfossil groups) and the biological relationships of some important extinct organisms (such as graptolites, conodonts, and acritarchs) were completely unknown for more than a century. That didn't stop them from being useful for biostratigraphy.

Of course, once the idea of evolution came along, it explained *why* fossils change through time. But to recapitulate my main point, the succession of fossils through time was established by devoutly Christian geologists decades *before* Darwin published his ideas about evolution. There is no possibility of the alleged fraud of arranging fossils to prove evolution, as creationists claim.

The Surreal World of "Flood Geology"

If the system of "flood geology" can be established on a sound scientific basis, and be effectively promoted and publicized, then the entire evolutionary cosmology, at least in its present neo-Darwinian form, will collapse. This, in turn, would mean that every anti-Christian system and movement (communism, racism, humanism, libertinism, behaviorism, and all the rest) would be deprived of their pseudo-intellectual foundation.
—Henry Morris, *Scientific Creationism*

Creationists like to dismiss evolution as only a theory. My favorite rejoinder is that creationism isn't even a theory. When examined in the light of well-known and thoroughly researched scientific phenomena, creationist "flood geology" fails the most basic and simple test known to forensic science: bodies don't pile up the way creationists insist they must.
—Walter F. Rowe, "Bobbing for Dinosaurs: A Forensic Scientist Looks at the Genesis Flood"

If the creationists cannot claim that the fossil record has been fraudulently shuffled by geologists to prove evolution, then they must acknowledge that it shows changing fossil faunas through time, not instantaneous creation. As we saw with the early fundamentalists of the late nineteenth and early twentieth centuries (chapter 2), most had come to terms with the idea that the change in fossil faunas through time did support the idea of evolution, even if they were unhappy with that notion. But we must never underestimate the wild imaginations of religious fanatics who will bend the truth in whatever way they need to win their battles. If the fossil record does show a sequence of faunas through time, then they believe that there *must* be a biblical explanation for it.

The first detailed attempt at such an explanation came from a Seventh-Day Adventist schoolteacher named George Macready Price, who published a series of books starting in 1902. Price had no formal training or experience in geology or paleontology and in fact attended only a few college classes at a tiny Adventist college. But inspired by Ellen G. White, the prophetess and founder of the Seventh-Day Adventist movement, he dreamed up an explanation called *flood geology* and aggressively promoted it for more than 60 years until his death in 1963. According to Price, the flood accounted for all of the fossil record, with the helpless invertebrates being buried first, and the larger land animals floating to the top to be buried in higher strata, or fleeing the floodwaters to higher ground. Price also originated the lie we just debunked about geologists dating rocks by their fossil content while simultaneously determining the age of fossils by their position in the geologic column. Ignorant of history or geology, Price was unaware of the fact that religious geologists had believed in a Noachian deluge explanation of the fossil record in the seventeenth and eighteenth centuries but abandoned it when their own work showed it to be impossible—long before evolution came on the scene. The most famous geologic treatise of the seventeenth century, *The Sacred Theory of the Earth*, by Reverend Thomas Burnet, dealt with the problem of the Noachian deluge explaining the rock record. Burnet, unlike the modern creationists, did not fall back on the supernatural. Although others urged him to resort to miracles, Burnet declared, "They say in short that God Almighty created waters on purpose to make the Deluge. . . . And this, in a few words, in the whole account of the business. This is to cut the knot when we cannot loose it."

In Price's later years, his bizarre ideas about geology were generally ignored as embarrassments by most creationists (see Numbers 1992:89–101). Most subscribed to the "day-age" idea of Genesis, where the "days" of scripture were geologic "ages," and did not try to contort all the evidence of geology into a simplistic flood model. Some disciples of Price actually tried to test his ideas and look at the rocks for themselves, which Price apparently never bothered to do. In 1938, Price's follower Harold W. Clark "at the invitation of one of his students visited the oil fields of Oklahoma and northern Texas and saw with his own eye why geologists believed as they did. Observations of deep drilling and conversations with practical geologists [none of whom were trying to prove evolution, but simply using biostratigraphy to find oil] gave him a 'real shock' that permanently erased any confidence in Price's vision of a topsy-turvy fossil record" (Numbers 1992:125). Clark wrote to Price,

> The rocks do lie in a much more definite sequence than we have ever allowed. The statements made in the New Geology [Price's term for flood geology] do not harmonize with the conditions in the field. . . . All over the Middle West the rocks lie in great sheets extending over hundreds of miles, in regular order. Thousands of well cores prove this. In East Texas alone are 25,000 deep wells. Probably well over 100,000 wells in the Midwest give data that have been studied and correlated. The science has become a very exact one, and millions of dollars are spent in drilling, with the paleontological findings of the company geologists taken as the basis for the work. The sequence of microscopic fossils in the strata is very remarkably uniform. . . . The same sequence is found in America, Europe, and anywhere that detailed studies have been made. This oil geology has opened up the depths of the earth in a way that we never dreamed of twenty years ago. (quoted in Numbers 1992:125)

Clark's statement is a classic example of a reality check shattering the fantasy world of the flood geologists. Unfortunately, most creationists do not seek scientific reality. They prefer to speculate from their armchairs and read simplified popular books about fossils and rocks rather than go out in the field and do the research themselves or do the hard work of getting the necessary advanced training in geology and paleontology.

In the 1950s, the young seminarian John C. Whitcomb tried to revive Price's ideas yet again. When Douglas Block, a devout and sympathetic friend with geologic training, reviewed Whitcomb's manuscript, he "found Price's recycled arguments almost more than he could stomach. 'It would seem,' wrote the upset geologist, 'that somewhere along the line there would have been a genuinely well-trained geologist who would have seen the implications of flood geology, and, if tenable, would have worked them into a reasonable system that was positive rather than negative in character.' He assured Whitcomb that he and his colleagues at Wheaton [College, an evangelical school] were not ignoring Price. In fact, they required every geology student to read at least one of his books, and they repeatedly tested his ideas in seminars and in the field. By the time Block finished Whitcomb's manuscript, he had grown so agitated he offered to drive down to instruct Whitcomb on the basics of historical geology" (Numbers 1992:190).

In 1961, Whitcomb and hydraulic engineer Henry Morris published *The Genesis Flood*, where they rehashed Price's notions with a twist or two of their own. Their main contribution was the idea of hydraulic sorting by Noah's flood, where the flood would bury the heavier shells of marine invertebrates and fishes in the lower levels, followed by more advanced animals such as amphibians, reptiles (including dinosaurs) fleeing to intermediate levels, and finally the "smart mammals" would climb to the highest levels to escape the rising floodwaters before they were buried.

The first time a professional geologist or paleontologist reads this weird scenario, they cannot help but be amazed at its naiveté. Price, Whitcomb, and Morris apparently never spent any time collecting fossils or rocks. What their model is trying to explain is a cartoon, an oversimplication drawn for kiddie books—not any real stratigraphic sequence of fossils documented in science. Those simplistic diagrams with the invertebrates at the bottom, the dinosaurs in the middle, and the mammals on top bear no resemblance to any local sequence on earth. In fact, those oversimplified cartoons show only the *first appearance* of invertebrates, dinosaurs, and mammals, not their order of fossilization in the rock record (since invertebrates are obviously still with us and are found in all strata from the bottom to the top; see page 1). This diagram is an abstraction based on the complex three-dimensional pattern of rocks from all over the world. In a few extraordinary places, such as William Smith's England, or the Grand Canyon, Zion, and Bryce National Parks in Utah and Arizona, we have a fairly continuous sequence of a long stretch of geologic time (fig. 3.4), so we know the true order in which rocks and fossils stack one on top of another. But even in that sequence, we have "dumb" marine ammonites, clams, and snails from the Cretaceous Mancos Shale found *on top of* "smarter, faster" amphibians and reptiles (including dinosaurs) from the Triassic and Jurassic Moenkopi, Chinle, Kayenta, and Navajo Formations.

Just to the north, in the Utah-Wyoming border region, the middle Eocene Green River Shale yields famous fish fossils quarried by commercial collectors for almost a century. The Green River Shale produces fossils of not only freshwater fish but also freshwater clams and snails, frogs, crocodiles, birds, and land plants. The rocks are finely laminated shale diagnostic of deposition in quiet water over thousands of years, with fossil mudcracks and salts formed by complete evaporation of the water. These fossils and sediments are all characteristic of a lake deposit that occasionally dried up, not a giant flood. These Green River fish fossils lie above the famous dinosaur-bearing beds of the Upper Jurassic Morrison Formation in places such as Dinosaur National Monument and above many of the mammal-bearing beds of the lower Eocene Wasatch Formation as well, so once again the fish and invertebrates are found *above* the supposedly smarter and faster dinosaurs and mammals.

If you think hard about it, why should we expect that marine invertebrates or fish would drown at all? They are, after all, adapted to marine waters, and many are highly mobile when the sediment is shifting. As Stephen Jay Gould put it,

> Surely, somewhere, at least one courageous trilobite would have paddled on valiantly (as its colleagues succumbed) and won a place in the upper strata. Surely, on some primordial beach, a man would have suffered a heart attack and been washed into the lower strata before intelligence had a chance to plot a temporary escape. . . . No trilobite lies in the upper strata because they all perished 225 million years ago. No man keeps lithified company with a dinosaur, because we were still 60 million years in the future when the last dinosaur perished. (Gould 1984:132)

In addition to the examples just given, there are hundreds of other places in the world where the "dumb invertebrates" that supposedly drowned in the initial stages of the rising flood

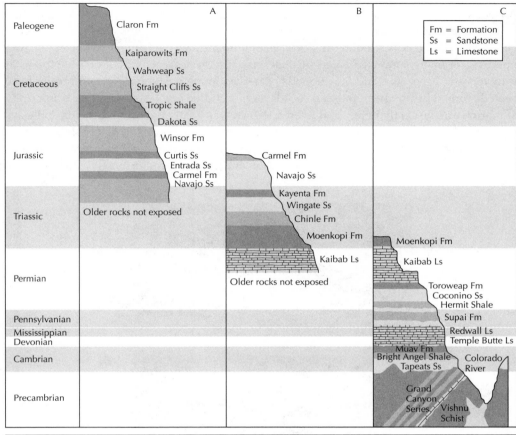

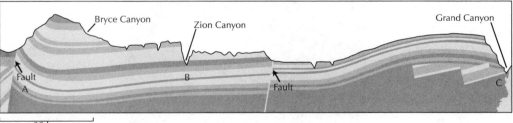

FIGURE 3.4. The sequence of rocks in the "Grand Staircase" and Grand Canyon country of Utah and Arizona shows that the superpositional order of formations and geologic time periods is no geologic fantasy but is instead based on empirical evidence.

are found on top of "smarter, faster land animals," including many places in the Atlantic Coast of the United States, Europe, and Asia, where marine shell beds overlie those bearing land mammals. In some places, like the Calvert Cliffs of Chesapeake Bay in Maryland (fig. 3.2A) or Sharktooth Hill near Bakersfield, California, the land mammal fossils and the marine shells are all mixed together, and there are also beds with marine shells above and *below* those containing land mammals! How could that make any sense with the "rising floodwaters" of the creationist model?

In a supreme twist of irony, the disproof of flood geology is found just beneath the Answers in Genesis creation "museum" in Kentucky. The museum is built upon the famous Ordovician rocks of the Cincinnati Arch, which span millions of years of the later Ordovician. If you poke around the slopes all around the area (as I often have), you will find hundreds of finely laminated layers of shales and limestones, each full of delicate fossils of trilobites and bryozoans and brachiopods preserved in life positions that could never have been disturbed by floodwaters—and each layer of hundreds represents another community of marine organisms that grew and lived and then was gently buried in fine silts and clays (fig. 3.5). There is no possibility these hundreds of individual layers of delicately preserved fossils were deposited in a single "Noah's flood." Over a century ago, paleontologists documented that these fossil communities change and evolve through time, so they can tell

FIGURE 3.5. The rocks of the Cincinnati Arch just beneath the "Creation Museum" debunk flood geology all by themselves. In many cases, you can find layer after layer of marine fossils in life position, undisturbed and buried by fine layers of mud, before another layer of organisms grew on the next layer, and so on. In these shots from road cuts in northern Kentucky just a few miles from the museum, you can see huge fossil coral heads in life position, each lying on a different layer of shallow seafloor mudstones. Each individual coral shows many years of growth bands, proving that it was not brought there by a flood but grew for many years on this shallow seafloor before it was finally buried. And there are many such examples, one layer after another. (A) Broad overview of one road cut, showing coral heads growing in place in multiple levels. (B) Close up of two large coral heads, showing their exposed growth bands, and also the fact that each grew on a different layer at a different time, so they were not dumped there by a single flood event. (Photos by the author)

exactly what part of the Ordovician each layer came from by its characteristic fossils. You could not ask for a better refutation of flood geology—yet the Answers in Genesis ministry that built the museum is so ignorant of geology and paleontology that they never noticed that the foundation of their own showcase building falsifies their ideas.

We could go on and on about the endless list of absurdities of flood geology (for a point-by-point demolition of Whitcomb and Morris's fantasy world, see McGowan 1984:58–67), but I will sum it up with one more example. I did some of my graduate dissertation work in the Big Badlands of South Dakota, one of the richest vertebrate-bearing fossil deposits in the world (fig. 3.6). The sequence of fossils there is very well known, and we can now establish the precise ranges of organisms through several hundred feet of sandstones and mudstones. Indeed, establishing the biostratigraphic sequence of the mammal fossils was a major part of

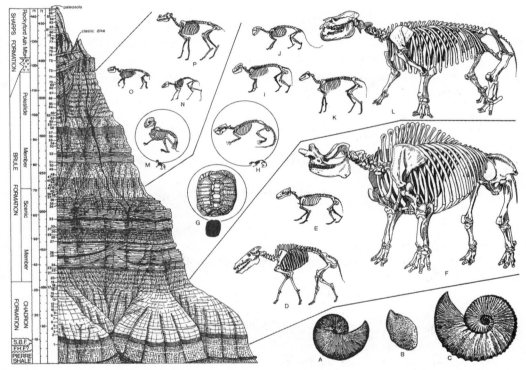

FIGURE 3.6. The sequence of fossils in the White River Group in Badlands National Park shows that even slow turtles are fossilized above supposedly smarter, faster mammals, so the "flood geology" model makes absolutely no sense whatsoever. The fossils are as follows: A–C: mollusks from the Cretaceous interior seaway: (A) the ammonite *Hoploscaphites nicolleti*; (B) the clam *Tenuipteria fibrosa*; (C) the ammonite *Discoscaphites cheyennensis*; D–F: mammals from the upper Eocene Chadron Formation; (D) the piglike enteledont *Archaeo-therium mortoni*; (E) the predatory creodont *Hyaenodon horridus*; (F) the giant titanothere *Megacerops*; G–L: fossil vertebrates from the early Oligocene Scenic Member of the Bruel Formation; (G) the tortoise *Stylemys nebrascensis*; (H) the squirrel-like rodent *Ischyromys typus*; (I) the larger oreodont *Merycoidodon culbertsoni*; (J) the "false sabertooth" *Hoplophoneus primaevus*; (K) the three-toed horse *Mesohippus bairdi*; (L) the hippo-like rhinoceros *Metamynodon planifrons*; M–P, middle Oligocene fossil mammals from the Poleslide Member of the Brule Formation; (M) the rabbit *Palaeolagus haydeni*; (N) the tiny deerlike *Leptomeryx evansi*; (O) the oreodont *Leptauchenia decora*; (P) the horned *Protoceras celer*. (Drawing by G. J. Retallack; reprinted with permission of the author and the Geological Society of America)

the research I did in my doctoral dissertation. At the base of the sequence are marine fossils, but right above them are the late Eocene fossils of the Chadron Formation, which include many large and spectacular mammals, including the huge rhino-like brontotheres. Above these in the overlying Brule Formation is a different assemblage of fossil mammals, none of which look like they could have outrun the huge brontotheres. Many of these are rodents. It's hard to imagine them doing a better job at "scrambling for higher ground" than the bigger, longer-legged animals. The clincher, however, is the fact that the most abundant fossils in the Brule Formation are tortoises! We have a new version of Aesop's fable of the tortoise and the hare, although here the dumb tortoises not only beat the hares to higher ground but also nearly all the rest of the smarter, larger, longer-legged mammals as well. If there ever was a clear-cut falsification of the flood geology model, this alone should be enough!

In addition to this stratigraphic fantasy world, Price and later flood geologists were particularly obsessed with overthrusts, places on earth where older rocks are shoved on top of younger ones along a fault plane. Price claimed that these overthrust faults were imaginary. Because they put the fossils in the wrong order, evolutionists had to explain away this anomaly by claiming that older rocks were thrust on top of younger rocks. Many creationists have repeated this claim (often verbatim from Price). For example, Whitcomb and Morris (1961:187) lifted the following partial quote from Ross and Rezak (1959) on the Lewis thrust in Glacier National Park:

> Most visitors, especially those who stay on the roads, get the impression that the Belt [the oldest rocks, of Precambrian age] are undisturbed and lie almost as flat today as they did when deposited in the sea which vanished so many million years ago.

Whitcomb and Morris fail to give the rest of the citation, which reads,

> Actually, they are folded, and in certain places, they are intensely so. From points on and near the trails in the park, it is possible to observe places where the Belt series, as revealed in the outcrops on ridges, cliffs, and canyon walls, are folded and crumpled almost as intricately as the soft younger strata in the mountains south of the park and in the Great Plains adjoining the park to the east.

So much for the supposed evidence that the fossils were deposited out of order! If the creationists were at all interested in real geology, they would spend the time and effort to see the rocks for themselves and realize that there is good independent evidence for the overthrusting. At the very least, they should not resort to deceptive quoting out of context, when the complete quotation clearly denies their claim.

In summary, the flood geology model constructed by Price and modified by Whitcomb and Morris bears no relation to any actual sequence of rocks or fossils on earth but was dreamed up to explain oversimplified cartoons. If these authors had any real experience with rocks or fossils, they would never have considered the model remotely reasonable. Indeed, religious geologists who have done their homework on fossils (as in the quotations from Clark and Block above) admit that the flood geology model does nothing to explain the real fossil record. Any consideration of a real sequence of fossils and rocks (such as the Big Badlands or the Grand Canyon) immediately demolishes the notion that a single Noachian deluge can account for the rock record and the actual sequence of fossils contained in those rocks.

The Grand Canyon Through the Looking Glass

The main reason for insisting on the universal Flood as a fact of history and as the primary vehicle for geological interpretation is that God's Word plainly teaches it! No geological difficulties, real or imagined, can be allowed to take precedence over the clear statements and necessary inferences of Scripture.

—Henry Morris, *Biblical Cosmology and Modern Science*

In *Alice's Adventures Through the Looking-Glass*, Alice steps through a mirror into a world in which all the rules are backward or reversed, and everything is the opposite of reality. A practicing geologist gets the same sensation when he or she reads about flood geology: the photographs of rocks are the same, and some of the same words are used, but the thinking is entirely alien to this planet. Nowhere is this more apparent than the creationist attempts to explain the geology of the Grand Canyon as a product of a single Noah's flood.

The reasons for the creationists' laser-like focus on the Grand Canyon (and almost no other geologic feature or national park) are obvious. People all over the world have seen pictures of and often visit this legendary place for its spectacular scenery; they cannot help but be impressed by the evidence it presents for millions of years of geologic history. The creationists are trying to show their followers that they can explain all of geology with the Noah's flood myth, so naturally they spend their energies on the most spectacular national park that best shows that earth has a long history. They've even managed to get one of their books (edited by a river guide who had a religious conversion experience, not a real geologist) offered for sale at the visitors' centers on the rim of the Grand Canyon. This has remained true for years now, despite the fact that the rangers and geologists at Grand Canyon National Park have repeatedly protested the sale of the book. The fact that this tract pushing the view of a specific religious minority is sold by a federal facility seems to be a violation of the separation of church and state and is probably unconstitutional.

Creationist flood geologists have made up their minds that the ancient flood myths of a sheep-herding culture must be literal truth. Then they bend and twist and special plead the entire history of the Grand Canyon into their preconceived notions that somehow this immense and magnificent pile of rocks must have been produced by one supernatural flood. If flood geologists were real scientists, they would look at real flood deposits and ask what they should look like. Because they never bother to do this, let's do it for them.

Geologists who study sedimentary rocks (known as sedimentologists) have become very sophisticated forensic detectives, looking at the clues found in sandstone or limestone and discovering amazing evidence of the source of the sediments, how the sediments were transported, what environment they were formed in, how they were deposited, and then how they were turned into rock. Sedimentology is also the principal skill required to find nearly all the oil, natural gas, coal, groundwater, uranium, and many other economically important natural resources, so we ignore their expertise at our own peril (for a basic background in the subject, see Prothero and Schwab 2013). These same sedimentologists who have found all the oil and coal and groundwater we require also have studied actual flood deposits and know exactly what they should look like. If the Noah's flood story were actually true, we would expect to find that the geology *around the world* (not just in the Grand Canyon) would begin with coarse-grained poorly sorted deposits of sand and gravel and boulders from the fast-water stage of a flood. Once a flood recedes, it can leave only one kind of deposit: a single layer of mud. Most floods inundate an area, and then the mud slowly settles out of suspension from the standing floodwaters until it accumulates in a thin layer. Even a worldwide flood would produce only one relatively thin layer of mudstone (not shale, because that requires burial and compaction over millions of years).

How does this compare to the real Grand Canyon? It's not even remotely close! Even a cursory glance at the sequence of layers in the Grand Canyon (fig. 3.4) shows that it is highly complex and cannot be explained by a single superflood (or even many floods, if that were an option). For one thing, there is no great deposit of coarse gravel and boulders and sand near the base representing the high-energy phase of rapidly moving water. For another, the upper part of the Grand Canyon sequence is not just a single thin layer of mud. Instead, it is a complex sequence of shales (*not* mudstones), sandstones, and limestones that alternate in a sequence that resembles no known flood deposits.

Let's start at the very bottom of the Canyon. Instead of coarse gravel, sand, and boulder deposits that flood geologists might expect, we have the ancient rocks of the Grand Canyon Series (fig. 3.7A and B). These are mostly quiet-water shales, plus sandstones and even some limestones. Many of these limestones contain stromatolites (figs. 3.7B, 6.1, and 7.1), dome-like mounds of layered sediment formed by algal mats that can only grow in the quiet waters of a sunny coastal lagoon. The individual layers in these stromatolites testify to hundreds of years of growth on each one—and there are multiple layers of stromatolites, each representing a separate episode of slow growth followed by burial, and then another phase of growth on a new surface. And this was supposedly formed during a single huge flood event only 40 days in duration?

A clear refutation of the flood geology model is the abundant mudcracks (fig. 3.7A) found in many of the shale units of the Grand Canyon Series. We've all seen mud dry up and form cracks. Common sense should tell even the creationists that the entire muddy surface was deposited and then dried up, not formed during the inundation of a flood. There's not just one layer of mudcracks, but hundreds, sometimes stacked in a long sequence. Clearly, these rocks represent dozens of small episodes of mud deposition and then complete drying, not a single catastrophic flood. A few creationist books and websites try to squirm out of this by mentioning unusual features such as syneresis cracks. What they don't mention is that even syneresis cracks still require drying and evaporation and shrinkage, so they are completely inconsistent with the flood geology model.

Even more strongly falsifying the flood geology model is that in the middle of this Grand Canyon Series sedimentary sequence are the Cardenas lava flows, dozens of individual flows totaling almost 1,000 feet in thickness. If these rocks had erupted into the floodwaters, they would be entirely composed of blobs of lava known as pillow lavas, which we can see erupting from undersea lava flows today. Instead, the Cardenas lavas show clear signs that they are normal subaerial eruptions and flowed downhill from their nearest volcano, much like the lavas erupting from Mount Kilauea in Hawaii. The very top of the lava flows shows evidence that they had completely cooled and were even weathered and eroded by wind and rain before the next sequence of sedimentary rocks was deposited on top of them. This is hardly consistent with the idea of lavas that erupted underwater during a major flood!

Finally, the clincher is that all these ancient Grand Canyon Series rocks at the base of the Grand Canyon are now found tilted on their sides and eroded on the edges, and then the rest of the Grand Canyon strata are deposited on top of them (fig. 3.4). A flood geologist simply cannot explain this. If these rocks were all soft soupy sediments deposited by Noah's flood,

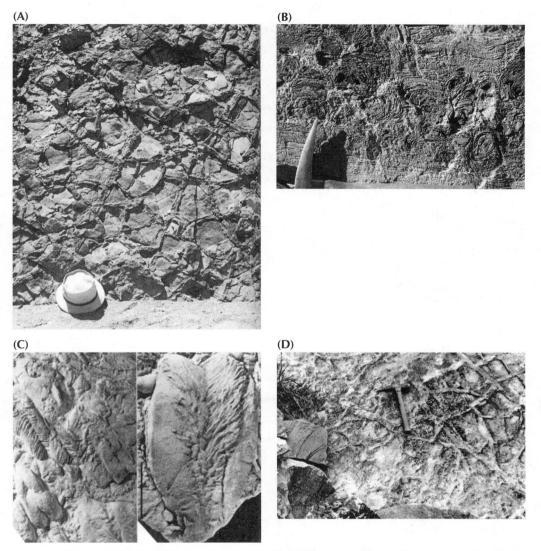

FIGURE 3.7. Close examination of the actual rocks in the Grand Canyon makes the "flood geology" hypothesis completely absurd. (A) Large mudcracks in the Precambrian Grand Canyon Series, in the lowest tilted sequence in the Grand Canyon. There are layer after layer of cracks like these in these shales, showing that there were hundreds of individual drying events—not possible with a single flood. (Photo by the author) (B) In other places, there are layered algal mats known as stromatolites, which were formed by daily fluctuations of sediment and algal growth. Some actually record decades or centuries of growth. These are abundant in the tilted late Precambrian limestones beneath the Paleozoic rocks of the Grand Canyon. (Photo courtesy U.S. Geological Survey) (C) The lower Cambrian (left) Tapeats Sandstone and (right) Bright Angel Shale are full of layer after layer of sediments with complex burrows and trackways, showing that each layer had been part of another sea bottom that was crawled upon and burrowed into and then buried again and again. (Photo courtesy L. Middleton) (D) The Pennsylvanian-Permian Supai Group and Hermit Shale are also full of layer upon layer of mudcracks, showing that they went through hundreds of episodes of drying, completely falsifying the flood geology model. (Photo courtesy U.S. Geological Survey) (E) The Permian Coconino Sandstone is composed entirely of huge cross-beds that could only have formed in desert sand dunes, not underwater. (Photo courtesy U.S. Geological Survey) (F) The Coconino dune faces also are covered with the trackways of reptiles that could never have been formed underwater. (Photo courtesy U.S. Geological Survey)

(E)

(F)

FIGURE 3.7. (*Continued*)

then as soon as some supernatural force rapidly tilted them on their sides, they would have all slumped downhill and left big gravity slump folds, a feature well known to sedimentologists. Instead, the entire sequence is undisturbed and full of stromatolites, mudcracks, and lava flows that belie the entire flood geology model right then and there. We have evidence of deposition of the Grand Canyon Series sediments (along with long erosion between them, when the Galeros lavas flowed across the landscape), then the hardening of these soft sediments into sedimentary rock layers, then tilting, then erosion, then another long *sequence* that makes up the higher part of the Grand Canyon. All of this is supposedly formed in a single large flood event?

And so it goes, layer by layer, right up through the rest of the Grand Canyon. The first unit above the tilted Grand Canyon Series rocks is the Tapeats Sandstone (figs. 3.4 and 3.7C), a classic beach and nearshore deposit. It is chock-full of trackways and burrows of trilobites, worms, and other invertebrates, layer after layer. When would these animals have had any time to crawl across the bottom and leave tracks or burrow through the sediment if it had been rapidly dumped by a flood? Above the Tapeats is the Bright Angel Shale, which real geologists interpret as deposited on a shallow marine shelf below the action of storm waves. It too is full of tracks and burrows, but of the types that today occur in the deeper part of the ocean. How did these tracks and burrows get there, layer after layer, if all the deposits of the Grand Canyon are a single flood deposit that drowned and buried all the marine life before it had a chance to begin burrowing?

The Bright Angel Shale has a complex interfingering relationship with the next unit above, the Muav Limestone. These types of relationships, where a thin layer of limestone alternates with a thin layer of shale, are very typical of deposits we find today when sea level slowly fluctuates back and forth—but it is impossible to explain such a complex relationship by a single flood dumping these sediments in a flat "layer cake." The Muav Limestone is one of three consecutive limestones forming the steepest cliffs in the Grand Canyon. Above the Muav is a sharp erosional surface with deeply eroded collapse features (from ancient collapsed caves that slowly dissolved out of the Muav), into which the much younger Temple Butte Limestone is deposited. The Temple Butte was then eroded away in most places (except the remnant fillings of those collapse features), and above it is deposited the big cliff of the Redwall Limestone. All three limestones have the features typical of modern limestones, made largely of the delicate remains of fossils. Today, we find such sediments forming in tropical, clear-water lagoons or shallow seas, such as those in the Bahamas or Yucatan or the South Pacific. In no case do these sediments form where there is the huge energy of floods or lots of mud stirred up by floodwaters. Particularly diagnostic is the fact that many of the fossils are extremely delicate (such as the lacy "moss animals," or bryozoans), yet they are intact and undisturbed, which proves the flood cannot have occurred. Even more evocative are the delicate animals such as the sea lilies (crinoids) and lamp shells (brachiopods), which are sitting just as they sat in life, layer after layer growing over and over, undisturbed by high-energy currents and buried by lime mud (not flood-type mud) that gently filtered in around them without disturbing them. This is true of limestones like this around the world and in the geologic past, so the Grand Canyon is not a special case—and definitely not evidence of a supernatural flood. The same is true of the Toroweap and Kaibab Limestones, which form the rim of the Grand Canyon.

Above the Redwall Limestone are the alternating sandstones and shales of the Supai Group, followed by the red Hermit Shale. The sandstones of the Supai Group are full of small ripples and small cross-beds, features of gentle deposition in rivers, not raging floodwaters or muds settling out after flood movement has stopped. The Supai Group and Hermit Shale not only contain layer after layer of mudcracks (fig. 3.7D), clearly demonstrating that they dried out repeatedly, but also delicate ferns and other plant fossils preserved intact, which is hard to explain if they were created by energetic floodwaters.

One of the best lines of evidence is the distinctive white band that is visible just below the rim on both sides of the Canyon, known as the Coconino Sandstone. This unit has huge cross-beds (fig. 3.7E) that are only known to form in large-scale desert sand dunes, *not* underwater. They also have small pits that are characteristic of the impacts of raindrops. How did raindrops land on these surfaces if they were immersed in a great flood? Even stronger proof is that many of the dune surfaces are covered with trackways of land reptiles (fig. 3.7F). How does a creationist reconcile these dry sand dune features and dry land reptile trackways with a huge flood event? I've read the creationists' attempts to explain these features, and they are classic examples of special pleading, twisting, and distorting scientific evidence as they thrash around in their completely unconvincing scenarios.

And the most impossible thing the creationists ask you to believe is this: the entire pile of sediments of the Grand Canyon sequence, soft, soupy, and supposedly deposited during a single flood event, was then eroded down to form the present-day Grand Canyon by the recession of the floodwaters. Wait a minute—didn't the creationists just use the recession of the floodwaters and the settling out of still water to deposit the thick piles of postflood shales, sandstones, and limestones in the first place? Or if that's not their scenario, then how did a soft pile of wet mud, sand, and lime hold up without slumping down and sliding into the gorge as the torrential retreating floodwaters rushed through? Did the supernatural flood also suspend the laws of gravity, too? Anyone with common sense can watch the Grand Canyon as it erodes today, with the long-hardened sediments (now sedimentary rocks) slowly weathering and eroding, dropping into the canyon by the action of gravity or by rains and small local canyon floods, and then slowly carried away by the Colorado River.

Even more revealing is the fact that if you trace the rock units of the Grand Canyon across distance, they change *facies* (gradually transform and intergrade sideways from one rock type to another). For example, in the Grand Canyon, the Pennsylvanian period (323–290 million years ago) is represented by the brick-red ledges and slopes of the Supai Group, which is composed of mudstones and sandstones deposited in broad rivers and plains. But, if you travel just 80 miles (130 km) to the west to the Arrow Canyon Range, just east of Las Vegas, Nevada, the same interval (between the Mississippian limestones below, and the Permian Coconino-Toroweap-Kaibab Formations above) is represented by a marine Bird Spring Limestone, full of the shells of foraminifera and brachiopods. About 300 miles (480 km) to the northwest, that same interval is represented by boulder conglomerate and sandstones shed from an ancient mountain range of the Antler orogeny that eroded away long ago and no longer exists. The fossils, and their position between nearly identical Mississippian limestones and Permian rocks, show that these Pennsylvanian rocks in Arizona and Nevada are all correlative and roughly the same age. Yet they look entirely different. These things are completely inexplicable if these rocks were all deposited as a worldwide uniformly consistent layer cake by a single Noah's flood.

Or let's follow the Permian rocks (Hermit Shale, Coconino Sandstone, and Toroweap and Kaibab Limestones) to the east of the Grand Canyon (fig. 3.8). By the time we get as far as Monument Valley, these units have all changed facies. The Toroweap and Kaibab Limestones have become thinner and thinner until they vanish completely near the Utah border. In their place is a thick sandstone (much thicker than the Coconino) known as the Cedar

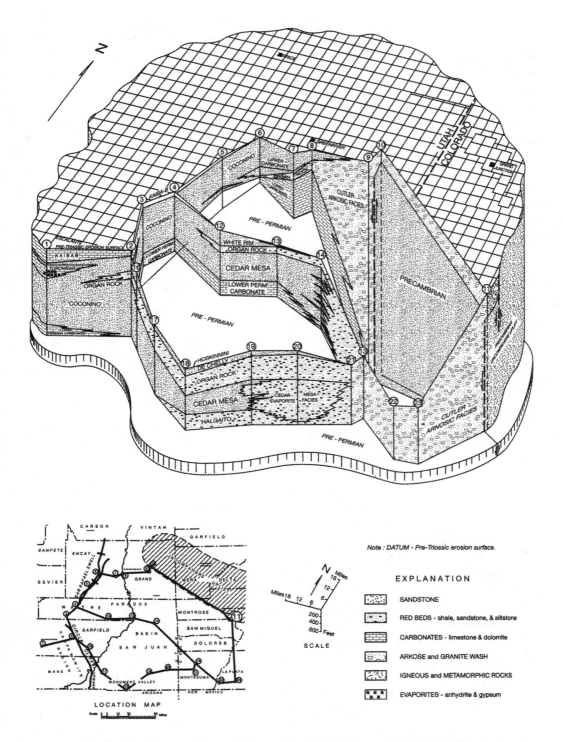

FIGURE 3.8. The flood geology model completely ignores the fact that rock formations, like the Permian sequence of the Grand Canyon, gradually change appearance, or facies, over distance. This panel diagram shows how the characteristic rocks of the Permian of the Grand Canyon are replaced by entirely different rock units as you move north and east through Utah, Arizona, Colorado, and New Mexico. (From Kunkle 1958; reprinted by permission of the Utah Geological Association)

Mesa Sandstone, which forms the spectacular cliffs and spires of Monument Valley. This unit also interfingers in a complex non–layer cake way with several red shales known as the Organ Rock and Halgaito Formations, which are full of ancient mudcracks indicating repeated drying events. These are the soft units that erode away quickly and have caused the sandstone cliffs and spires of Monument Valley to keep collapsing over the centuries. If you follow those rocks to the northeast, you find that the Cedar Mesa Sandstone has been largely replaced by thick deposits of salt and gypsum. Today, we can find only one place where such deposits form: in dry lakes and salty lagoons where the high rate of evaporation removes the water and concentrates the salts into a brine that eventually dries up completely. There's absolutely no way you can deposit hundreds of feet of salt and gypsum in a single flood event (especially since the same unit becomes mostly sandstones to the west). If a creationist tried to argue that it was formed as the flood finally dried up, then what about the hundreds of feet of sandstone, shale, and other rocks that lie above the salt and gypsum?

The final clincher comes even farther to the northeast, into southwestern Colorado. The sandstones and shales of Monument Valley and the salt and gypsum deposits of the Four Corners change laterally into coarse pebbly sediments known as the Cutler Arkose. Today these materials are only found in thick alluvial deposits eroding out of mountains. They are clear-cut evidence that during the Permian there was a range of mountains in the area known to modern geologists as the Ancestral Rockies. But they look nothing like a flood deposit nor does their lateral transformation into salt and gypsum deposits to the south, or dune sands and mudcracked shales to the west, make any sense in the world of "layer cake" flood geology.

Creationists have always focused on the Grand Canyon because it seems to fit their "layer cake" notions of how sediments would settle out after Noah's flood. But the Grand Canyon is also practically the only place in the world that looks this simple and undeformed. A more typical situation is the outcrops in the Basin and Range province of Nevada and Utah, just to the northwest of the Grand Canyon (fig. 3.9). There you do not have any horizontally layered rocks that might be called "Noah's flood deposits." Instead, you find ancient layers of fossiliferous Paleozoic rocks, cut up by faults, then buried by more layers of Mesozoic rocks, which are then cut by more faults. Which layers were deposited in Noah's flood? Clearly the faulted Mesozoic layers are different in age from the Paleozoic ones that they cut across on fault planes. Where in the Bible is this complex sequence of older faults and folds cut by younger faults and still younger faults mentioned? Finally, the youngest deposits fill the basins in between the fault block ranges—and they are full of fossils of extinct mammals and plants from the Miocene and older beds. Where do they fit in the flood geology model? This is the kind of geology that is typical in most parts of the world and that real geologists deal with all the time. It is no surprise that real geologists don't even bother to explain any of this with a simplistic flood model.

We could go on and on with this point, but this should be sufficient for anyone with an open mind and common sense not blinded by religious dogmatism. In the 1600s and 1700s, long before Darwin, religious scientists like Thomas Burnet, Abraham Gottlob Werner, William Buckland, and others tried to explain the world's rocks with the Noah's flood story. In the 1830s, also before Darwin's book was published, they gave it up completely, as the complexity of the real geologic record became apparent. The Noah's flood model predicts a simple layer cake worldwide sequence of coarse flood gravels and sands, overlain by a worldwide mudstone deposit. By contrast, the real geologic record is highly

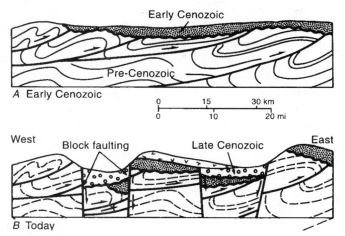

FIGURE 3.9. Creationists try to align relatively simple flat-lying sequences like the Grand Canyon with Noah's flood and ignore the vast majority of geologic settings around the world that in no way resemble a "layer cake" that could be deposited by a flood. For example, in the Basin and Range province of Utah and Nevada, just north of the Grand Canyon, the geologic relationships are extremely complex. (A) Paleozoic and Mesozoic beds are faulted and folded many times, with Paleozoic beds full of marine shells overthrust above Mesozoic dinosaur-bearing beds (contrary to the idea that dinosaurs could outrun the marine invertebrates in the rising flood). These older beds are then eroded off and unconformably overlain with early Cenozoic beds containing fossil mammals. (B) The early Cenozoic beds were then cut by Miocene normal faults, and the basins were filled with late Cenozoic sediments containing extinct horses, camels, mastodonts, and other Miocene land mammals. None of this complex geometry could be explained by simplistic "Noah's flood" models. (Modified from Prothero and Dott 2010)

complex and variable from region to region, with complex intertonguing contacts between units and facies that change dramatically over relatively short distances within the same part of the sequence. It is full of thousands of individual mudcracked layers and many different layers of salt and gypsum that simply cannot be explained by a single flood. There are thousands of different layers with delicate fossils in life positions, undisturbed by a single flood event.

All of this adds up to a simple conclusion: 200 years of mainstream geology (largely done by scientists who were very religious) has shown that the geologic record is far too complex for simplistic Bible myths. If the creationists were intellectually honest, they would face this fact, instead of imagining fantastic explanations for the Grand Canyon alone, and ignoring the remaining 99 percent of geology that cannot be twisted to fit their peculiar ideas. (For a blow-by-blow discussion, see www.talkorigins.org/faqs/faq-noahs -ark.html#georecord.) Real scientists are not allowed to twist and torture data to fit their preconceived conclusions or to ignore 99 percent of the data that cannot be made consistent with the flood geology model.

The most significant implication of flood geology and its fantasy view of the earth is a practical problem. Without real geologists doing their work, none of us would have the oil, coal, gas, groundwater, uranium, and most other natural resources that we extract from the earth. There are lots of devout Christians in oil and coal companies (I know many of them personally), but they all laugh at the idea of flood geology and would never attempt to use it to find what they're paid to find. Instead, like the Clark and Block quotes above demonstrate, they have seen the complexity of real geology in hundreds of drill cores spanning whole continents and don't even begin to try to interpret these rocks in a creationist mold (even though they may be devout Christians and believe much of the rest of the fundamentalists' credo). If they tried, they'd find no oil and lose their jobs! As creationists keep trying to get their bizarre notion of flood geology inserted into classrooms and places like the Grand Canyon, we have to ask ourselves: Are we willing to give up the oil, gas, coal, groundwater, and uranium that our civilization requires? That would be one of the steepest prices we would pay if we listened to the creationists.

Ship of Foolishness?

Alice laughed: "There's no use trying," she said: "one *can't* believe impossible things."

"I daresay you haven't had much practice," said the Queen. "When I was younger, I always did it for half-an-hour a day. Why, sometimes I've believed as many as six impossible things before breakfast."
—Lewis Carroll, *Through the Looking-Glass*

Most creationists believe that the Noah's ark story is historical fact. Never mind that there are actually two different stories in Genesis 6 and 7 from different sources that don't even agree with one another, or that large parts of both flood myths are cribbed almost word for word from the much older accounts in the *Epic of Gilgamesh* (see chapter 2). Never mind that creationists must explain why one verse has seven pairs of clean animals on the ark, while another only has one pair. Creationist books are full of incredible mental gyrations needed to make the Noah's ark story remotely believable. However, as I found out from my encounter with Gish, creationists will avoid discussing these points if they are brought up in debate, because they sound foolish and ruin the creationists' credibility with most audiences. A number of expeditions have been sent to Mount Ararat in Turkey (the supposed landing site) and made fantastic claims that they have found evidence of the ark, but none of these have stood up to scrutiny, despite the claims made by some creationist books and TV shows.

First, let us start with what the Bible says and delve into the world of arkeology. McGowan (1984: chap. 5) and Moore (1983) discuss the logistical details of the Noah's ark story at length, so I will not repeat their entire analysis here. A whole series of questions and problems come up when you look at the ark story in detail. The biggest problem is that a wooden boat the size of the ark would break up under the smallest stresses. This is a well-known problem in nautical engineering. Beyond a certain size, wooden vessels are simply not strong enough or flexible enough and begin to rip apart or leak catastrophically. The strength of wood as a vessel increases in size is not enough to hold the vessel rigid and intact. If Noah's ark were indeed 137 meters (450 feet) long, as in most estimates, it would be larger than any wooden ship that has ever been built in history. The sailing ship *Wyoming* was 100 meters (329 feet) long, leaked constantly due to the problems with the flexing of the hull, and sank 14 years after its launch in 1910. So did the 99 meter (324 foot) barge *Santiago*, which sank in 1918. Two wooden British warships, the HMS *Orlando* and HMS *Mersey*, were

102 meters (335 feet) long, but were scrapped only a few years after they were built because they were not seaworthy. Yet Noah's ark was allegedly about 30 percent larger than these boats that could not stay afloat and intact, despite the most advanced technology of wooden shipbuilding ever devised. Creationists sometimes mention legendary Chinese treasure ships of the fifteenth century that may have approached 137 meters, but if they were real and actually this big, they were just barges that barely moved, and none of them are known to have floated outside quiet harbors and rivers.

McGowan (1984:55) calculates that the biblical dimensions give a boat with about 55,000 cubic meters of internal volume. As we discussed earlier, there are at least 1.5 million species on earth today, which gives us only about 0.0367 cubic meters per species, or about one-third the capacity of a domestic oven—and these animals would have to be packed like shoeboxes stacked on top of one another to make this solution work. Clearly this is not enough space for most large animals. The pairs (or is it seven pairs?) of elephants, rhinos, and hippos would take up much of the ark all by themselves. The problem gets even more complicated if we consider that the true estimate is about 4 or 5 million species on earth.

The creationists, of course, are aware of this problem. When the flood myths were written, most ancient Middle Eastern cultures recognized only a handful of animals (domesticated plus wild). They paid no attention to insects, different kinds of fish, or many other less conspicuous forms of life, so they saw no problem in accounting for all living things that were important to them in a single boat. But the modern-day creationists must account for all of the millions of life forms on earth or else admit that some things have evolved from others since the days of Noah. They do this by claiming that Noah only took the "created kinds" (their translation of *baramin* in Hebrew) on the boat and that these kinds have since evolved into many more forms (a concession that evolution occurs!). By this method, they claim that there were only about 30,000 to 50,000 created kinds on board, but then that only gives each "kind" about a cubic meter to live in—still not much of an improvement.

Let's look closer at the term *baramin*. It was created out of nothing by Seventh-Day Adventist Frank Marsh in 1941 by tacking two words together from a Hebrew glossary (*bara*, "created"; *min*, "kind") without any idea how Hebrew actually works. Since almost none of the creationists read the Old Testament in the original Hebrew (or they would spot the problems and inconsistencies that make literal interpretation absurd), they don't realize how ridiculous this term is, and why it doesn't mean what they think it means. As I learned when I studied Hebrew, the Semitic root "b-r-a" (vowel points were not invented until centuries later) is translated "he created" or "he conjured," so it is a past-tense verb, not a past participle of a verb, as Marsh used it. And *min* can be used to mean not only a "kind" but also a species, or even a sex. Slapped together in Marsh's construction, the object *min* replaces the original subject *Elohim* (one of the names for the gods), so literally translated, *baramin* means "the species created," not "god created"—and certainly not "created kinds" in any sense the scriptures use. If Marsh had known any Hebrew and wanted to create a grammatically correct translation of "created kind," it would have been *min baru* (past participle). But given the consistently incompetent scholarship of creationists, I would never expect them to get this part right.

Leaving aside their ignorance of Hebrew, the whole topic of "baraminology" reminds one of a laughably poor imitation of science—science as imagined by kids at play or amateurs who are parroting the forms without understanding any of the principles or protocols or implications of the actual research, or the silly imitation of science in the movies and TV, where they spout scientific-sounding words that make no sense whatsoever. The focus of their "research" is to skim over the entire field of modern animal classification and then imagine ways to shoehorn hundreds of individual species and genera into the smallest possible number of categories. They don't bother to work with actual animals, or get their hands dirty with the dissections and anatomical work that established the modern taxonomy of organisms, or spend the years in graduate school to obtain the kind of training necessary to understand and analyze molecular phylogenetic data, or wade into the gigantic literature of modern systematic theory since the days of George G. Simpson and Ernst Mayr and cladistics. No, that would require that they be trained in actual science, and confront the evidence for evolution that runs throughout life. Instead, they do superficial, high school–level "book report" types of analyses, where they cherry-pick ideas here and there from highly simplified Internet sources and Wikipedia articles. They know just enough science to pick up a stray factoid here and there without any understanding of the caveats and methods behind the data or the relative significance or importance of one kind of data versus another that only comes with years of graduate study in a field.

In reality, their *baramin* "solution" to minimizing the number of animals on the ark creates a whole new set of problems. Not only does it concede evolution from the created kinds, but the kinds have no basis in biology. When you examine the creationist literature or try to pin them down, sometimes "the kinds" are species, sometimes they are genera, and sometimes they are whole families, orders, or even phyla of animals (Siegler 1978; Ward 1965)! Creationists are so wildly inconsistent and their theories are so completely out of line with the known taxonomy of organisms that it is clear that a created kind is one of those slippery words that people use to weasel out of difficult spots. As Humpty Dumpty said to Alice (in *Through the Looking-Glass*), "When *I* use a word, it means just what I choose it to mean." Nevertheless, a lot of the creationist "research" focuses on just this fruitless, unscientific version of chasing their own tails, and they even have a name for it: baraminology.

Some creationists try to squirm out of the problem by claiming that the fish and marine invertebrates stayed outside the ark and lived through the flood. But this reveals their complete lack of understanding of basic biology. To a creationist, apparently, if it lives in the water, it's all the same, but marine fish and invertebrates are highly sensitive to changes in salinity, so if the oceans were flooded by freshwater, these organisms would die immediately. If, on the other hand, these supernatural clouds rained marine seawater (a physical impossibility, because salt is mostly left behind when water evaporates), then the salty worldspanning waters would have killed all the freshwater fish and invertebrates, which cannot tolerate high salinities. Of course, pushing the aquatic forms off the boat and into the water doesn't begin to solve the space or numbers problem because these forms account for only a few hundred thousand species.

To this point we have only addressed the issue of cramming thousands of species into shoebox-sized spaces stacked to the top of the ark. Where would they put all the food for so many animals? How did the carnivores survive without eating their neighbors? Finally, the most unpleasant thought of all: so many animals produce a lot of dung. Did Noah and his sons spend most of their 40 days and nights shoveling it out of the boat? Instead of evaluating a reasonable and testable hypothesis, the special pleading and twisting of the facts of nature makes it clear that we're dealing with an explanation that is a load of dung (fig. 3.10).

FIGURE 3.10. Flood geology just doesn't hold water. (Cartoon courtesy Los Angeles Times Syndicate)

Dates with Rocks

We now have literally thousands of separate analyses using a wide variety of radiometric techniques. It is an interlocking, complex system of predictions and verified results—not a few crackpot samples with wildly varying results, as creationists would prefer to believe.

—Niles Eldredge, *The Monkey Business*

The principle that one rock layer is normally deposited on another older one (*superposition*) and the principle of faunal succession laid the framework for the relative geologic time scale by the 1840s, but geologists still could not attach numerical ages to these rock units or say whether they took thousands or millions or billions of years to deposit. It was clear by looking at the immense thicknesses of rock and their complexity (such as in the Grand Canyon example just discussed) that the rock record could not be explained by Noah's flood. But there was no reliable "clock" in the eighteenth or nineteenth century that could tell us whether the earth was only 20 million years old (as physicist Lord Kelvin argued) or billions of years old (as most geologists since Hutton had estimated).

Then, in 1895, the discovery of radioactivity by Henri Becquerel provided the first mechanism to produce reliable numerical ages (formerly called "absolute" dates) on rocks. By 1913, geologists such as Arthur Holmes had developed the radioactive decay method and found rocks on earth that were at least 2 billion years old. Since then, numerous additional dates have been established for older and older rocks. The oldest known rocks on earth currently date to 4.28 billion years ago, and there are individual mineral grains that give dates of 4.3 to 4.4 billion years old (Dalrymple 2004). Rocks from the moon and meteorites are even older, with many samples giving ages of 4.5 to 4.6 billion years. Although we have no earth rocks that old yet, this is not surprising, because the earth has a dynamic crust that is constantly melting and recycling older material. It is amazing that earth materials as old as 4.4 billion years still survive. The moon and the meteorites, by contrast, have changed very little since their formation, so it would be expected that they preserve very ancient dates (for further details, see Dalrymple 2004). The earth, moon, and meteorites apparently formed at the same time from the primordial solar system, which is why we say the earth is also 4.6 billion years old.

The basic principles of radiometric dating are relatively straightforward and very well understood. Certain isotopes of elements, such as potassium-40, rubidium-87, uranium-238, and uranium-235, spontaneously break down or "decay" into atoms of different "daughter" elements (argon-40, strontium-87, lead-206, and lead-207, respectively, for the "parent" material just listed) by emitting nuclear radiation (alpha and beta particles and gamma radiation) plus heat. The rate of this radioactive decay is well known for all these elements and has been checked and double-checked in the laboratory hundreds of times. Geologists obtain a fresh sample of the rock, break it down into its component mineral crystals, and then measure the ratio of parent atoms to daughter atoms within the mineral. That ratio is a direct mathematical function of the age of the crystal.

Like any other technique in science, there are limitations and pitfalls that have to be avoided. Because dating is a measure of the time since a crystal cooled and locked in the radioactive parent atoms, it only works in rocks that cool down from a molten state, known as igneous rocks (such as granites or lava rocks). Creationists mock scientists over the fact that we cannot directly date the crystals in sandstone or any other sedimentary rock (the crystals are recycled from older rocks and have no bearing on the age of the sediment). But geologists long ago circumvented this problem by finding hundreds of places all over earth where datable volcanic lava flows or ash falls are interbedded with fossiliferous sediment or where intruding granitic magma bodies cut across the sedimentary rocks and provide a minimum age. From settings such as these, the numerical ages of the geologic time scale are derived, and their precision is now so well resolved that we know of the age of most events that are millions of years old to the nearest 100,000 years or less.

If the crystal structure has somehow leaked some of its parent or daughter atoms or allowed new atoms in to contaminate the crystal, the parent/daughter ratio is disturbed and the date is meaningless. But geologists are always on the lookout for this problem, running dozens of samples to determine whether their age data are reliable and cross-checking their dates against other sources for determining age. The newest techniques and machinery are so precise that a skilled geologist can spot the error in almost any date and quickly reject those that don't meet very high standards. Creationists will mention a specific date that proved to be wrong as evidence that the entire field of geochronology is unreliable, when in fact it was the geologists themselves who spotted the erroneous date, and quickly rejected it. As Dalrymple (2004) points out, it's as if we had a variety of watches and clocks, a few of which don't keep accurate time. But that fact doesn't mean we completely ignore clocks and watches altogether, as creationists are doing by rejecting *all* radiometric dating out of hand. We simply keep checking them against one another to determine which ones are reliable and which ones are not.

In another commonly repeated claim, creationists mock geologists over the example of a living clam that gave a radiocarbon date of many thousands of years old. But this is a well-understood anomaly. Normal radiocarbon dating works when nitrogen-14 from the atmosphere is transformed by cosmic radiation into carbon-14, which is then incorporated into living tissues. When the organism dies, the carbon-14 begins to decay, making the bone, shell, or wood (or anything bearing carbon) datable, as long as it is less than 80,000 years old (since the radiocarbon decay rate is relatively fast). These peculiar clams live in water covering ancient limestone that releases radioactively dead carbon into the water. That ancient carbon (instead of the normal carbon derived from the atmosphere) then becomes part of the mollusk shells, where it throws the ratio off. Radiocarbon specialists have long been aware of this minor problem and never rely on dates where this kind of contamination could be an issue.

Nearly every time you read a creationist account of radiometric dating, they find some way to misuse or misinterpret the system, then accuse the system of being unreliable when they have only demonstrated that they are incompetent and have no idea how it works. Take, for example, the familiar creationist trope about a volcanic lava in Australia dated at 45 million years old which flowed around trees which were radiocarbon dated at 40,000 years old. They point to this example and say, "Aha, this proves that all radiometric methods are unreliable."

So why did the Australian anomaly occur? The thing to remember is that each radioactive "clock" ticks at a different rate, but all of them keep good time (if they are used properly). Radiocarbon has a very short half-life of 5,370 years, so it decays extremely rapidly. By 40,000 years (some labs can push it to 80,000 years now), it is radiocarbon dead—no more decay is occurring, and you cannot use it to measure *anything* anymore. Thus, radiocarbon is primarily used by scientists who work on really young events of the last ice age and Holocene: archeologists and those who work on the last glacial cycle. A real scientist would never even *consider* using it for anything older, and any fools who do so show that they have no clue what they are doing. Other isotopic systems are useful in different age ranges. U-Pb (both isotopic pairs) and Rb-Sr decays over billions of years, so it is only used on the oldest earth rocks, plus moon rocks and meteorites. K-Ar is the system used by most geologists, since it can date rocks as young as 1 million years, and as old as the oldest rocks we have, so its useful age range covers the vast majority of common geologic settings. To go back to our clock analogy, radiocarbon is like a clock that ticks really fast and runs down quickly. U-Pb and Rb-Sr are like a big grandfather clock that ticks very slowly but doesn't run down except over a very long time. The Australian trees-in-lava scenario only demonstrates the complete incompetence of creationists. They cited the K-Ar analysis to date the lava at 45 million years old, which means that no real geologist would waste their time dating the radiocarbon-dead trees. But the creationist ran the radiocarbon on them anyway, and sure enough, he got dates around 40,000—*which only means the sample is radiocarbon dead* and older than 40,000, *not* that the age of the sample is 40,000 years. It's like looking at the time on a fast-running clock that has stopped and comparing it with the slow grandfather clock. The creationist says that since they are different, *no* clocks can be trusted, when in fact he was foolishly looking at a clock that has run down and stopped.

Creationists don't give scientists any credit for being skeptical and self-critical about their own data. But anyone who deals with geochronology knows that the dates are subject to constant scrutiny by multiple labs, and anything that is fishy is quickly challenged and rejected. The result is an extremely robust set of data, where multiple independent radioactive atomic systems (for example, potassium-argon, uranium-lead, and rubidium-strontium) are used on the same samples, so if any one of them is giving problems, it clearly can be thrown out. The creationists point to one or two examples of supposedly unreliable dates, but when three or more independent dating methods are run in different competing labs on the same rock and give the same answer, there is no chance that this is accident. After nearly a century of analyses, with thousands of dates checked and rechecked like this, geologists are as confident about the reliability of radiometric dating as they are about gravity or any other well-established principle of science. The earth is about 4.567 billion years old; this is as much a fact as the observation that it is round!

Creationists cannot abide the idea that the earth is more than a few thousand years old. Ironically, the idea that the earth is 6,000 to 10,000 years old or that it was created in 4004 B.C.E. (originally calculated by Archbishop Ussher and still found in the margins of some Bibles) is not based on the scriptures but on much later theological extrapolations. There are too many gaps and unrecorded intervals of time in the scriptures between the "begat" verses, such as "and Methuselah lived after he begat Lamech seven hundred eighty and two years, and begat sons and daughters" (Genesis 5:26) to allow a precise calculation of the age of the earth from biblical texts. Nevertheless, young-earth (but not day-age) creationists will not concede the millions of years that geologists have documented and seek to deny any kind of evidence for the great age of the earth or the universe. We have already seen (in chapter 1) how they resort to Gosse's untestable and unscientific *Omphalos* hypothesis to explain how starlight reaching us appears to have traveled billions of years already ("it was created that way").

The creationists' insistence that the earth was created only 6,000 years ago runs into all sorts of problems, even if you don't accept radiometric-dating methods. There are bristlecone pines in the White Mountains of California that record over 10,000 years of tree rings, and individual trees more than 5,000 years old! There are clonal trees older than 13,000 years in California and 9,500 years in Norway. In fact, there are ice cores like the EPICA-1 core in Antarctica that have more than 680,000 annual cycles of winter and summer recorded in their layers. That is 100 times the age that creationists accept. Are they saying that the Antarctic had 100 winter/summer cycles every year?

A creationist named Thomas Barnes argued that the earth's magnetic field is decaying, and if you extrapolate the field strength back in time, it suggests that the earth is very young. But the long-term behavior of the earth's magnetic field is very well known. It fluctuates in strength and direction over thousands of years (which we can measure by measuring the magnetic intensity recorded in ancient rocks). Barnes assumes a simple linear change through time, but he is apparently not aware of the abundant scientific evidence of the increases as well as decreases in the earth's field strength.

The other creationist "proofs" of a young earth are just as simpleminded and show their complete lack of understanding of how to do science and math. For example, Henry Morris attempted to calculate the age of the earth by estimating how long it would take for 3.5 billion people (now more than 7.4 billion) to arise from Adam and Eve. But this completely ignores the fact that human populations (especially in the distant past) didn't increase exponentially like bacteria in a Petri dish. For most of human history, there is very solid archeological evidence that human populations remained roughly constant for hundreds of thousands of years because of the constraints of death and disease. Only in the past 50 years has the population explosion come to resemble that of bacteria. Morris's extrapolation is entirely unsupported by the data.

Morris is fond of citing figures that indicate that about 5 million metric tonnes of cosmic dust falls on the earth each year. He calculates that over 5 billion years, we should have accumulated a layer of dust 55 meters (182 feet) thick! But if you do the calculation correctly (which Morris does not), it amounts to only a shoebox full of dust over an entire square kilometer. This is so minuscule it can barely be detected even in the best sedimentary records from deep-sea cores, which are undisturbed and have extremely low sedimentation rates (the only place so little dust could be detected). In shallow marine or terrestrial sediments, which are highly mixed and weathered, this tiny amount of dust would be homogenized quickly.

Creationists are so wildly inconsistent in their attempts to discredit all the lines of evidence of an ancient earth and universe that sometimes they stumble into self-contradiction. Chris McGowan (1984:89) recounts the following example:

> During a recent encounter I had with Dr. Gish, he was careless enough to make some reference to astronomical distances in terms of millions of light-years. When I asked him how he could rationalize this statement with a belief in a ten-thousand-year-old universe, he was unable to give a reply.

Punk Eek, Transitional Forms, . . . and Quote Miners

> Since we proposed punctuated equilibria to explain trends, it is infuriating to be quoted again and again by creationists—whether through design or stupidity, I do not know—as admitting that the fossil record includes no transitional forms. Transitional forms are generally lacking at the species level, but they are abundant between larger groups. Yet a pamphlet entitled "Harvard Scientists Agree Evolution Is a Hoax" states: "The facts of punctuated equilibrium which Gould and Eldredge . . . are forcing Darwinists to swallow fit the picture that Bryan insisted on, and which God has revealed to us in the Bible."
>
> —Stephen Jay Gould, "Evolution as Fact and Theory"

Like most people in Victorian England, Charles Darwin was part of a culture that believed in progress and gradual change. Shaken by the revolutions in France and America, the British instead subscribed to the notion that slow, steady change was the best model for societal change and reform. As a close friend and disciple of Charles Lyell, Darwin wanted to extend Lyell's approach of gradual, uniformitarian transformations of the earth to biology. As he began to work on his ideas about evolution, the notion of gradual change was deeply embedded in his thinking. As he wrote in 1859, the fossil record should yield "infinitely numerous transitional links" that demonstrated the slow, steady work of natural selection. "Natural selection is daily and hourly scrutinizing, throughout the world, every variation, even the slightest; rejecting that which is bad, preserving and adding up all that is good; silently and insensibly working. . . . We see nothing of these slow changes in progress until the hand of time has marked the long lapse of ages. . . . Why then is not every geological formation and every stratum full of such intermediate links? Geology assuredly does not reveal any such finely graduated organic chain; and this, perhaps, is the gravest objection which can be urged against my theory" (Darwin 1859:280). In reviewing his manuscript, however,

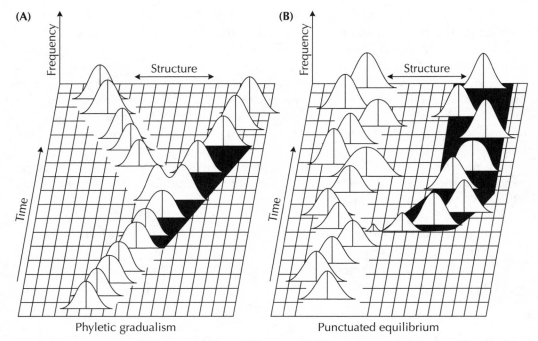

(A) Phyletic gradualism

(B) Punctuated equilibrium

FIGURE 3.11. (A) The classic notion (inherited from Darwin) was that evolution should show gradual transformations of species (shown by bell-shaped frequency distributions) through time, or "phyletic gradualism." (B) Eldredge and Gould (1972) pointed out that according to Mayr's allopatric speciation model, most speciation should happen too rapidly to be seen in the fossil record. Instead, fossil samples will show apparent abrupt speciation from populations outside the sampled area, followed by long periods of stability or stasis.

Darwin's friend and supporter Thomas Henry Huxley warned him that, "You have loaded yourself with an unnecessary difficulty in adopting *Natura non facit saltum* [Nature does not make leaps] so unreservedly." To Huxley, gradualism was not a necessary part of evolutionary theory.

As the century after Darwin progressed, paleontologists set out to try to document what Darwin expected. A few examples of apparently gradually evolving fossil sequences were documented (fig. 3.11A) but generally not a lot were found. As we shall discuss in chapter 4, however, evolutionary theory quickly became the domain of geneticists by the early and middle twentieth century, and paleontologists were pushed into the background.

In the 1940s and 1950s, another development occurred in evolutionary biology: the development of modern speciation theory. Darwin had assumed that all one needed to explain the origin of species (as in his title) was the transformation of lineages. But by the middle twentieth century, it was clear that *the real issue* in speciation was the splitting of lineages into two or more new species. Field biologists who observed species interacting in nature noticed that species are defined by reproductive isolation; that is, different species cannot successfully interbreed. In particular, ornithologist and evolutionary biologist Ernst Mayr (1942) found from studying birds in New Guinea that most species had distinct non-overlapping geographical ranges, and islands often harbored their own species. From this, he proposed the *allopatric speciation model*.

By the 1940s, studies of breeding and gene flow had shown that large populations are resistant to evolutionary change, because the unusual genes of any new variant are quickly diluted by interbreeding with the normal members of the rest of the large population. But small populations can change dramatically in relatively short periods of time. For example, when a population (or just a single pregnant female) reaches an isolated place like an island, all of their rare genetic mutations will soon become dominant, because all of the island's inhabitants will be their descendants. This effect is not restricted to islands. Small populations on the mainland can be genetically distinct if they do not interbreed with other populations. Religious sects such as the Amish in Pennsylvania, for example, have high frequencies of unusual genes, because they are highly inbred and very few non-Amish convert and marry into the Amish population to replenish the gene pool.

From these studies, Mayr (1942) concluded that small isolated populations on the fringe of the main population (*peripheral isolates*) would be the likeliest source of new species. Their ranges are usually separated from the main population (*allopatric*), usually by water or mountain barriers. Once these populations have become distinct, they can then come into contact with the main population (becoming *sympatric*), but they will no longer be able to interbreed. This, in a nutshell, is the allopatric speciation model. By the mid-1950s, it was widely accepted among nearly all biologists (with modifications as more exceptions were encountered), and soon they began to modify their definition of species to incorporate the interbreeding criterion.

Surprisingly, paleontologists seemed to be unaware of the implications of the allopatric speciation model, perhaps because apparently very few of them thought about fossils in a biological way or kept up with the literature on modern speciation theory. It was not until almost 30 years later in 1972 that two young scientists, Niles Eldredge of the American Museum of Natural History in New York and Stephen Jay Gould of Harvard University, first integrated modern concepts of biological species with the fossil record. If the allopatric speciation model applied to the fossil record, then we should not expect to see speciation in the fossils from the main population. Instead, speciation should occur in small, peripherally isolated populations that have little chance of being fossilized. In addition, all of the data from biology showed that this process of speciation typically takes place in tens to hundreds to thousands of years, which is a geologic instant as far as paleontologists are concerned. The age difference between two bedding planes is often many thousands of years. Thus we would not expect to see the gradual transitions between species preserved very often; instead, we expect to see new species when they immigrate back into the main population after their isolation and speciation events. In other words, they would suddenly appear in the fossil record. Once they were established, speciation theory would predict that the main population would remain stable and not change gradually through time but that new species would continually arise on the periphery and migrate back to the homeland. Eldredge and Gould (1972) called their idea *punctuated equilibrium*, because the fossil record seem to show species stability without change (equilibrium or *stasis*) except when it is punctuated by the arrival of new species from elsewhere (fig. 3.11B).

When the original Eldredge and Gould (1972) paper appeared, it caused a storm of controversy in paleontology. Gradualism was a deeply embedded concept, and many paleontologists had been trying to study it for their entire careers. Example after example of possible gradualism was raised, but they all suffered from some problem in the analysis or the data, which Gould and Eldredge (1977) quickly pointed out. Eventually it became clear that gradualism was extremely rare among multicellular animal fossils. (Microfossils, on the other hand, show a lot of gradualism, but they are not strictly sexual but largely clonal or asexual and so are not bound by the interbreeding criterion of Mayr's allopatric speciation theory.) Fossil species *do* show an incredible stability over many millions of years of strata, which Gould and Eldredge (1977) called stasis. Some biologists tried to explain away this stasis with mechanisms such as stabilizing selection (selection against the extremes of a population, reinforcing the mean tendency), but this does not explain how some fossil populations persist unchanged through millions of years of well-documented climatic change (surely, a strong selection pressure), as documented by Prothero and Heaton (1996), Prothero (1999), and Prothero et al. (2012). As Gould (1980a, 2002) pointed out, the persistence of fossil species through millions of years of intense selection pressure suggests that they are not infinitely malleable by selection but instead have an integrity or some sort of internal homeostatic mechanism that resists most external selection. This is a radical notion for evolutionary biology and is still hotly controversial. Most paleontologists argue that the fossil record shows things that can't be seen in fruit flies or living populations, but many biologists are unconvinced that the fossil record can't be explained by some neo-Darwinian mechanism (see chapter 4).

But this was not as startling for paleontologists as it was for biologists. Even though paleontologists had been trying to find Darwinian gradualism in the fossil record for a century, it was the stasis and abrupt appearance of fossil species that made biostratigraphy work so well. If everything evolved gradually, there would be a major problem of how to break up a gradually transforming lineage into nonarbitrary species segments. As Gould wrote, "All paleontologists knew that the practical world of fossil collecting rarely imposed such a dilemma. The oldest truth of paleontology proclaimed that the vast majority of species appear fully formed in the fossil record and do not change substantially during the long period of their later existence (average durations for marine invertebrate species may be as high as 5 to 10 million years). In other words, geologically abrupt appearance followed by subsequent stability." But instead of being an embarrassment of absence of evidence for evolution, the prevalence of stasis is a powerful message for evolutionary biologists that there is something to be explained that still has not been explained by experiments or observations of living species. Gould (1993) put it this way:

> Stasis, or nonchange, of most fossil species during their lengthy geological lifespan was tacitly acknowledged by all paleontologists, but almost never studied explicitly because prevailing theory treated stasis as uninteresting nonevidence for nonevolution. . . . The overwhelming prevalence of stasis became an embarrassing feature of the fossil record, best left ignored as a manifestation of nothing (that is, nonevolution).

We have seen that punctuated equilibrium is simply an application of modern biological speciation theory to the fossil record, an application that happened to explain and highlight the long-known fact of stasis in fossil species. This stasis, in turn, is now causing discomfort among many evolutionary biologists, because there is not yet any good mechanism in neo-Darwinian theory for it, suggesting we still have a lot to learn about evolution and speciation. *But this is a good thing!* If we had all the answers, and paleontology provided no new or interesting facts and ideas, science would be very boring.

Through all this intense debate within evolutionary biology, the creationists are constantly on the lookout for some tidbit they can quote of out context to say just the opposite of the author's meaning. Sure enough, many of the quotations about punctuated equilibria are misconstrued to indicate that Gould and Eldredge claim there are no transitional forms or that the fossil record doesn't show evidence of evolution! Typically, these "quote miners" pull a single short section out of a longer quotation that gives exactly the opposite impression of what the author really said. Such a practice suggests that the creationists either can't read, don't understand the entire quote, or are intentionally trying to deceive their own readers by claiming that Gould and others have said something that is actually the opposite of what was meant (which means the creationists are dishonest and deceitful)! (For a complete archive of corrections of the commonest creationist misquotes, see www.talkorigins.org/ faqs quotes.) For example, Gould (1980b:181) writes,

> The extreme rarity of transitional forms in the fossil record persists as the trade secret of paleontology. The evolutionary trees that adorn our textbooks have data only at the tips and nodes of their branches; the rest is inference, however reasonable, not the evidence of fossils.

This is quoted again and again in creationist books and websites. But if the creationists read closer and tried to understand the context of the argument, it is clear that Gould is not claiming that there are *no* transitional forms *between species* but only that they are rare (as expected from the fossil record and allopatric speciation theory). More importantly, there are *many* transitional fossils *between larger groups*, as we shall document through the entire second part of this book. If a series of stable fossil species shows an overall trend linking one major group to another, each of those fossil species is a "transitional form," even though we rarely get all the fossils of the transitions *between* those species (fig. 3.12).

An important concept here is the distinction between a *lineal ancestor* and a *collateral ancestor* (terms borrowed from genealogy). A lineal ancestor is one of your direct ancestors: your father and mother, grandparents, great-grandparents, and so on. A collateral ancestor shares an ancestor with you, but you are not their direct descendant: your uncles and aunts, great-uncles and great-aunts, and so on. In the fossil record, we often talk about a specific fossil not as being directly ancestral to some other organism but as showing anatomical features that indicate it is almost certainly a collateral ancestor (or a sister group, or closest relative, if you prefer). The creationist Jonathan Wells (2000) frequently confuses this point in his highly muddled and misleading book *Icons of Evolution*. For example, Wells (2000:138) attacks the use of the fossil *Archaeopteryx* by arguing that it is not an "ancestor" because modern birds are not descended from it and that paleontologists no longer consider it "transitional" and have "quietly shelved" it while looking for other "missing links." First of all, there is no such thing as a missing link (as discussed in chapter 5), and paleontologists have not quietly shelved it but continue to debate and discuss it along with the huge number of other Mesozoic bird fossils discussed in chapter 12. *Archaeopteryx* has many *transitional features* between living birds and Mesozoic dinosaurs, so if it was not a direct ancestor, it was certainly a collateral ancestor. Actually, *Archaeopteryx* has no derived features that would exclude it from the ancestry of later birds, so it very well could be the ancestor—but that is not a testable scientific statement, so paleontologists are cautious about using the term "ancestor" in this instance.

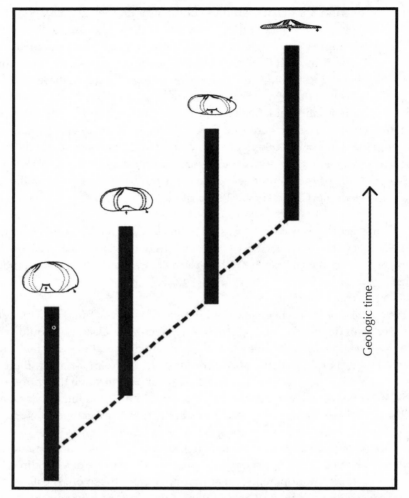

FIGURE 3.12. We do not need to have every single transitional fossil connecting several species to show evolutionary transformation. For example, a sequence of related species that are stable through time nevertheless forms an evolutionary transformation series, even though not every transitional fossil has been preserved. The evolution of sand dollars from sea urchins is an excellent example (see figure 8.13).

In the *Dictionary of Science and Creationism*, Ecker (1990:195–196) defines the issue with transitional forms this way (capitalized words indicate other entries in the dictionary):

> There are nevertheless all sorts of transitional forms in the fossil record to belie the creationist argument that they do not exist. They exist not only at the species level (creationists consider these only "variant forms of the same basic kind") but between major groups: There are intermediates between FISH and AMPHIBIAN (*Ichthyostega*), amphibian and REPTILE (*Seymouria*), reptile and MAMMAL (the mammal-like reptiles), reptile and BIRD (see *ARCHAEOPTERYX*), extinct ape forms and MAN (see AUSTRALOPITHECINES).

But creationists quibble to no end, their basic argument, as paraphrased by biologist Kenneth Miller, being that "the intermediates are not intermediate enough." The creationists consider the reptile-bird intermediate *Archaeopteryx*, for example, to be "100 percent bird" because it had wings and feathers and flew, when in fact *Archaeopteryx* was basically a flying, feathered dinosaur. What creationists challenge evolutionists to show them, it seems, is a "perfect 10" transitional form, exactly halfway between, say, fish and amphibian. But no such "fishibian," says the Institute for Creation Research (ICR), has ever been found in the fossils.

The creationists through such arguments exhibit no understanding of the nature of transitional forms. There is no general conversion of all parts of a transitional form at the same time. Genetics would not produce a smooth gradation of all features of an intermediate such as the creationists with their fishibian require; rather, it is to be expected that the characteristics of an intermediate will be mixed, a pattern called *mosaic evolution*. Nor does a fossil form need to be in the direct line of descent between two groups to be considered transitional. *Archaeopteryx*, for example, was doubtless not the direct ancestor of birds but rather one of that ancestor's cousins. Similarly, the fishlike amphibian *Ichthyostega* was probably a dead-end collateral branch of the fish-to-amphibian transition. The point is that a cousin of an ancestor is the more likely paleontological find, given the multiple splitting off of species and the general spottiness of the fossil record, and is evidence enough that a transition occurred.

The fact is, however, that not even a direct ancestral "10" would make any difference to creationists. No such form could be accommodated to their preconceived belief system. Thus creationist leader Henry Morris states that even the discovery of a fossil intermediate between men and apes—Morris believes that no such intermediate has been found, the australopithecines being "merely extinct species of apes"—would not be proof of human evolution. "An extinct ape," says Morris, "could have certain man-like features and still be an ape," and a man could have some ape-like features and "still be a man." In other words, no conceivable ape-man transitional form could be anything other than either true ape or true man. Creationists simply cannot allow transitional forms to exist, for to do so would be to admit that evolution has occurred.

Since the remainder of this book after chapter 5 will document transitional fossils in almost every major group, we will not dwell on the concept any longer in this chapter. But the point is clear: if you want to talk about transitional forms and the fossil record, *you go do the basic research on fossils; you don't quote people out of context*. Not only is doing research by quote mining lazy, unscientific, and deceitful, but it is also trying to prove your case by argument from authority not by actual scientific data or experiment (which is the only real evidence in science).

In his classic article on evolution, the great geneticist Theodosius Dobzhansky (1973:125) wrote,

> This is not to imply that we know everything that can and should be known about biology and about evolution. Any competent biologist is aware of a multitude of problems yet unresolved and of questions yet unanswered. After all, biologic research shows no sign of approaching completion; quite the opposite is true. Disagreements and clashes of opinion are rife among biologists, as they should be in a living and growing science. Antievolutionists mistake, or pretend to mistake, these disagreements as indications of dubiousness of the entire doctrine of evolution. Their favorite

sport is stringing together quotations, carefully and sometimes expertly taken out of context, to show that nothing is really established or agreed upon among evolutionists. Some of my colleagues and myself have been amused and amazed to read ourselves quoted in a way showing that we are really antievolutionists under the skin.

As we saw from the Gould quotes just given and Whitcomb and Morris's (1961) deceptive use of the quote about the Lewis overthrust earlier in the chapter, quote mining is dishonest and unscientific. In his book *Abusing Science: The Case Against Creationism*, Philip Kitcher (1982:185) said it all,

So it goes. One scientist after another receives the Creationist treatment. Any qualifying comment, any deviation from orthodoxy is a potential target. Ripped from its context, it can be made to serve the Creationists' purpose, namely, to convince the uninitiated that Creationist theses are sometimes advanced by scientists in scientific debates. But anyone can play the same game. In conclusion, I cannot resist turning the weapon against the Creationist who has used it to its greatest effect. Referring to the controversy about transitional forms, Gish writes, "There should be no room for question, no possibility of doubt, no opportunity for debate, no rationale whatsoever for the existence of the Institute for Creation Research" (Gish, 1981, ii).

How true.

For Further Reading

Berra, T. 1990. *Evolution and the Myth of Creationism*. Stanford, Calif.: Stanford University Press.

Beus, S., and M. Morales, eds. 1990. *Grand Canyon Geology*. Oxford: Oxford University Press.

Dalrymple, G. B. 1991. *The Age of the Earth*. Stanford, Calif.: Stanford University Press.

Dalrymple, G. B. 2004. *Ancient Earth, Ancient Skies: The Age of Earth and Its Cosmic Surroundings*. Stanford, Calif.: Stanford University Press.

Eldredge, N. 1982. *The Monkey Business: A Scientist Looks at Creationism*. New York: Pocket Books.

Eldredge, N. 1985. *Time Frames*. New York: Simon & Schuster.

Eldredge, N. 2000. *The Triumph of Evolution and the Failure of Creationism*. New York: Freeman.

Eldredge, N., and S. J. Gould. 1972. Punctuated equilibria: An alternative to phyletic gradualism. In *Models in Paleobiology*, ed. T. J. M. Schopf. San Francisco: Freeman Cooper, 82–115.

Gould, S. J. 1980. Is a new and general theory of evolution emerging? *Paleobiology* 6:119–130.

Gould, S. J. 1992. Punctuated equilibria in fact and theory. In *The Dynamics of Evolution*, ed. A. Somit and S. A. Peterson. Ithaca, N.Y.: Cornell University Press, 54–84.

Gould, S. J. 2002. *The Structure of Evolutionary Theory*. Cambridge, Mass.: Harvard University Press.

Gould, S. J., and N. Eldredge. 1977. Punctuated equilibria: The tempo and mode of evolution reconsidered. *Paleobiology* 3:115–151.

Kitcher, P. 1982. *Abusing Science: The Case Against Creationism*. Cambridge, Mass.: MIT Press.

Mayr, E. 1942. *Systematics and the Origin of Species*. New York: Columbia University Press.

McGowan, C. 1984. *In the Beginning: A Scientist Shows Why the Creationists Are Wrong*. Buffalo, N.Y.: Prometheus.

Numbers, R. 1992. *The Creationists: The Evolution of Scientific Creationism*. New York: Knopf.

Prothero, D. R. 1990. *Interpreting the Stratigraphic Record*. New York: Freeman.

Prothero, D. R. 1992. Punctuated equilbria at twenty: A paleontological perspective. *Skeptic* 1(3):38–47.

Prothero, D. R. 2013. *Bringing Fossils to Life: An Introduction to Paleobiology*. 3rd ed. New York: Columbia University Press.

Prothero, D. R., and R. Dott. 2010. *Evolution of the Earth*. 8th ed. New York: McGraw-Hill.

Prothero, D. R., and F. Schwab. 2013. *Sedimentary Geology*. 3rd ed. New York: Freeman.

Winchester, S. 2002. *The Map That Changed the World: William Smith and the Birth of Modern Geology*. New York: Harper.

FIGURE 4.1. We are familiar with pictures of Darwin as a sickly, bearded old man, but when he traveled the world on the *Beagle* voyage, climbed the Andes, and came up with natural selection, he was a vigorous young man in his twenties. Darwin's 5-year research expedition would have been equivalent to a Ph.D. project, because there were no modern doctoral programs back in 1836. The original caption reads, "How Charles Darwin might have looked as a modern graduate student just back from five years of field work. The picture is intended to fix in readers' minds that Darwin was at his most innovative at this age, and readers should remember that Darwin might now be denied admission to a good graduate school because of his deficiencies in languages and math." (From P. R. Darlington Jr., 1980, *Evolution for Naturalists: The Simple Principles and Complex Reality*, frontispiece. Reprinted with permission of John Wiley & Sons, Inc.)

THE EVOLUTION
OF EVOLUTION

4

Before Darwin

Would it be too bold to imagine, that in the great length of time since the earth began to exist, perhaps millions of ages before the commencement of the history of mankind, . . . that all warm-blooded animals have arisen from one living filament, which THE GREAT FIRST CAUSE endued with animality, with the power of acquiring new parts, attended with new propensities, directed by irritations, sensations, volitions and associations, and thus possessing the faculty of continuing to improve by its own inherent activity, and of delivering down those improvements by generations to its posterity!

—Erasmus Darwin, *Zoonomia*

Although Charles Darwin (fig. 4.1) deserves most of the credit for bringing about the scientific revolution in biology, he was by no means the first to suggest that life had changed through time. As early as the fifth century B.C.E., Greek philosophers such as Empedocles promoted the idea that life is constantly transforming. In 50 B.C.E., the Roman philosopher Lucretius wrote the poem *De rerum naturae* ("On the nature of things"), which postulated the existence of atoms and argued that everything in nature was in flux. With the fall of the Roman Empire, however, this bold style of thinking was suppressed by church orthodoxy, and the Genesis account of earth and life history ruled for almost 1,300 years. By the early 1700s, most people in Europe and North America still believed in a literal interpretation of the Bible and in the idea that the earth had been formed about 6,000 years ago, with no further change except for decay and corruption due to the sins of Adam and Eve.

During the next century, however, it became increasingly difficult to reconcile the Genesis accounts of the origin of earth and life with the expanding knowledge of nature. As we discussed in chapter 3, the discovery of faunal succession between 1795 and 1805 made Noah's flood explanations of geology more and more implausible. By 1840 flood geology had been abandoned altogether by devout Christian geologists. By the mid-1700s, natural historians like Linnaeus and his successors had already recognized over 6,000 species of animals and many more plants, far too many to squeeze into Noah's ark. At the same time, exploration of exotic places—Africa, South America, Australia, and Southeast Asia— was producing still more new species previously unknown to Europeans, and the ark story became ludicrous. Instead of showing a pattern that reflected dispersal from Mount Ararat in Turkey, the distribution of animals showed a completely nonbiblical pattern. There were many biogeographic puzzles, like the fact that many islands (such as New Zealand, Madagascar, and Hawaii) were inhabited by unique species not found anywhere else on earth or the fact that Australia was dominated by peculiar beasts such as the egg-laying platypus and the pouched marsupials but had no native placental mammals. Why had nothing but marsupials migrated from Mount Ararat to Australia?

In addition to all these new facts from natural history, new attitudes were prevailing in the 1700s. Often called "the Enlightenment," it was a period of skepticism and questioning of dogmatic authority (especially royal authority and religious dogmas), as scholars and philosophers sought rational, nonsupernatural explanations for the world. Inspired by the breakthroughs in science and explaining nature pioneered by Bacon, Newton, Leibniz, Pascal, and Galileo in previous centuries, Enlightenment scientists and philosophers explored new and daring ideas, unfettered by the shackles of the powers that be. In France, the Enlightenment was led by Rousseau, Diderot, and Voltaire, who questioned the authority of the king of France and the church, and by Lavoisier and Buffon, who made scientific breakthroughs in chemistry (Lavoisier) and proposed daring nonbiblical explanations of nature (Buffon). In England, John Locke and Thomas Hobbes examined political and economic systems and laid the foundation for the American experiment in democracy; George Berkeley broke new ground in philosophy (as did Spinoza and Kant in Holland and Germany); and the Industrial Revolution began with steam engines, textile manufacturing, and canal systems transforming England from an agrarian nation to an industrialized one. In 1764, the Lunar Society (so named because they met for dinner on the Monday night nearest the full moon; they called themselves the "Lunatics") was formed in Birmingham and promoted new scientific and technological ideas. The original founders included Erasmus Darwin (Charles Darwin's grandfather), William Small (Jefferson's mentor), and the industrialist Matthew Boulton. Soon the "Lunatics" included many of the great minds in Britain (including Benjamin Franklin when he visited). In Scotland, the brilliant men who met in Edinburgh and Glasgow pubs included the philosopher David Hume (discussed in chapter 2), the pioneering economist Adam Smith (the "father of capitalism"), the inventor of the first practical steam engine, James Watt, and the father of modern geology, James Hutton. Eventually, these new ideas and challenges to royal authority helped lead to the American Revolution against Britain in 1775. American patriots such as Jefferson were inspired largely by British and French political philosophers, such as Locke and Rousseau. These ideas also triggered the French Revolution in 1787, which overthrew the centuries-long domination of France by the Bourbon monarchy and the church.

The Enlightenment had the greatest effect on natural history in France, where Buffon had proposed daring nonbiblical ideas as early as 1749. The foremost French naturalist of the late 1700s was Jean-Baptiste Antoine de Monet, the chevalier de Lamarck (1744–1829). He began his career as a Linnaean-style botanist, but when the French Revolution swept through the king's gardens (*les Jardins du Roi*), he was reassigned to study the less glamorous "insects, shells, and worms," which had been neglected by naturalists for centuries. Lamarck soon revolutionized invertebrate zoology, laying out the foundation of our modern understanding of these "spineless wonders." Because he had started as a botanist, he soon recognized that zoology and botany made an integrated whole, which he named "biology." His ideas culminated in 1809 with the publication of *Philosophie Zoologique*.

In this work, Lamarck points out how all of life is highly variable and interconnected, not composed of discrete fixed species from the original Creation. He arranged animals in the traditional "scale of nature" (*scala naturae*) from the primitive sea jellies and corals to worms to mollusks and insects to vertebrates, with humans at the top of the scale. In some versions, the upper rungs of the "ladder of nature" were occupied by divine beings, with the ranks of cherubim, seraphim, angels, and archangels culminating in God. Like many people of his time, Lamarck believed that life was constantly being spontaneously generated out of the mud. Once formed into sea jellies, life would be transformed and move up the ladder, so that lineages that had originated long ago had already climbed to the higher rungs of fish or mammals or humans.

This "ladder of nature" notion from the late 1700s is still a common misconception that leads to all sorts of philosophical and biological absurdities. We will talk more about this topic in the next chapter, but the main point is that the concept of the ladder has been replaced by the metaphor of the bushy, branching "family tree of life."

Although Lamarck's ideas were evolutionary, they were not very similar to modern concepts of evolution. Instead of the bushy family tree of life we now recognize, Lamarck's concept was that of many different "blades of grass" independently arising by spontaneous generation out of the mud and climbing the ladder of complexity and not interconnected by common ancestry. Typical of the works of its time, *Philosophie Zoologique* was highly speculative and philosophical and not supported by much hard evidence from nature or experimental data. A minor idea in Lamarck's book was the *inheritance of acquired characters*. Like most people of that time (including Charles Darwin 50 years later), Lamarck believed that the characteristics that you developed during your lifetime (such as the muscles of a blacksmith or a bodybuilder) could be passed directly to your offspring. According to Lamarck, the giraffe would keep stretching and stretching its neck and pass those improvements to its descendants until they all had long necks. After Lamarck's death, this idea was lampooned by his enemies as "Lamarckism" or "Lamarckian inheritance" (even though Charles Darwin believed it, too). Consequently, most of Lamarck's great achievements are largely forgotten, and his name is now attached to a minor part of his ideas that has become disreputable.

The Evolution of Darwin

In the survival of favoured individuals and races, during the constantly-recurring struggle for existence, we see a powerful and ever-acting form of selection.
—Charles Darwin, *On the Origin of Species*

Charles Darwin (fig. 4.1) was born on the same date as Abraham Lincoln (but about 15 hours earlier)—February 12, 1809. Like Lincoln, he was a liberating force for humankind, but instead of freeing people from slavery, he freed biology from the bondage of supernaturalism. Philosophers of science have long pointed to Darwinian evolution as the greatest scientific revolution within biology, comparable to the role of Isaac Newton's or Albert Einstein's revolutionary ideas in physics or the plate tectonics revolution in geology. Before Darwin, it was possible (although increasingly difficult) to see nature as divinely created as we see it, unchanged over thousands of years. After Darwin, all of life was subject to natural law, just as Newton had shown that the stars and planets followed natural laws and did not require God to move them. As Sigmund Freud (1917) commented,

In the course of centuries the naïve self-love of men has had to submit to two major blows at the hands of science. The first was when they learnt that our earth was not the center of the universe but only a tiny fragment of a cosmic system of scarcely imaginable vastness. This is associated in our minds with the name of Copernicus, though something similar had already been asserted by Alexandrian science. The second blow fell when biological research destroyed man's supposedly privileged place in creation and proved his descent from the animal kingdom and his ineradicable animal nature. This revaluation has been accomplished in our own days by Darwin, Wallace and their predecessors, though not without the most violent contemporary opposition.

For such a revolutionary idea, it is surprising that it came from such a conventional, outwardly conservative man. Charles Darwin was born of a family of doctors, including his grandfather Erasmus, who was the king's physician, and his father, Robert, also a distinguished physician. But evolution was in Charles's blood as well. Erasmus wrote a 1794 poem entitled *Zoonomia*, which contained some of the most advanced evolutionary speculation of its time but was made less threatening by its poetic format and its religious references. Young Charles could not help but be influenced by his grandfather's ideas. When Charles was a teenager, his father sent him to medical school in Edinburgh. Charles proved to have no stomach for dissecting rotting corpses robbed from graves or watching people scream through surgery (as was common in medicine then, because there were no anesthetics). He was, however, exposed to the ideas of Robert Grant (the man who proved that sponges are animals), who was in turn influenced by the French evolutionists like Lamarck and Geoffroy. After Charles had dropped out of medical school, his father next sent him to Cambridge University, where he could become a clergyman and do something useful with his life while he pursued his mania for natural history collecting. There he neglected his theological studies and was instead influenced by botanist John Stevens Henslow and the geologist Adam Sedgwick (first professor of geology anywhere in the world). Several years later, Darwin had the opportunity to sail on the oceanographic voyage of the HMS *Beagle*. At first his father was horrified at the prospect, but eventually he relented.

The voyage lasted 5 years (1831–1836) and the *Beagle* traveled completely around the world. Darwin collected fossils of huge extinct animals in Argentina, climbed the Andes in Chile, and was one of the first to study the anthropology of the natives of Tierra del Fuego, in the freezing southern tip of South America. When the ship stopped at the Galapagos Islands for water and food, Darwin ended up making many observations and collections, yet not until he returned did he appreciate the importance of what he saw. The *Beagle* continued on its journey to Australia and South Africa before finishing its survey of the Brazil-Argentina coast and then returning home. When Darwin returned home, his father said, "Why, even the shape of his head has changed!"

Even more remarkable is what had changed *inside* his head. As a Cambridge-educated, wealthy gentleman, Darwin didn't need to work for a living; he could conduct research and access the elite scientific societies of the time. He set about publishing a book about his *Beagle* voyage and a book about coral atolls in the Pacific, which immediately made his reputation. He also made arrangements to have his specimens studied by the eminent specialists of the day. He married his first cousin Emma Wedgwood (of the Wedgwood pottery family), settled in the town of Down (just southeast of London), and began to raise a family. As he began to work, he read the writings of the Reverend Thomas Malthus, who pointed out that human populations were prone to exponential size increase unless held in check by death and disease. Darwin also became a fancier of domestic pigeons and soon developed a network of other gentlemen who raised and bred exotic domestic animals to develop unusual varieties, practicing a form of artificial selection. These ideas, and many others, combined to give Darwin his first notions of evolution by *natural selection*.

Only 6 years after returning from the *Beagle* voyage, Darwin had written his first sketch of his evolutionary ideas but put it under lock and key with instructions to his wife to open it and publish it only after his death. Darwin had good reason to be worried about his reputation if he published ideas about evolution. The entire concept was associated with radical French thinking and revolutionary politics of the lower-class medical schools in London. It was an anathema to the powerful conservative elites of the wealthy, the nobility, and the Anglican Church. In 1844, a Scottish publisher named Robert Chambers wrote an anonymous tract called *Vestiges of the Natural History of Creation*, which caused a national furor at its evolutionary thinking. Although the science in *Vestiges* was amateurish and easy to discredit, clearly evolutionary thinking was in the air in the 1840s. But it was still too controversial for a respectable Cambridge-educated gentleman like Darwin to touch.

So Darwin sat on his dangerous idea for 15 years, working quietly at home on barnacles. At that time, these animals were poorly understood, but Darwin discovered that they were highly modified crustaceans related to shrimp, showing amazing adaptations. He published several scientific books on barnacles, which gave him an impeccable scientific reputation. Darwin accomplished this despite working only a few hours a day because he was wracked by a mysterious illness, which may have been a psychosomatic response to the dangerous ideas he was considering. All the while, he was amassing notes for a gigantic book on the "species problem." He would have procrastinated for many more years had he not received a letter in the mail from a younger British naturalist, Alfred Russel Wallace, who was collecting specimens in what are now Malaysia and Indonesia. During a bout of malarial fever, Wallace had come up with a similar concept of natural selection and had sent his ideas to Darwin, of all people. Horrified, Darwin went to his friends Charles Lyell (the famous geologist) and Joseph Hooker (a famous botanist), searching for an honorable solution to the dilemma of who would get credit for the idea. They arranged to have Darwin's 1842 abstract and Wallace's letter read at an 1858 meeting of the Linnean Society. The president of the Linnean Society, Thomas Bell, while summarizing the discoveries for the year 1858, wrote: "The year which has passed has not, indeed, been marked by any of those striking discoveries which at once revolutionize, so to speak, the department of science on which they bear." But Darwin now had to work quickly or he would be scooped. He abandoned his planned gigantic work and wrote a shorter (155,000 words, which was short by Victorian standards) summary of his ideas, *On the Origin of Species by Means of Natural Selection*. It sold out all 1,250 copies on the day it was published and eventually went through six editions while Darwin was alive.

The argument of *On the Origin of Species* is very simple yet powerful. First, Darwin drew an analogy to the artificial selection of domesticated animals practiced by animal breeders. He argued that if they could modify the ancestral wolf into dogs as different as a Chihuahua and a Great Dane, then species were not as fixed and stable as commonly believed. Then he borrowed the idea from Malthus that natural populations are capable of exponential growth, yet they remain stable in nature because of high mortality rates. From this he deduced that *more young are born than can survive*. Darwin next described the variability of natural populations and pointed to the evidence from domesticated animals that these variations are highly heritable. He concluded that *organisms that inherit favorable variations are more likely to survive and breed*, and he called this process *natural selection* (called by others "survival of the fittest").

As we saw, Darwin was not the originator of the concept of evolution, and at least two others proposed something like natural selection. Why, then, does Darwin deserve most of the credit? For one thing, Darwin was the right man at the right time. In 1844, the idea was still too controversial and Chambers's amateurish efforts only made people scoff at the idea of evolution. By 1859, however, the time was right and many people were thinking along these lines (as Wallace's independent inspiration shows). In addition, Darwin had worked hard to build a sterling scientific reputation and was also a member of the Oxford-Cambridge elite. He was not a radical from the lower-class medical schools of London. Most importantly, Darwin put all the pieces together in one book and provided two important concepts: the *evidence* that life had changed through time (the "fact" of evolution) and a *mechanism* for how it occurred, natural selection (the "theory" of evolution). He overwhelms the reader with example after example, so that by the end the conclusion is inescapable.

Darwin's ideas were controversial at first, but by the time he died in 1882, the fact that life had evolved was universally accepted in all educated parts of the world (including most of the United States). When he died, Darwin was hailed as one of Britain's greatest scientists. He was buried in "Scientists' Corner" of Westminster Abbey, next to Isaac Newton and the rest of Britain's scientific geniuses. However, his mechanism of natural selection initially did not fare so well. Many of Darwin's critics could not imagine how it was sufficient to shape organisms. Some argued that if favorable variations occurred, they would be blended out of existence in a few generations by backcrossing with the normal strains of animals. Darwin never solved this problem by the time he died in 1882.

Ironically, the solution had already been discovered in 1865 by an obscure Czech monk named Gregor Mendel. He found that by breeding strains of pea plants in his garden, he could produce very simple and mathematically predictable inheritance patterns. More importantly, he showed that inheritance does not *blend* the genes of both parents but is *discrete*, so that rare genes from one parent can seem to vanish for a generation but then reappear fully functional in the next generation if the genes are recombined in a certain way. Mendel's work remained unknown until it was independently rediscovered by three different lab groups in 1900, when the time was ripe for appreciating his insights. Genetics made enormous strides over the next 50 years, culminating with the discovery of the DNA molecule and its role in inheritance in 1953.

The Neo-Darwinian Evolutionary Synthesis

Evolution is a change in gene frequencies through time.
—Theodosius Dobzhansky, *Genetics and Origin of Species*

Evolution is merely a reflection of changed sequence of bases in nucleic acid molecules.
—John Maynard Smith, *The Theory of Evolution*

Although the world had accepted the fact that life has evolved by the time of Darwin's death in 1882, it was less convinced that his mechanism, natural selection, was sufficient to explain all of evolution. Natural selection slowly lost favor as several different genetics labs rediscovered Mendelian inheritance and then built new ideas of how evolution occurred. Paleontologists, meanwhile, became even more heterodox, with some following Darwin's mechanism, others subscribing to some sort of inheritance of acquired characters ("neo-Lamarckians"), and still others completely agnostic as to what mechanism drove evolution. Although the fossil record continued to provide more and more evidence for how life had evolved, paleontologists were not at the forefront for theoretical mechanisms of how evolution occurred. Meanwhile, systematists (biologists who study the naming and relationships of organisms) were busy describing new species, but few thought of the evolutionary implications of their work. There was simply no common thread among them, and there appeared to be no way to show that Darwinian natural selection was compatible with genetics, paleontology, and systematics.

The breakthrough occurred in the 1930s, when three scientists introduced a set of mathematical models known as *population genetics*. Two were British (Sir Ronald Fisher and J. B. S. Haldane) and one was an American, Sewall Wright, whom I had the great fortune to meet in 1983 while he was still active and alert at age 94. These mathematical simulations allowed evolutionists to describe changing gene frequencies through many generations and simulate the effects of mutation and selection. Population genetics clearly showed that even slight selection pressure can quickly change gene frequencies and made evolution by Darwinian natural selection plausible again. By the late 1930s, several books tied population genetics to the various subdisciplines of evolutionary biology. In 1937, geneticist Theodosius Dobzhansky published *Genetics and the Origin of Species*, which updated and synthesized all the genetics known at the time, and showed how selection in fruit fly experiments gave amazing demonstrations of evolution in action. In 1942, ornithologist Ernst Mayr published *Systematics and the Origin of Species* (discussed in chapter 3), which dealt with the problem of speciation in nature and showed how the allopatric speciation model was consistent with natural selection. And in 1944 (written in 1941, but delayed by World War II), paleontologist George Gaylord Simpson published *Tempo and Mode in Evolution*, which attempted to show that nothing in the fossil record was inconsistent with Darwinian natural selection. Together, these works brought the major threads of evolutionary biology—genetics, systematics, and paleontology—back into the Darwinian fold. By the late 1940s, biologist Julian Huxley was referring to this new consensus as the "modern synthesis" or the "neo-Darwinian synthesis." By the 1959 centennial of the publication of *On the Origin of Species*, the neo-Darwinian synthesis completely dominated evolutionary biology, and most biologists thought that all the major problems had been solved and questions had been answered. After almost 60 years, most current textbooks on evolution still reflect this dominance of neo-Darwinism.

What are the main ideas of the neo-Darwinian synthesis? Its central core comes from genetics (with its focus on the *genotype* of the organism), which has shown just how effective natural selection can be in changing the frequencies of genes in populations. From this, neo-Darwinists define evolution in an extremely reductionist manner as a change in gene frequencies through time, without regard to the embryology or development of the organism or the influence of the body (*phenotype*). Some extreme neo-Darwinists argue that the body is simply a device for genes to make more copies of themselves. Population genetics and fruit fly experiments showed that most variation is due to the recombination of genes from both parents but that additional variation is the result of slight mutations. These random variants are then weeded out by natural selection, and the stronger the selection, the more rapid the genetic change. In some extreme versions of neo-Darwinism, natural selection was treated as an all-powerful, all-pervasive force that, in Darwin's words, "is daily and hourly scrutinizing throughout the world every variation, even the slightest; rejecting all that which is bad, preserving and adding up all that is good; silently and insensibly working." Some even argued that all changes are adaptive in some way, even if we can't detect how. To them, there are no features of an organism unaffected by natural selection. Such views are often called *panselectionism*.

Along with the reductionist attitude that organisms are nothing more than vessels to carry their genes came the extrapolation that the tiny genetic and phenotypic changes observed in fruit flies and lab rats were sufficient to explain all of evolution. This defines all evolution as *microevolution*, the gradual and tiny changes that cause different wing veins in a fruit fly or a slightly longer tail in a rat. From this, neo-Darwinism extrapolates all larger evolutionary changes (*macroevolution*) as just microevolution writ large. These central tenets—reductionism, panselectionism, extrapolationism, and gradualism—were central to the neo-Darwinian orthodoxy of the 1940s and 1950s and are still followed by the majority of evolutionary biologists today.

As we shall see later in this chapter, the evidence for microevolutionary change is abundant throughout nature, and we see evolution in action all the time. Neo-Darwinian evolutionary biology has had many great successes (detailed in all the textbooks), so there is no reason to doubt that natural selection is the most important engine for evolution. But is it the only factor involved? Is evolution truly reducible to changes in gene frequencies through time?

Challenges to Neo-Darwinism

The variations detected by electrophoresis may be completely indifferent to the action of natural selection. From the standpoint of natural selection they are neutral mutations.
—Richard Lewontin, *The Genetic Basis of Evolutionary Change*

Unfinished Synthesis

When neo-Darwinism swept through the profession in the 1940s and 1950s, it achieved almost a complete consensus. Many evolutionary biologists thought that the major problems were all solved; only the details needed to be worked out. But it is not a good thing when a field in science seems to have all the answers and is no longer questioning its assumptions. A continuing critical attitude, new unsolved problems, and skepticism and controversy are essential to the health of good science. If a science does not continue to test its ideas and views all essential problems as solved, then it soon stagnates and dies.

Fortunately, the neo-Darwinian synthesis has been continually scrutinized and challenged by legitimate biological and paleontological data, so the field is rife with healthy controversy. Most of these challenges only question some of the more extreme neo-Darwinian tenets or argue that natural selection is not the only mechanism by which life evolves. Despite the misleading misquotations of creationists, *none* of these ideas challenges the well-established fact that *life has evolved* or that natural selection is an important (if not exclusive) mechanism for evolution. For example, creationists frequently quote Stephen Jay Gould saying that "neo-Darwinism is dead" and suggest that Gould does not believe in evolution! In fact, Gould is arguing for the importance of non–neo-Darwinian mechanisms for evolution. A typical Gould quote in context reads:

I well remember how the synthetic theory beguiled me with its unifying power when I was a graduate student in the mid-1960's. Since then I have been watching it slowly unravel as a universal description of evolution. The molecular assault came first, followed quickly by renewed attention to unorthodox theories of speciation and by challenges at the level of macroevolution itself. I have been reluctant to admit it—since beguiling is often forever—but if Mayr's characterization of the synthetic theory is accurate, then that theory, as a general proposition, is effectively dead, despite its persistence as textbook orthodoxy. (Gould 1980b:120)

As the full quotation clearly shows, Gould is talking about the neo-Darwinian synthesis and is *not* expressing doubts that evolution occurred. Creationists either can't tell the difference or are deliberating misquoting Gould to mislead their readers.

Now that we have cleared up the creationist misunderstanding that a challenge to neo-Darwinism is *not* a denial that evolution has occurred, let us look at the legitimate scientific issues that are raised by Gould's words.

Lamarck Revisited

As we discussed already, most naturalists of the nineteenth century, including Lamarck and Darwin, concluded that features acquired during one's lifetime could be passed directly to descendants (the inheritance of acquired characters). This type of inheritance is unfortunately known as "Lamarckian inheritance"; as we mentioned already, it was an old notion from before Lamarck and only a minor part of Lamarck's ideas and was accepted by Darwin as well. It is obvious why this idea is appealing. Instead of the wasteful Darwinian mechanism of the death of many offspring, just so a few favorable variants can survive, Lamarckian inheritance allows new variations to be passed on directly in a single generation and allows organisms to adapt more readily.

By the 1880s, however, some geneticists began to doubt whether Lamarckian inheritance was real. Instead, they asserted the primacy of Darwinian natural selection. The German biologist August Weismann performed a series of experiments that seemed to discredit the idea of the inheritance of acquired characters. He cut off the tails of twenty generations of mice, but each new generation developed a tail, despite this rather extreme form of selection pressure. From this, Weismann concluded that anything that happens to our bodies ("soma" in Weismann's terminology) during our lifetimes does not get back into the genome ("germ line"). This became known as "Weismann's barrier" or the *central dogma* of genetics: the flow of information is a one-way flow, from genotype to phenotype, but not the reverse. When James Watson and Francis Crick discovered DNA, the central dogma was redefined to mean the one-way flow of information from DNA to RNA to proteins to phenotype—but never back to the DNA.

For decades, the central dogma seemed to work, and even the hint of Lamarckism was considered highly controversial and unorthodox. But as early as the 1950s, embryologist Conrad Waddington showed that repeated environmental stresses could cause abrupt genetic change without direct selection, or what he called genetic assimilation. The best evidence, however, comes from immunology. When we are born, our immune system is functional but does not yet recognize all the foreign germs and pathogens it must defend against. We acquire immunity through our lifetimes each time our immune system is exposed to a germ and develops an antibody to defend against it. However, a series of experiments showed that laboratory mice could pass on their immunity directly to their offspring (Steele et al. 1998). It is hard to see how this is explained by anything other than Lamarckian inheritance.

More recently, molecular biologists have found that acquired inheritance is the norm, rather than the exception, in most microorganisms. Viruses work entirely this way, inserting their DNA into the cell of a host and making more copies of themselves. Many bacteria and some other organisms (including plants such as corn) seem to have "jumping genes" that exchange gene fragments between strains of organisms without sex or even recombination. One group of viruses (the retroviruses that cause HIV, among other infections) copies their own genetic information from host to host and may be capable of carrying the DNA of one organism into another.

All of these new mechanisms of inheritance suggest the genome is not as simple and "one-way" as we thought only 40 years ago. John Campbell (1982) summarized a full range of genetic interactions, starting with the simple "structurally dynamic" genes that respond to a certain environmental stimulus by producing a particular response. At a more sophisticated level are genes that apparently sense their environment and change their response. Automodulating genes change their future responsiveness to stimuli when stimulated. The most Lamarckian of all are "experiential genes," which transmit specific modifications induced during their lifetimes into the genome of their descendants. The example from immunology may fit this, as does bacterial and viral DNA swapping.

Clearly, the simplistic "central dogma" no longer applies to microorganisms, which are remarkably promiscuous in swapping DNA around. It may also not apply to many multicellular organisms either, if the immunologic experiments are correctly interpreted.

Neutralism, Junk DNA, and Molecular Clocks

One of the first challenges to neo-Darwinism came when molecular biology began to understand the details of the genome in the 1960s. Prior to this, geneticists had assumed that each gene in the chromosome coded for only one protein (and the structures built from them), so that inheritance would be simple (the "one gene, one protein" dogma). They also asserted that every gene was under the constant scrutiny of natural selection (panselectionism) and no gene was selectively neutral (even if we can't detect how selection operates). But in the 1960s, a series of discoveries shattered this simplistic idea of the genome. Using a newly developed technique called electrophoresis, Lewontin and Jack Hubby (1966) found that organisms had far more genes than they actually use or that can be expressed in the phenotype. Soon, geneticists were discovering that as much as 85–97 percent of the DNA in some organisms (including about 90 percent of human DNA) is not critical for expression of a phenotypic feature and is either "silent" DNA or "junk" DNA left over from the distant past when it had some function. If it is not expressed, it cannot be detected by natural selection and is neutral with respect to selective advantages or disadvantages. This new idea of *neutralism* completely shattered the old belief in panselectionism. In recent years, a few geneticists have tried to salvage the idea that there is less "junk DNA" than once thought, and Project ENCODE made the claim that most of the DNA was minimally functional. However, these claims have been debunked by many different lines of evidence. Most of our DNA is indeed "junk" that is never read or used in any real functional sense.

At the most basic level, the fundamental structure of the genetic code guarantees that a high percentage of mutations will be invisible to natural selection. The genetic code (fig. 4.2) consists of a three-letter "triplet" sequence of nucleotides (adenine, cytosine, guanine, and

The genetic code, which specifies by three letters in the genome (A = adenine; C = cytosine; G = guanine; U = uracil) any one of 20 amino acids, or a stop command.

First base in the codon	Second base in the codon				Third base in the codon
	U	C	A	G	
U	Phenylalanine	Serine	Tyrosine	Cysteine	U
	Phenylalanine	Serine	Tyrosine	Cysteine	C
	Leucine	Serine	Stop	Stop	A
	Leucine	Serine	Stop	Tryptophan	G
C	Leucine	Proline	Histidine	Arginine	U
	Leucine	Proline	Histidine	Arginine	C
	Leucine	Proline	Glutamine	Arginine	A
	Leucine	Proline	Glutamine	Arginine	G
A	Isoleucine	Threonine	Asparagine	Serine	U
	Isoleucine	Threonine	Asparagine	Serine	C
	Isoleucine	Threonine	Lysine	Arginine	A
	Methionine	Threonine	Lysine	Arginine	G
G	Valine	Alanine	Aspartic acid	Glycine	U
	Valine	Alanine	Aspartic acid	Glycine	C
	Valine	Alanine	Glutamic acid	Glycine	A
	Valine	Alanine	Glutamic acid	Glycine	G

FIGURE 4.2. The genetic code. Each protein is specified by a three-letter "triplet" codon combination of adenine, guanine, cytosine, or uracil. Note how most amino acids can be specified by just the first two letters, and the third letter makes no difference—it is adaptively neutral, and most mutations at this locus are silent and nonselective.

uracil). As the DNA is transcribed by tRNA, it interprets each three-letter sequence as the code for one of the 20 amino acids (plus a few codes are used to stop the transcription of DNA). Notice in figure 4.2 that of the 64 possible combinations of three letters, many of them specify the same amino acid. It is usually the first two letters of the triplet that count, and the third letter makes no difference. For example, if the first two nucleotides are cytosine and uracil, it produces the amino acid leucine, no matter what the third letter is. Clearly, most mutations in the third-letter position (every third nucleotide in the DNA) are invisible to natural selection and must be neutral as a result.

From these discoveries, geneticists have come to realize that many mutations are adaptively neutral and continue to occur without interference from natural selection. This has led to the discovery of the *molecular clock*. When molecular biologists began to compare the DNA of closely related organisms, they found that there seemed to be a regular, predictable amount of change in their DNA that depended only upon how long ago the two lineages had been separated. When they calibrated their divergence points on the molecular family tree with the fossil record, they found that they could determine how long ago various lineages branched off, even in the absence of fossil evidence. All of this works because so much of the genome is invisible to selection and can constantly change by random mutation without interference. Although the molecular clock has had some great successes, mutation rates can vary unpredictably, so scientists are cautious about putting too much weight on molecular clock estimates for the age of a lineage when all the other evidence disagrees.

More importantly, the fact that 80–97 percent of the DNA in most organisms codes for nothing at all (so far as we know) says that evolution and selection must work entirely on the remaining few percent of the DNA that *does* code for something. Those remaining genes are known as *regulatory genes*. They are the master switches that control the reading of the rest of the DNA, some of which is used to make the basic structures of life (*structural genes*) and therefore does not differ between organisms. So from the assertion in the 1950s that *every* gene codes for one protein, we now know that most genes don't code for anything, and only a few regulatory genes exert almost complete control over every other gene in the DNA. By tiny changes in those "switches" or regulatory genes, the organism can make big evolutionary leaps.

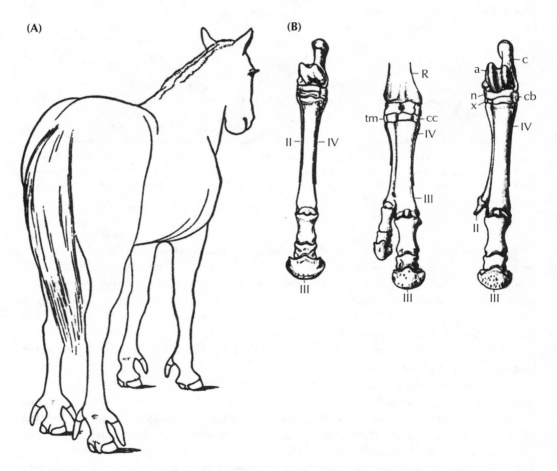

FIGURE 4.3. (A) A famous example of a rare mutant horse that has three toes, rather than one. (B) Bone structure of the feet of the mutant horses. On the left is a normal horse foot; in the middle is an extra toe formed by duplicating the central toe; on the right is an extra toe formed by enlarging the reduced side toes (splint bones), which were functioning side toes in earlier horses. (From Marsh 1892)

We can see the importance of these regulatory genes when something goes wrong, and a bizarre atavistic mutant, or "evolutionary throwback" occurs. Humans still have the genes for the long tail of our monkey ancestors, and every once in a while, the suppression of those genes fails and a human is born with an external tail (fig. 15.9). A simple failure in the transcription in the genes of a horse, and you get a horse with three toes (fig. 4.3). The side toes are poorly developed, but they still resemble the condition of the ancestral horses, which had two functional side toes. This experiment shows that the genes for the ancestral side toes are not lost in modern horses, only suppressed by the regulatory genes, and when there is a mistake in regulation, these ancient features reappear. Such freakish "horned horses" were thought to have great powers, and Julius Caesar rode one into battle.

The most striking example was an experiment that showed that birds still have the genes for teeth, even though no living bird has teeth. The embryonic mouth tissues of a chick were grafted into the mouth area of a developing mouse. When the mouse grew teeth, they were not normal mouse teeth, but conical peg-like teeth similar to those of the earliest toothed birds, or the dinosaurian ancestors of birds. All it took was the removal of the regulatory genes that a chick would normally have (by grafting tissues into a mouse) and the long-suppressed genes for reptilian teeth carried by all birds finally emerged. Other embryonic studies have managed to change the genes that code for the birds' short, stumpy, bony tail, resulting in the development of a long bony tail like a dinosaur. Another genetic modification experiment in developing chickens gave rise to chickens with dinosaur-like feet, not bird feet. Yet another experiment produced a bird with a dinosaurian snout with teeth instead of a normal beak. Birds have nearly all their old dinosaurian genes residing in their genome, just not expressed.

Macroevolution and Evo/Devo

You have loaded yourself with an unnecessary difficulty in adopting *Natura non facit saltum* [Nature does not make leaps] so unreservedly.
—Thomas Henry Huxley, in an 1859 letter to Charles Darwin

The importance of regulatory genes goes far beyond neutralism and junk DNA. It raises the question again of whether microevolution, which is so successful at making small changes (such as the number of bristles or wing veins in a fruit fly or the length of the beak of a Galapagos finch) is sufficient to explain macroevolution (the development of large-scale changes in evolution, such as new body plans). If you just keep accumulating tiny microevolutionary changes through time, would this produce wholly novel organisms?

This debate goes back to the earliest days of evolutionary biology. Darwin was a convinced gradualist, but his friend and defender Huxley warned him (in the quotation at the start of this section) that he need not tie his evolutionary ideas to gradualism or rule out evolutionary "leaps" to new body forms. When neo-Darwinism became dominant in the 1940s and 1950s, Richard Goldschmidt, a German-born geneticist at Berkeley, protested the strict gradualist position. He argued from his studies of gypsy moths that the changes required to build new body plans and new species were not the same as those he found within the normal variation within a species. Goldschmidt argued that some sort of large-scale genetic change was needed (a "systemic mutation" in his words) to jar species out of their normal range of variation and into new body plans. These changes were due to slight changes in "controlling genes" (what we now call regulatory genes). According to Goldschmidt, speciation was a discontinuous, rapid process that was caused by changes in controlling genes, not by accumulation of small microevolutionary changes. If a new macromutation appeared that gave the individual a big advantage, it might produce a "hopeful monster" that could establish a new species or a new adaptive zone.

Naturally, such opinions were highly unorthodox with respect to the gradualistic ideas of the newly dominant neo-Darwinians, and they subjected Goldschmidt to ridicule and scorn. When I was taking evolution classes from hard-core neo-Darwinians in graduate school, they would scoff "How does the hopeful monster find a mate?" Without more than one hopeful monster, there is no possibility of breeding or establishing a new population, and thus there would be no chance of a new species forming.

Ironically, the past 20 years have vindicated Goldschmidt to some degree. With the discovery of the importance of regulatory genes, we realize that he was ahead of his time in focusing on the importance of a few genes controlling big changes in the organism, not small-scale changes in the entire genome as neo-Darwinians thought. In addition, the hopeful monster problem is not so insurmountable after all. Embryology has shown that if you affect an entire population of developing embryos with a stress (such as heat shock), it can cause many embryos to go through the same new pathway of embryonic development, and then they all become hopeful monsters when they reach reproductive age (Rachootin and Thomson 1981).

The more we learn about regulatory genes, the more we realize their primary importance to evolution. A common example is the study of *heterochrony*, where organisms change the sequence of their developmental timing. This allows evolution to take advantage of the changes already encoded in our embryology and development. For example, nature frequently makes changes through *neoteny*, where an organism retains its juvenile body form while achieving reproductive maturity. The most famous case involves the salamanders (such as the Mexican axolotl) that do not complete their metamorphosis into lunged salamanders but hold on to their juvenile gills and body form, yet they can breed like adults (fig. 4.4). Whenever these salamanders are exposed to stagnant water conditions, they can complete their metamorphosis into lunged adults and walk to the next fresh pool of water. Thus, this ability to choose to breed either as the juvenile or adult body form gives them great ecological flexibility, all with a few tiny changes in the regulation of their development.

As Stephen Jay Gould pointed out in his book *Ontogeny and Phylogeny* (1977), this mechanism is extremely common in nature, especially when the juvenile and adult body forms have radical differences in shape and ecology, and allows the organism to "switch-hit" for whatever works best. Those pesky aphids that invade your flowers each spring are a classic example. When the food resources are abundant (in the spring and summer), they multiply rapidly, with each female giving birth to immature daughters as asexual clones (no males are born at all). Those offspring, in turn, also reproduce asexually as juveniles, so they can make literally hundreds of daughters in a short period of time (which is why they can infest your flowers so quickly). When the fall comes and the food resources dry up and cold weather approaches, they switch to sexual reproduction. A few males are born and mature into adults and then quickly mate with adult females. These lay normal eggs that can survive the winter and hatch out next spring to start the process all over again. All of this evolutionary flexibility does not require big changes in the genome, just small changes in regulating the normal sequence of embryonic development already encoded in the organism.

FIGURE 4.4. The Mexican salamander *Ambystoma*, known to the Aztecs as the axolotl. In normal conditions, it retains its larval gills into sexual maturity, enabling it to remain aquatic. However, if the water becomes stagnant, it completes its metamorphosis into an adult salamander with lungs, so it can crawl out and find a new lake to live in. By using different parts of its natural developmental cycle, it has great evolutionary flexibility. (After Dumeril 1867)

The most important recent development, however, has been the discovery of the master regulatory genes, known as the *homeotic genes* (especially the "Hox" genes). These genes are found in nearly all multicellular organisms and regulate the fundamental development of the body plan and how major organ systems develop. They were first discovered with experiments on fruit flies that had unusual mutations. Some had legs growing on their heads instead of antennae (fig. 4.5A); this is known as the "antennipedia" mutation. Some flies developed two pairs of wings instead of the usual single pair (fig. 4.5B). Normal flies have tiny knob-like balancing organs called halteres where the second pair of wings would be, but these mutant flies have apparently changed their regulatory genes so that ancestral wings appeared instead of the halteres.

From these early discoveries, molecular biologists have identified most of the Hox genes in a number of organisms and found that nearly all animals (including flies, mice, and humans) use a very similar set of Hox genes, with slight variations, subtractions, and additions. Each Hox gene is responsible for the development of part of the organism and all its normal organ systems (fig. 4.6). Small changes in the Hox genes can put different appendages on a segment of a fly (like the leg where the antenna would go or the wing where a haltere belongs) or even multiply the number of segments. Clearly, then, a tiny change in Hox genes can make a big evolutionary difference. In the arthropods (the "jointed legged" animals, such as insects, spiders, scorpions, and crustaceans), for example, a small

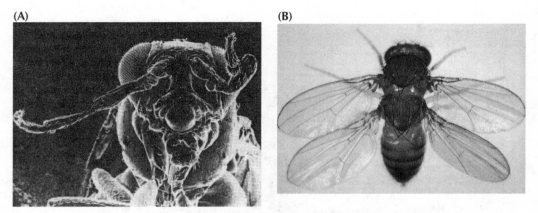

FIGURE 4.5. Homeotic mutants show that big developmental changes can result from small genetic mutations, giving rise to dramatic differences in body plan. (A) The antennipedia mutation, where a leg grows instead of an antenna. (Photo courtesy F. R. Turner, Biology Department, Indiana University) (B) The bithorax mutant fly, which has a second pair of wings instead of the halteres normally found behind the front pair of wings. (Photo courtesy W. Gehring and G. Backhaus)

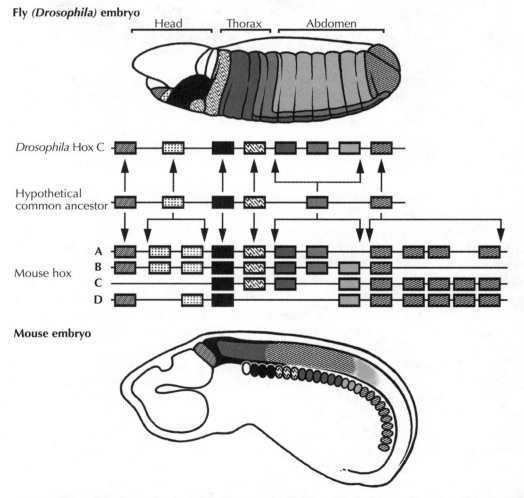

FIGURE 4.6. Map of the locus of action of the Hox genes in the fly and in the mouse. Note that the basic Hox genes are similar in almost all bilaterally symmetrical animals, so the system goes back to the very origin of complex animals. Small changes in any of these Hox genes make big differences in body plans. (Illustration by Carl Buell)

change in the Hox genes can multiply the number of segments or reduce them and switch one appendage (e.g., a leg) on each segment with another (e.g., a crab claw or an antenna or mouth parts). Arthropods are a classic example of this modular development with interchangeable parts; with a small change in Hox genes, whole new body plans can easily evolve to exploit new resources.

All of these ideas are part of the exciting new research field known as "evolutionary development" (nicknamed "evo/devo"), and it is now the hottest topic in evolution. From the neo-Darwinian insistence on every gene gradually changing to make a new species, we now realize that only a few key regulatory genes need to change to make a big difference, often in a single generation. This circumvents many of the earlier problems with ideas about macroevolution and makes it entirely possible that the processes that build new body plans and allow organisms to develop new ecologies are not the small-scale microevolutional changes extrapolated upward. Some evolutionary biologists still see evo/devo as just an extension of the neo-Darwinian synthesis (e.g., Carroll 2005), but others argue that it is an entirely different type of process than neo-Darwinists envisioned in the 1950s (e.g., Gould 1980a, 2002).

When Gould wrote his 1980 article, "Is a New and More General Theory of Evolution Emerging?," he was pointing to these provocative new ideas (heterochrony, regulatory genes, homeotic mutants, and the stability of species through time as shown by punctuated equilibrium). He argued that the neo-Darwinian synthesis, with its emphasis on tiny changes in the genotype adding up to new species by microevolutionary change, was not sufficient to explain macroevolution but that these new developments showed how macroevolution could occur.

There are many hard-core neo-Darwinians who do not agree, of course, so evolutionary biology is in an interesting, controversial time where new ideas are being intensely debated. It may turn out that we understand less about how evolution works than we thought we did back during the heyday of the synthesis in the 1950s and 1960s. But the important point is that *this is how normal science operates*. Even if we knew nothing about the mechanisms that drive evolution, it would not change the factual data that show it has occurred and still is occurring (as discussed in the next section). We still don't know exactly how gravity works, but it does not change the fact that objects still fall to the ground. We may never know completely how evolution works, but life keeps evolving. And to repeat our earlier point: even if "neo-Darwinism is dead" (as the creationists like to misquote Gould), that's only one possible explanatory mechanism for evolution. *Neo-Darwinism is not all there is to evolution. Evolution happened in the past and is still happening right now.*

The Evidence of Evolution

Well, evolution is a theory. It is also a fact. And facts and theories are different things, not rungs in a hierarchy of increasing certainty. Facts are the world's data. Theories are structures of ideas that explain and interpret facts. Facts do not go away when scientists debate rival theories to explain them. Einstein's theory of gravitation replaced Newton's, but apples did not suspend themselves in mid-air, pending the outcome. And humans evolved from apelike ancestors whether they did so by

Darwin's proposed mechanism or by some other, yet to be discovered. . . . Scientists regard debates on fundamental issues of theory as a sign of intellectual health and a source of excitement. Science is—and how else can I say it?—most fun when it plays with interesting ideas, examines their implications, and recognizes that old information might be explained in surprisingly new ways. Evolutionary theory is now enjoying this uncommon vigor. Yet amidst all this turmoil no biologist has been lead to doubt the fact that evolution occurred; we are debating how it happened. We are all trying to explain the same thing: the tree of evolutionary descent linking all organisms by ties of genealogy. Creationists pervert and caricature this debate by conveniently neglecting the common conviction that underlies it, and by falsely suggesting that evolutionists now doubt the very phenomenon we are struggling to understand.
—Stephen Jay Gould, "Evolution as Fact and Theory"

How can we talk about "the fact that evolution has occurred"? On what evidence can we make that statement? As we saw in chapter 1, scientists must use the word *fact* cautiously, as a description of nature, an observation, or hypothesis that has accumulated so much overwhelming evidence without falsification that it is a fact in common everyday parlance. As Gould (1981) put it,

Moreover, "fact" does not mean "absolute certainty." The final proofs of logic and mathematics flow deductively from stated premises and achieve certainty only because they are not about the empirical world. Evolutionists make no claim for perpetual truth, though creationists often do (and then attack us for a style of argument that they themselves favor). In science, "fact" can only mean "confirmed to such a degree that it would be perverse to withhold provisional assent." I suppose that apples might start to rise tomorrow, but the possibility does not merit equal time in physics classrooms.

In this sense, the idea that life has evolved and is evolving is "confirmed to such a degree that it would be perverse to withhold provisional assent." We see life evolving all around us, and we have abundant evidence that it has done so in the past. Creationists may put religious blinders on and refuse to face reality, but to the unbiased observer, the fact of evolution is as clear as the fact that the sun rises in the east or that objects fall to the earth. Some "evolution deniers" cannot live with this reality, but their denial won't stop viruses and bacteria from evolving new ways to attack us.

Ironically, the evidence that life had evolved was accumulating long before Darwin, as we saw with our discussion of faunal succession (chapter 3), Chambers's premature efforts to document evolution in 1844 (this chapter), or the evidence that Philip Henry Gosse tried to explain away with his *Omphalos* hypothesis (chapter 1). But the great strength of Darwin's book is that it accomplished two different functions: it laid out a huge amount of evidence that life had evolved, thereby *establishing the fact of evolution*; and it proposed a *mechanism* for how it had occurred (the "theory" of evolution), which was Darwinian natural selection. As we have just discussed, Darwin's mechanism is still being debated as to whether it explains everything, but Darwin's evidence that life has evolved is still valid. Much new evidence has accumulated in the past 150 years that Darwin could only have dreamed about. What was Darwin's evidence? Why does it demand an evolutionary explanation? How does creationism fail to explain it?

The Family Tree of Life

The first line of evidence had been emerging ever since the days of the Linnaean classification of animals in 1758, a full century before *On the Origin of Species*. The purpose of Linnaeus's classification scheme was to document God's handiwork by discovering the "natural system" of classification that God had used. Inadvertently, Linnaeus stumbled on an obvious fact of nature: each group (such as a species) of animals and plants clusters with other groups into larger groups (called *taxa* in the plural, *taxon* in the singular), such as a genus or order, and those higher-level supergroups cluster into even larger groups (such as classes or phyla) with additional taxa. For example, humans are part of a taxon (the family Hominidae) that also includes chimps, gorillas, orangutans, and gibbons. The apes, in turn, cluster together with the Old World monkeys (family Cercopithecidae) and New World monkeys, lemurs, and bush babies into a larger group, the order Primates. The primates are clustered with cows, horses, lions, bats, and whales in the class Mammalia. The mammals are clumped with fishes, birds, reptiles, and amphibians in the subphylum Vertebrata. Together with sponges, corals, mollusks, and other invertebrates, the vertebrates are part of the kingdom Animalia. The natural system for arranging and classifying life is a hierarchical system of smaller groups clustered into larger groups, which is best represented as a branching tree of life.

By Darwin's time, this branching pattern of life was even more strongly supported and led many people toward the notion that life had undergone a branching pattern of evolution (although not as boldly as Darwin suggested it). All of this was deduced by comparison of features visible to the naked eye or simple magnifier, primarily in the anatomy of the organism. But not even Darwin could have dreamed that the genetic code of every cell in your body also shows the evidence of evolution. Whether you look at the genetic sequence of mitochondrial DNA or nuclear DNA or cytochrome *c* or lens alpha crystallin or any other biomolecule, the evidence is clear: the molecules show the same pattern of nested hierarchical similarity that the external anatomy reveals (fig. 4.7). Our molecules are most similar to those of our close relatives, the great apes, and progressively less similar to those more distantly related to us.

Teasing out the details of this molecular similarity shows us a simple fact: every molecular system in every cell reveals the fact that life has evolved! If we were specially created and unrelated to the apes, why would we share over 98 percent of our genome with the chimpanzee and progressively less shared genome with primates who are less closely related to us? If God created it to *look* that way, then we are back on the "deceptive God" problem that faced Gosse's *Omphalos* hypothesis. No, the simplest interpretation is that the molecules tell the truth: life has a common origin and exhibits a branching pattern of ancestry and descent.

Homology

As comparative anatomy became a science in the early 1800s, anatomists were struck by how animals were constructed. Organisms with widely differing lifestyles and ecologies used the

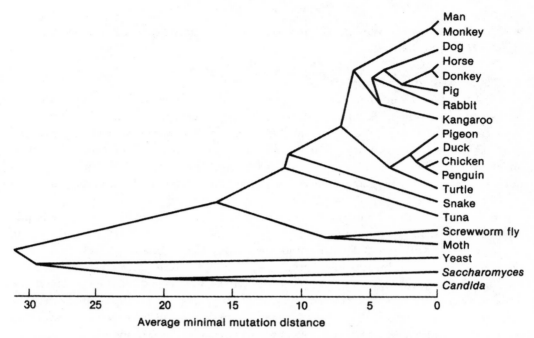

FIGURE 4.7. Branching diagram of the similarities in cytochrome *c* among various organisms. Nearly every biochemical system shows a similar branching pattern, which is identical to the branching pattern of life during its evolution. (From Fitch and Margoliash 1967. Copyright © 1967 American Association for the Advancement of Science. Reprinted with permission.)

same basic building blocks in their anatomy but had modified those parts in remarkably different ways. For example, the basic vertebrate forelimb (fig. 4.8) has the same basic elements: a single large bone in the upper arm (the humerus), a pair of two long bones in the forearm (the radius and ulna), a number of wristbones (carpals and metacarpals), and multiple bones (phalanges) supporting five digits (fingers). But look at the wide array of ways that some animals use this basic body plan! Whales have modified them into a flipper, while bats have extended the fingers out to support a wing membrane. Birds also developed a wing, but in an entirely different way, with most of the hand and wrist bones reduced or fused together, and feather shafts providing the wing support instead of finger bones. Horses have lost their side toes and walk on one large finger, the middle finger. None of this makes any sense unless these animals inherited a standard body plan in place from their distant ancestors and had to modify it to suit their present-day function and ecology. These common elements (bones, muscles, nerves) that serve different functions despite being built from the same basic parts are known as *homologous structures*. For example, the finger bones of a bat wing are homologous with our finger bones, and so on.

If the system had been divinely created by an "intelligent designer," why would there be this underlying similarity? A good engineer would create all wings the same best possible way from scratch, rather than jury-rig the structure using bones that the animal inherited from its ancestors. In fact, nature uses a variety of nonhomologous ways to build a wing. We have already seen how vertebrates build wings in two completely different ways, even though bats and birds started out with the same bones from a common ancestor—and neither

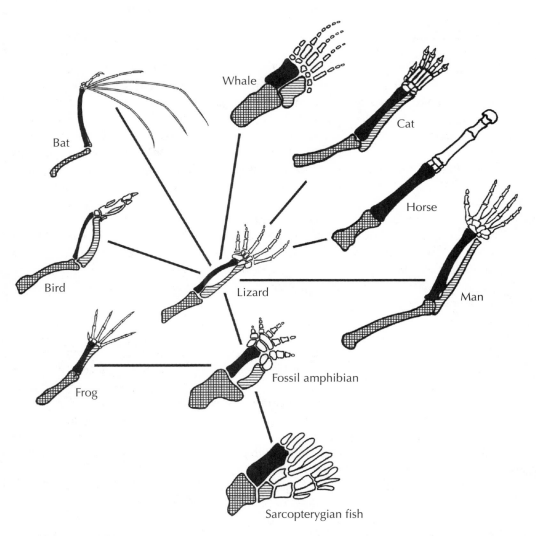

FIGURE 4.8. The evidence of homology. All vertebrate forelimbs are constructed on the same basic plan with the same building blocks, even though they perform vastly different functions. The basic vertebrate forelimb has been modified into flippers in whales, a wing in bats, and a one-fingered running hand in horses, yet the basic bony structure remains the same. (Drawing by Carl Buell)

of their solutions resembles the wing of a pterodactyl (which is supported only by the bones of the fourth finger), let alone the completely different structure of the wings of an insect. The flippers of a whale may perform the same function as the paddles of a marine reptile or the fins of a fish, but all three structures with a common function have completely different bony structures. These different types of wings and flippers found in unrelated organisms are *analogous* organs that perform the same function but have a fundamentally different structure. The fact is that organ systems are jury-rigged with whatever bones the animal inherited from its ancestors and not built from scratch in the optimal shape for its current use. This only makes sense if life had evolved to use what anatomy is already available.

Vestigial Structure and Other Imperfections

In chapter 2, we already alluded to the fact that organs that are remnants of past structures but no longer serve a function were evidence against an intelligent designer. The list of such vestigial organs (fig. 4.9) is overwhelming. They include not only the appendix, tonsils, and tailbones of humans (none of which have a function now), but the tiny splint bones in the feet of horses, which are remnants of the time when horses had three toes. When these bones break, the horse is crippled for life. Whales and snakes both have tiny hips and thigh bones buried deep in their bodies, with no function whatsoever. Why would they have these features unless they evolved from ancestors that did have hind limbs? Any creationist attempt to explain these remnants of ancient history falls into the same trap that Gosse did: a God who would plant these vestiges of the past is a deceiver, tricking us by making it *look* like life evolved.

A related issue is all the organ systems that are poorly or suboptimally designed, or jury-rigged so they work just well enough for the organism to survive. We discussed these in chapter 2 (especially the panda's thumb and the fishing lures used by the anglerfish and the clam *Lampsilis*). Once again, these organ systems do not look like they were intelligently designed and only make sense if the organism must use whatever building blocks its ancestors provided.

Embryology

Even before Darwin, the studies of embryos began to provide important evidence for evolution. In the 1830s, the great German embryologist Karl Ernst von Baer documented that the embryos of all vertebrates show a common pattern (fig. 4.10). Whether they develop into fish, amphibians, or humans, all vertebrate embryos start out with a long tail, well-developed gill slits, and many other fishlike features. In adult fish, the tail and gills develop further, but in humans, they are lost during further development. Von Baer was simply trying to document how embryos developed, not to provide evidence of evolution, which had not yet been proposed.

Darwin used this evidence in *On the Origin of Species*, and embryology soon developed into one of the growth fields of evolutionary biology. One of the foremost advocates of evolution was the flamboyant German embryologist Ernst Haeckel. He not only promoted Darwinism in Germany, but he went so far as to argue that we could see all details of evolutionary history in embryos and reconstruct ancestors from embryonic stages of living animals. His most famous slogan, the "biogenetic law," was "Ontogeny recapitulates phylogeny." This is simply a fancy way of saying embryonic development ("ontogeny") repeats ("recapitulates") evolutionary history ("phylogeny"). To the limited extent that von Baer had shown 40 years earlier, this is true. But embryos also have many unique features (yolk sac, allantois, amniotic membranes, umbilical cords) that have nothing to do with the

FIGURE 4.9. The evidence from vestigial organs. (A) Both whales and snakes retain tiny remnants of their hind legs and hip bones in their bodies, although they are normally not visible externally, nor do they have any function. These facts only make sense if whales and snakes had four-legged ancestors. Horses also retain vestiges of their ancestral side toes, known as splint bones. (B) Close-up of the hip regions of a mounted fin whale skeleton, showing the tiny vestigial hip bones and thigh bones. (Photo by the author) (C) In 1921, Roy Chapman Andrews documented a specimen of a humpback whale that actually had atavistic hind limbs that extended from its body. These are the bones of those hind limbs. (From Andrews 1921)

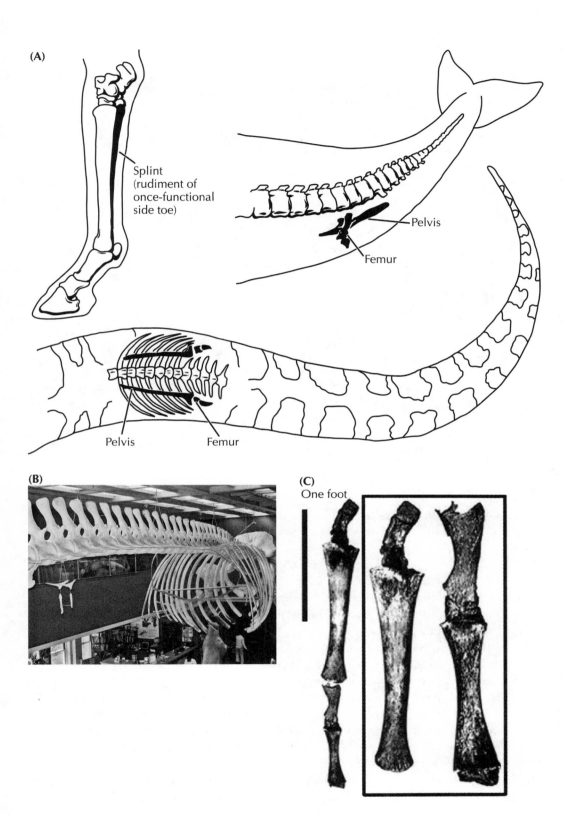

(A)

Splint
(rudiment of
once-functional
side toe)

Pelvis

Femur

Pelvis   Femur

(B)

(C)

One foot

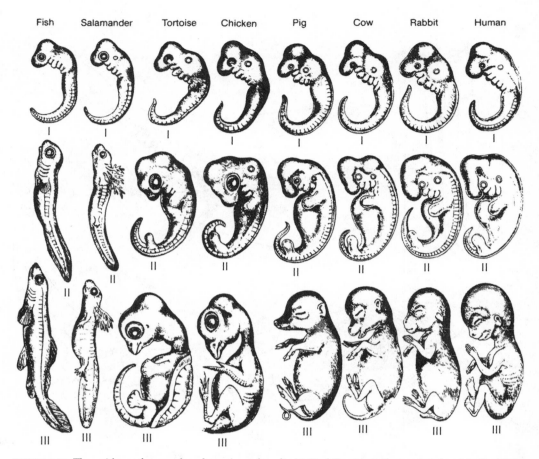

FIGURE 4.10. The evidence from embryology. As embryologist Karl Ernst von Baer pointed out in the 1830s, long before Darwin published his ideas about evolution, all vertebrates start out with a very fishlike body plan early in embryology, including the predecessors of gills and a long tail. As they develop, many lose their fishlike features on their way to becoming reptiles, birds, and mammals. (From Romanes 1910)

evolutionary past and are adaptations to their developmental environment. Thus it is dangerous to overextend the evolutionary implications of the stages in an embryo, but they are useful guides nonetheless.

Creationists such as Jonathan Wells (2000), in their eternal effort to mislead the uninitiated and miss the forest for the trees, will crow about how the biogenetic law has been discredited. But Haeckel's overenthusiasm does not negate the careful embryological work of von Baer that shows that many features of our past evolutionary stages are preserved in our embryos. Wells, in particular, nags about how some of Haeckel's original diagrams had errors and oversimplifications, but this does not change the overall fact that the sequence of all vertebrate embryos shows the same patterns in the early stages, and all of them go through a fishlike stage with pharyngeal pouches (which become the gill slits in fishes and amphibians) and a long fish-like tail, then some develop into fishes and amphibians and others lose these features and develop into reptiles, birds, and mammals.

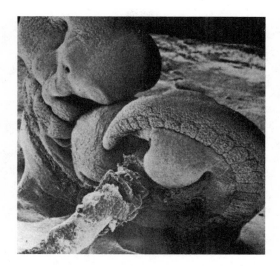

FIGURE 4.11. This is what you looked like 5 weeks after conception. You still had many fishlike features, such as a well-developed tail and the embryological precursors of gill slits, both of which are lost in most human embryos as they develop. (From IMSI Photo Images, Inc.)

If you had any doubts that you once had ancestors with fish-like gills and a tail, figure 4.11 shows what you looked like 5 weeks after fertilization. Why did you have pharyngeal pouches (predecessors of gills) and a tail if you had not descended from ancestors with those features?

Biogeography

As was pointed out in chapter 3, the great expeditions of discovery in the 1700s and 1800s soon produced a huge diversity of animals and plants that were unfamiliar to Europeans and not anticipated by the authors of either the Noah's ark story or Linnaeus in 1758. Not only do these animals and plants render any version of the Noah's ark story impossible, but they created another problem as well. They are not distributed around the earth in pattern from Mount Ararat in Turkey (the supposed landing site of the ark) but instead have their own unique distribution patterns that only make sense in light of evolution.

Darwin got a hint of this on the Galapagos Islands, where each island had a slightly different species of giant tortoise or finch. Instead of populating the islands with the same species as occurred on the mainland, apparently God had seen fit to put new, unique species on each island (and this phenomenon is true of islands in general). Later studies of the unusual animals of exotic places further confirmed the fact that these far-flung locations were largely populated with unique animals found nowhere else, and their distribution patterns made no sense in the context of migration from the ark. For example, Australia is home to its own unique fauna of native pouched mammals, or marsupials (fig. 4.12). These include not only the familiar kangaroo and koala, but many other creatures that evolved to fill the same ecological niches that placental mammals occupy on other continents. There are marsupial equivalents of placental wolves, cats, flying squirrels, groundhogs, anteaters, moles, and mice. If the animals had all migrated away from the ark, why had nothing but marsupials arrived in Australia and then apparently evolved to fill the niches left vacant by the lack of placentals?

Other patterns are equally persuasive. For example, many of the southern continents have at least one flightless bird species, all members of the primitive group known as ratites.

Placentals

Marsupials

Wolf
(Canis)

Tasmanian wolf
(Thylacinus)

Ocelot
(Felis)

Native cat
(Dasyurus)

Flying squirrel
(Glaucomys)

Flying phalanger
(Petaurus)

Ground hog
(Marmota)

Wombat
(Phascolomys)

Anteater
(Myrmecophaga)

Anteater
(Myrmecobius)

Mole
(Talpa)

Mole
(Notoryctes)

Mouse
(Mus)

Mouse
(Dasycercus)

FIGURE 4.12. The evidence from biogeography. The native fauna of Australia consists mainly of pouched marsupials, which have converged remarkably on their placental counterparts from other continents, even though the two groups are not closely related. In Australia, there are marsupials that look vaguely like wolves, cats, flying squirrels, groundhogs, anteaters, moles, and mice—but they are all pouched mammals. (Modified from Simpson and Beck 1965)

Africa has the ostrich, South America the rhea, Australia the cassowary and emu, and New Zealand the kiwi. This distribution makes no sense in the Noah's ark story but does fit the idea that they were closely related when all these southern landmasses were part of the great Gondwana supercontinent about 100 million years ago. Since the time that these continents have drifted apart, so too have their ratite natives diverged from one another.

Finally, the fossil record provides the details of how life has evolved and is now the strongest piece of evidence for evolution. The remaining chapters of this book will detail the incredible evolutionary stories revealed by fossils, so we will not discuss them further here.

These were all lines of evidence that Darwin mustered in 1859 and they have only grown stronger in the past 150 years with the accumulation of more details and examples. Each alone is strong evidence that life has evolved and is impossible to explain by creationism, and added together they make the case overwhelming. But we have even better evidence: we can see life evolving today, so evolution is as much an observed fact of nature as the fact that the sky appears blue.

Evolution Happens All the Time!

Nothing in biology makes sense except in the light of evolution.
—Theodosius Dobzhansky, 1973

Biologists finally began to realize that Darwin had been too modest. Evolution by natural selection can happen rapidly enough to watch. Now the field is exploding. More than 250 people around the world are observing and documenting evolution, not only in finches and guppies, but also in aphids, flies, graylings, monkeyflowers, salmon and sticklebacks. Some workers are even documenting pairs of species— symbiotic insects and plants—that have recently found each other, and observing the pairs as they drift off into their own world together like lovers in a novel by D. H. Lawrence.
—Jonathan Weiner, "Evolution in Action"

If the evidence mustered by Darwin and many other scientists in the past 150 years was not enough to prove that life has evolved, there is an even simpler test: watch life evolve right now! Creationists try to discredit evolution by saying that it all happened in the past; they seem unaware that it continues to happen all around us, even as I write this.

We can see natural selection operating on many different scales and on many different types of organisms. The details of many of these recent studies are provided in Jonathan Weiner's excellent book, *The Beak of the Finch: A Story of Evolution in Our Time* (1994) or David Mindell's *The Evolving World: Evolution in Everyday Life* (2006). Looking over the shoulders of the hundreds of hardworking, dedicated, self-sacrificing biologists who spend years enduring harsh conditions in the field to observe evolution in action inspires admiration in us real scientists. This is in sharp contrast with the creationists who sit in their comfortable homes and write drivel about subjects they have never studied and do not understand.

The classic example, of course, has long been the finches of the Galapagos Islands (fig. 4.13). Darwin himself collected many of them when he was there in 1835, but they were all so different that he did not notice that the diverse birds he had shot were all finches with

FIGURE 4.13. Although Darwin didn't notice this while he was in the Galapagos, the majority of the birds are finches, which have evolved from a generalized finch ancestor blown over from South America into birds with a wide variety of bills for nutcracking, probing for insects, picking up tiny seeds, and many other tasks performed by different families of birds on the mainland. (Modified from Lack 1947; used by permission)

highly modified beaks and different color patterns. It was not until he got back to England that ornithologist John Gould (recruited to study Darwin's specimens) pointed this out to him. In the twentieth century, David Lack did a much more detailed study published in 1947. In recent years, Darwin's finches have been the focus of research of Peter and Rosemary Grant of Princeton University. The Grants visited the islands year after year, documenting the changes in the finch populations. On one island (Daphne Major), the finch population changed dramatically from year to year. During a 1977 drought, the finches with strong beaks survived, because they could crack the toughest seeds and survive the shortage of food. The next few years, all of the finches on that island were their descendants and had the stronger nut-cracking bills found on other species of Galapagos finches. Since that time, the return of wetter conditions has changed the finches yet again, so forms that have more normal beaks for eating a wide variety of seeds could also survive. From this, it is easy to see how such strong selection pressures could transform the ancestral finches (which still live in South America) into a wide variety of specialized finches that perform the roles that other birds play on the mainland. Instead of nuthatches, there are thick-billed finches; instead of wood-peckers, there are finches with long bills for drilling wood and probing for grubs; instead of warblers, there are finches with similar bills called warbler finches. One finch has even learned to use a twig as a tool for fishing for insects in the hollows of trees! Recent research has identified the genes that control beak shape in these finches and artificially duplicated the pattern seen in nature by adding or subtracting those genes.

We don't have to live in the Galapagos Islands to see evolution happen. We can see evolution in action in our own backyards. The common European house sparrow is found all over North America today, but it is an invader, brought from Europe in 1852. The initial populations escaped and quickly spread all over North America, from the northern boreal forests of Canada down to Costa Rica. We know that members of the ancestral population were all very similar because they were introduced from a few escaped immigrants. Because they have spread to the many diverse regions of North America, they are rapidly diverging and on the way to becoming many new species. House sparrows now vary widely in body size, with more northern populations being much larger than those that live in the south. This common phenomenon, known as Bergmann's rule, is due to the fact that larger, rounder bodies conserve heat better than smaller bodies. House sparrows from the north are darker in color than their southern cousins, perhaps because dark colors help absorb sunlight and light colors are better at reflecting it in warm climates. Many other changes in wing length, bill shape, and other features have been documented.

New species can arise even faster than people once thought. A study by Andre Hendry at McGill University in Montreal analyzed the sockeye salmon near Seattle (fig. 4.14). These salmon tend to breed either in lakes or in streams and have different shapes dependent on the environment in which they breed. In the 1930s and 1940s, sockeye salmon were intro-duced to Lake Washington east of Seattle and rapidly became established in the mouth of the Cedar River. By 1957, they had also colonized a beach called Pleasure Point. In less than 40 years, these two populations have rapidly diverged. The males of populations that live in the swift-flowing waters of Cedar River are more slender to fight the strong currents; the females are bigger so they can dig deeper holes for their eggs to prevent the river from erod-ing them away. The populations that live in the warmer, quieter waters of the lakeshore near Pleasure Point have males with deeper, rounder bodies, which are better at fending off rivals for mating privileges, and females with smaller bodies because they do not have to dig deep holes for their eggs. These populations are genetically isolated and already show the differ-ences that would be recognized as separate species in most organisms. Hendry was able to show that this species split started in less than 40 years, and in just a few more generations, they might be genetically isolated and become distinct species.

Another rapidly evolving fish is the three-spined stickleback (fig. 4.15). Sticklebacks that live in the ocean have heavier body armor than those that live in lakes. In one pond near Bergen, Norway, biologists have been able to document this change in less than 31 years.

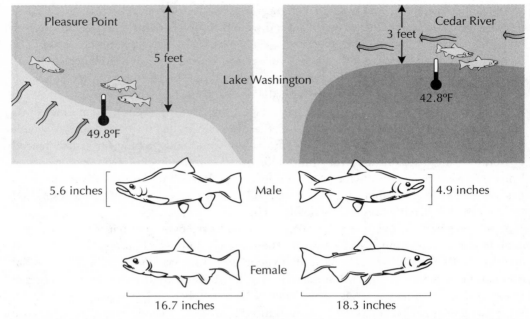

FIGURE 4.14. Evolution has happened in some groups over just a few decades. In the swift currents of Cedar River, Washington, the sockeye salmon introduced in the 1930s have adapted so that the males can swim in the strong currents and the females dig deeper nests in the sand to lay their eggs. But the salmon that invaded the shallow waters of Pleasure Point in 1957 have developed rounder, deeper bodies in males to help fight off rival males; the females dig shallower nests for their eggs because there are no strong currents. (Modified from Weiner 2005)

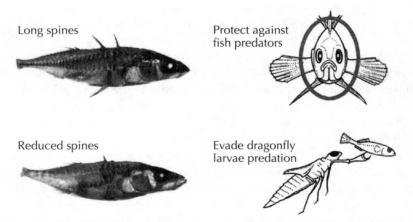

FIGURE 4.15. The three-spined stickleback fish also show rapid evolution. Species that live in the lakes or in the ocean have longer spines that make it harder for a predator to swallow them. But sticklebacks living in shallow streams have evolved shorter spines so predators like dragonfly larvae cannot catch them by their protruding spines. (Photos courtesy D. M. Kingsley and S. Carroll)

In Loberg Lake, Alaska, the change took only a dozen years, or just six generations. Stickle-backs also change their spines in response to local conditions. In open water, longer spines are an advantage because they protect against being swallowed by predators. But in shallow water, the long spines are a liability because they make it easier to be captured by dragonfly larvae, which have long pincers. A single Hox gene, *Pitx1*, turns off and on the switches that regulate spine length. Other studies have shown that when captive sticklebacks are artificially modified so they have unusual new combinations of spines, the females only mate with the males that have new traits, so sexual selection is a driving force in their evolution of novelty. Prior to these studies, ichthyologists would readily assign specimens with different spine counts and different body armor to different species, but these studies show just how easy it is for one stickleback population to transform to another species given the right conditions.

There are the classic cases of industrial melanism, so familiar from every textbook. The peppered moth, *Biston betularia*, normally has a speckled appearance that blends in well with mottled appearance of trunks and branches of trees. During the Industrial Revolution, soot in the air made the tree trunks black, and the normal form was conspicuous. Instead, a dark-colored mutant became dominant, because they were well camouflaged against the dark tree trunks, while the birds picked off the normal speckled varieties. When environmental regu-lations cleaned up the air and eliminated the sooty tree trunks, the normal speckled varieties returned, and the dark mutants were again selected against.

Examples like these could be multiplied endlessly. In New England, the periwinkles dramatically changed their shell shape and thickness in less than a century, probably due to predation pressure by newly introduced crabs. In the Bahamas, the anole lizards (the com-mon "chameleon" in the pet shops, which are not true chameleons) changed the proportions of their hind limbs after people introduced them to new islands with different vegetation. In Florida, the soapberry bug evolved a significantly longer beak in response to the invasion of its habitat by a nonnative plant with larger fruits. In Hawaii, the honeycreeper birds evolved shorter bills as their favorite food source, the native lobelloids, has disappeared, and the birds switched to another source of nectar. In Nevada, the tiny mosquito fish that lives in isolated desert water holes that were once connected during the last ice age quickly evolved major differences in less than 20,000 years. And in Australia, the introduced wild rabbits (brought by European settlers less than a century ago) modified their body weight and ear size in response to the different conditions of the outback.

Humans are often the strongest agents of selection for many wild animals. In popula-tions of bighorn sheep, trophy hunters killed off most of the rams with spectacular horns, so the smaller males with reduced horns had a better chance of breeding, and the population no longer has many large-horned rams. Rattlesnakes that are too nervous and buzz when humans approach are quickly killed, so in many regions, the rattlers no longer give any warning. Overfishing of the Atlantic cod led to a population crash during the 1980s, and large cod nearly vanished; those that bred quickly while they were small and immature had a better chance of survival.

But the most dramatic and rapid examples of evolution in action occur with microor-ganisms, especially viruses and bacteria. Every year doctors have new flu strains to battle, because last year's cold and flu strains have evolved new protein coats that make them unrecognizable to our immune systems and allow them to infect us again. This is why there will never be a cure for the common cold—the virus evolves too fast for any drug to keep up. The heavy use of antibiotics has selected for strains of bacteria that are resistant to every drug we throw at them. When sulfonamides were introduced in the 1930s, resistant strains evolved in only a decade. Penicillin was introduced in 1943, and there were resistant strains by 1946. For this reason, doctors are now much more cautious about issuing antibiotics to sick patients who want the drugs even though they are useless against the viruses that cause cold and flu. Similarly, the heavy use of antiseptic cleansers and wipes has led to strains of bacteria that can resist most antiseptics. Many medical researchers think that our excessive cleanliness in the Western world is to our disadvantage because young people are no longer exposed to many different kinds of germs, and they become vulnerable when a strong strain (that doesn't infect people in the "dirty" Third World) invades. Now hospitals are worried—when one of these drug-resistant strains appears in a hospital, it can quickly spread to many patients and nothing can stop it.

Likewise, many insects and weeds have evolved resistance to pesticides and herbicides, all within a few decades, causing enormous economic damage to people all over the world. Every modern housefly now carries the genes that make it resistant not only to DDT, but also to pyrethroids, dieldrin, organophosphates, and carbamates, so there are few pesticides left that can suppress them. The mosquitoes that evolved resistance to DDT and other organo-phosphate insecticides apparently evolved in Africa during the 1960s, spread on to Asia, then reached California by 1984, Italy in 1985, and France in 1986. As entomologist Martin Taylor describes it (in Weiner 1994:255),

> It always seems amazing to me that evolutionists pay so little attention to this kind of thing, and that cotton growers are having to deal with these pests in the very states whose legislatures are so hostile to the theory of evolution. Because it is the evolution itself they are struggling against in their fields every season. These people are trying to ban the teaching of evolution while their own cotton crops are failing because of evolution. How can you be a creationist farmer any more?

Evolution is happening all around us. It happens every time a new germ invades your body, a new pest or weed destroys our crops, or a new insecticide-resistant fly or mosquito bites you. Creationists may get some personal comfort from their beliefs, but they cannot change the fact that life is evolving all around us and threatens our survival if we don't come to terms with that evolution (see fig. 1.3).

For Further Reading

Campbell, J. 1982. Autonomy in evolution, in *Perspectives on Evolution*, ed. R. Milkman. Sunderland, Mass.: Sinauer, 190–200.

Carroll, S. 2005. *Endless Forms Most Beautiful: The New Science of Evo/Devo*. New York: Norton.

Desmond, A., and J. Moore. 1991. *Darwin: The Life of a Tormented Evolutionist*. New York: Warner.

Eldredge, N. 1985. *Unfinished Synthesis*. New York: Oxford University Press.

Gould, S. J. 1977. *Ontogeny and Phylogeny*. Cambridge, Mass.: Harvard University Press.

Gould, S. J. 1980. Is a new and more general theory of evolution emerging? *Paleobiology* 6:119–130.

Gould, S. J. 1982. Darwinism and the expansion of evolutionary theory. *Science* 216:380–387.

Gould, S. J. 2002. *The Structure of Evolutionary Theory*. Cambridge, Mass.: Harvard University Press.

Levinton, J. 2001. *Genetics, Paleontology, and Macroevolution*. 2nd ed. New York: Cambridge University Press.

Mindell, D. P. 2006. *The Evolving World: Evolution in Everyday Life*. Cambridge, Mass.: Harvard University Press.

Ridley, M. 1996. *Evolution*. 2nd ed. Cambridge, Mass.: Blackwell.

Schwartz, J. 1999. *Sudden Origins: Fossils, Genes, and the Emergence of Species*. New York: Wiley.

Stanley, S. M. 1979. *Macroevolution: Patterns and Process*. New York: Freeman.

Stanley, S. M. 1981. *The New Evolutionary Timetable*. New York: Basic.

Steele, E.. 1979. *Somatic Selection and Adaptive Evolution: On the Inheritance of Acquired Characters*. Chicago: University of Chicago Press.

Steele, E., R. Lindley, and R. Blanden. 1998. *Lamarck's Signature: How Retrogenes Are Changing Darwin's Natural Selection Paradigm*. Reading, Mass.: Perseus.

Weiner, J. 1994. *The Beak of the Finch: A Story of Evolution in Our Own Time*. New York: Knopf.

Weiner, J. 2005. Evolution in action. *Natural History* 115(9):47–51.

Wesson, R. 1991. *Beyond Natural Selection.*, Cambridge, Mass.: MIT Press.

Wills, C. 1989. *The Wisdom of the Genes: New Pathways in Evolution*. New York: Basic.

FIGURE 5.1. A parody of the classic "march from ape to man," showing the cartoonist's opinion of creationists. (Cartoon by Bill Day, Detroit Free Press; used by permission)

SYSTEMATICS AND EVOLUTION

5

What Is Systematics?

The extent to which progress in ecology depends upon identification and upon the existence of sound systematic groundwork for all groups of animals cannot be too much impressed upon the beginner in ecology. This is the essential basis of the whole thing; without it, the ecologist is helpless, and the whole of his work may be rendered useless.

—Charles Elton, *Animal Ecology*

In the previous chapter, we briefly introduced the concept of systematics. However, we will not be able to talk about most of the fossils or animals in the rest of the book if we do not review the basic concepts of systematics and the major breakthroughs in systematic thought that have occurred in the past decades. Of all the topics in biology, systematics is the least understood by the general public, yet it is one of the most essential.

Most people are vaguely aware that there is a scientific system for naming and classifying organisms. This field is known as *taxonomy.* Most scientists who name and describe new species of animals and plants have to be familiar with its rules and procedures. But systematics is broader than just taxonomy. Systematics is "the science of the diversity of organisms" according to Ernst Mayr (1966:2) or "the scientific study of the kinds and diversity of organisms and of any and all relationships among them" in the words of George Gaylord Simpson (1961:7). Systematics includes not only taxonomic classification but also determining evolutionary relationships (*phylogeny*) and determining geographic relationships (*biogeography*). The systematist uses the comparative approach to the diversity of life to understand all patterns and relationships that explain how life came to be the way it is. In this sense, it is one of the most exciting fields in all of science.

Taxonomists and systematists may not be as numerous or well funded as other kinds of biologists, but everything else in biology depends on their classifications and phylogenies. If the physiologist or doctor wants to study the organism most similar to humans, taxonomists point to the chimpanzee, our closest relative. If ecologists want to study how a particular symbiotic relationship developed, they depend on the systematist for accurate classification of their organisms. Systematics provides the framework or scaffold on which all the rest of biology is based. Without it, biology is just a bunch of unconnected facts and observations.

Today, taxonomists are becoming scarce as funding dries up or goes to more glamorous fields that use big expensive machines. But this starves science at its heart. The humble systematist, collecting specimens in the field on a shoestring budget or analyzing specimens in museum drawers and jars, may not be as famous as the behavioral ecologists, watching animals in the wild, or molecular biologists with the white lab coats and million-dollar machines, but their work is just as essential. One of the hottest topics today, *biodiversity*, is the fundamental domain of the systematist. Many people are worried about how rapidly we're destroying habitat and wiping out the species on earth, but without enough systematists to identify and describe these species, we have no idea how bad the problem is. Many ecologists who are working on this problem complain that there are no longer enough trained systematists to even begin to identify all the species in danger—yet the funding agencies continue to starve systematics, and most students stay away from it because there isn't as much glamour or money. Likewise, the systematist is necessary for many other functions, like correctly identifying which pest species is causing problems or describing and naming new species that may someday hold the cure for a deadly disease.

What Is Taxonomy?

God created, but Linnaeus classified.
Carolus Linnaeus

There are many ways to classify things. We do it all the time. We quickly identify cars on the road as "sedans," "SUVs," "minivans," "pickups," and the like, but our kids might just identify them as "red cars" and "silver cars" and so on. A car buff or police officer might be able to recognize the make and model of each car as it flashes by. Similarly, small libraries use the Dewey decimal system to classify their books by topics, but larger libraries use an entirely different system developed by the Library of Congress; the categories in each are entirely different. Both of these systems try to be as "natural" as possible, clustering books that belong in the same category, such as "Science" with subcategories like "Geology," "Biology," "Physics," and "Chemistry."

In nature, there are also many ways to classify things. Many native cultures use simple rules like "good to eat" versus "eat only in emergency" versus "inedible" versus "poisonous." Even our own culture uses simple ecological properties to crudely classify things. For example, some people call nearly all marine life "fish," including "shellfish" (which are mollusks), "starfish" (which are echinoderms), and "jellyfish" (which are cnidarians related to corals and sea anemones). In the 1600s and early 1700s, a number of different naturalists had proposed classification schemes for life, but they were arbitrary and highly unnatural. For example, they often lumped together everything with wings, including birds, insects, and flying fish, or things with shells, like armadillos, turtles, and mollusks. The eventual solution was developed by the Swedish botanist Carl von Linné, better known by his Latinized name, Carolus Linnaeus. From his experiences with plants, Linnaeus realized that the best classification was based on reproductive structures, primarily flowers, rather than in the confusingly similar leaves or trunks or roots. This "sexual system" for plant classification was published in 1752 and became the basis for modern plant taxonomy. Linnaeus also applied the same idea to animals, focusing on fundamental properties like the reproductive system and hair or feathers, rather than on superficial ones like flight or armor. His first classification, entitled *Systema Naturae* (The System of Nature) was published in 1735, and the tenth edition of 1758 is regarded as the starting point of modern taxonomy.

Although the original Linnaean classification became outdated as hundreds of new species were described, his basic principles are still used around the world. Every species on earth has a two-part name (*binomen*), consisting of the *genus name* (always capitalized and either underlined or italicized) and the *trivial name* (never capitalized, but also always either underlined or italicized). For example, we belong to the genus *Homo* (Latin for "human") and the trivial name of our species is *sapiens* (Latin for "thinking"), so our full species name is *Homo sapiens* or *H. sapiens*. The trivial name can never stand by itself, but must always be paired with the genus because trivial names are used over and over, but a generic name can never be used for another animal. The genus (plural, *genera*) can be composed of a single species or more than one species. Our genus *Homo* includes not only *H. sapiens*, but also *H. erectus*, *H. neanderthalensis*, *H. habilis*, *H. rudolfensis*, and several other extinct species, such as the newly discovered *H. naledi*. Genera are clustered into larger groups called *families*, which always end with the suffix "-idae" in the animals and "-aceae" in plants. Our family is the Hominidae, while the Old World monkeys are the Cercopithecidae, and the New World monkeys are the Cebidae. Families are clustered into *orders* (which have no standardized ending or format), such as the order Primates (which includes all monkeys, apes, lemurs, and humans), the order Carnivora (cats, dogs, bears, and their flesh-eating relatives), or the order Rodentia (the rodents, the largest order of mammals on earth). Orders are then clustered into *classes* such as the class Mammalia, or mammals, including all the orders just listed. The class Mammalia is clustered along with the classes for birds, reptiles, amphibians, and fish into the *phylum* Chordata (all animals with a backbone or its precursors). Finally, there are a number of phyla (mollusks, arthropods, worms, echinoderms, etc.) that cluster with the chordates in the *kingdom* Animalia.

Although this system is over 360 years old, it has powerful advantages. The Linnaean classification scheme is flexible, allowing groups and new taxa to be changed, shuffled, and inserted as the situation demands. Taxonomic names were traditionally based on Greek or Latin roots, or Latinized forms of other words, because Latin was the language of scholars in Linnaeus's time. Although most scholars around the world no longer read Latin, the fact that the names are accepted worldwide means that no matter what language a biologist speaks, the names of the animals are consistent. You can pick up a journal article in an entirely different orthography, like Russian Cyrillic or Mandarin Chinese, and still recognize the Linnaean names. By contrast, every language has its own local names for familiar animals. Even within the United States, "gopher" could mean a small burrowing rodent in some regions, or a tortoise in others—but the scientific names for the rodent genus *Geomys* and the tortoise genus *Gopherus* are universally recognized and unambiguous.

When Linnaeus set up his "natural system," his goal was to understand the mind of God by understanding how his handiwork was arranged. Ironically, the system he developed was hierarchical, showing that life has a branching structure like a tree or bush. That branching structure of life became one of Darwin's best arguments for the fact of evolution (discussed in chapter 4). Taxonomy shifted its goals from theology to understanding life's evolutionary history. Consequently, the practice of systematics has had to make some important decisions about what criteria are used in classification. Should taxonomy be based solely on evolutionary history or should other components (such as ecology) also be included? Ecologically speaking, many people lump all swimming vertebrates with fins together as "fish." But not all fish are the same. Lungfish are actually more closely related to amphibians, reptiles, and us than they are to a tuna fish. In taxonomic terms, lungfish do not belong with fish but with the four-legged land animals in a group known as the Sarcopterygii (the lobe-finned fish and their descendants). Even though that is an accurate picture of evolutionary relationships, many biologists have trouble with thinking about organisms this way and prefer groups that reflect some ecology as well.

Ladders, Bushes, Mosaics, and "Missing Links"

Evolution usually proceeds by speciation—the splitting of one lineage from a parental stock—not by the slow and steady transformation of these large parental stocks. Repeated episodes of speciation produce a bush. Evolutionary "sequences" are not rungs on a ladder, but our retrospective reconstruction of a circuitous path running like a labyrinth, branch to branch, from the base of the bush to a lineage now surviving at its top.

—Stephen Jay Gould, "Ladders, Bushes, and Human Evolution"

The realization that the classification of life forms a natural bushy or treelike pattern has other implications as well. As we saw in chapter 4, the older (pre-Linnaean) way of arranging nature was in the *scala naturae* or "ladder of creation," with "lower" animals at the base, humans near the top, and divine beings up to God completing the ladder (figs. 5.1 and 5.2).

FIGURE 5.2. Evolution is *not* about life climbing the "ladder of nature" or the "great chain of being" from "lower" to "higher" organisms. Instead, evolution is a "bush" with many lineages branching from one another, and ancestors living alongside their descendants. (Drawing by Carl Buell)

But life is not a ladder, and there are no such things as "higher" or "lower" organisms. Organisms have branched off the family tree of life at different times in the geologic past, and some have survived quite well as simple corals or sponges, while others have evolved more sophisticated ways of living. Corals and sponges, although simple compared with other organisms, are not "lower" organisms, nor are they evolutionary failures for not advancing up the ladder. They are good at doing what they do (and have been doing for over 500 million years), and they exploit their own niches in nature without any reason to change whatsoever.

Nevertheless, this antiquated and long-rejected view of life as a ladder of creation still seems to lurk behind many people's misunderstandings of biology and evolution. For example, it is common for creationists to ask, "If humans evolved from apes, why are apes still around?" The first time biologists hear this question, they are puzzled, because it seems to make no sense whatsoever—until they realize this creationist is still using concepts that were abandoned over 200 years ago. We now know that nature is not a ladder, but a bush (fig. 5.2). Lineages branch and speciate and form a bushy pattern, with the ancestral lineages living alongside their descendants. Humans and apes had a common ancestor about 7 million years ago (based on evidence from both the fossils and the molecular sequences), and both lineages have persisted ever since then. It is comparable to saying, "If you are descended from your father, why didn't your father die when you were born? Why didn't your grandfather die when your dad was born?" We all understand that we children branch off from our parents, and they do not have to die when we are born. Similarly, the human lineage branched off from the rest of the apes about 7 million years ago, but both are still around.

Likewise, the tendency to put things into simple linear order is a common metaphor for evolution—and also one of its greatest misperceptions. The iconic image is the classic "ape-to-man" sequence of organisms marching up the evolutionary ladder (fig. 5.1). This icon of evolution is so familiar that it is parodied endlessly in political cartoons and advertisements (for an extended discussion with many humorous examples, see Gould 1989:27–38). Most people think that this is an accurate representation of evolution. WRONG! *Evolution is a bush, not a ladder!* As we shall discuss in chapter 15, human evolution is quite bushy and branching, with multiple human species living side-by-side at certain times in the past 5 million years (fig. 15.3). The old-fashioned line of prehistoric humans marching "up the ladder" may be familiar and easy to visualize, but it is a gross oversimplification of the truth.

Another familiar example is the evolution of the horse, which we shall discuss in detail in chapter 14. About 100 years ago, when fossil horses were first discovered, it appeared that they were but a single lineage getting progressively larger and more advanced through time (fig. 14.2). But the past 100 years of collecting shows that horse history, too, was highly bushy and branching, with multiple lineages living at the same time (fig. 14.3). It may be convenient to visualize the general trend of horse evolution as a single linear sequence, but it is a very poor representation of their actual history.

Related to this concept is the misconception about "missing links." Two centuries ago when people believed in the ladder of life, another related metaphor was the "great chain of being." According to this idea, all life was linked into a great chain of increased complexity up the ladder to God. In his divine providence, God would not allow any link in this chain to vanish. As Alexander Pope wrote in *An Essay on Man* (1735),

Who sees with equal eye, as God of all
The hero perish, a sparrow fall . . .
Where, one step broken, the great scale's destroy'd;
From Nature's chain whatever link you strike,
Ten or ten thousandth, breaks the chain alike.

As Lovejoy (1936) showed, the concept goes back to the ancient Greeks and was widespread in the Middle Ages and the Renaissance. At that time, it was used to justify the inequalities in human society and the divine right of kings and nobility, as well as to place all of nature in a religious context. Even as late as the 1790s, most naturalists (including Thomas Jefferson) refused to accept the idea that this great chain of being could be broken or that God would allow any of his creations to become extinct. But in the early 1800s, the great anatomist and paleontologist Baron Georges Cuvier showed conclusively that the skeletons of mastodonts and mammoths represented giant animals that could no longer be alive on earth today and must be extinct.

Even though the great chain of being was ultimately discredited when evolutionary theory came around in the mid-1800s, the imagery was still very powerful. A century before Darwin's ideas of evolution, people saw the close similarity between apes and humans and postulated that there must be a "missing link" to complete the chain between us. This missing link metaphor then took on an evolutionary meaning after 1859, so that in the late nineteenth century, people wondered where the missing link fossil could be found that would connect humans to their ape ancestors. Eugène Dubois's discovery of "Java Man," "*Pithecanthropus*" (now *Homo*) *erectus* in 1891 was hailed as the first such discovery, although it is still a member of our own genus. Certainly, Raymond Dart's 1924 discovery of the skull of the "Taung Child," *Australopithecus africanus*, should have been sufficient to show that there were fossils that were truly intermediate between apes and modern humans and clearly not members of either group. As we shall discuss in chapter 15, the fossil record of extinct humans is now incredibly rich, so there are more "discovered links" than there are "missing" links. Nevertheless, the misconception about missing links leads some people to think that if a certain fossil hasn't yet been found, then evolution cannot be true.

Creationists are particularly shifty when this topic comes up. If they bring up the discredited concept of a missing link (knowing that their audience doesn't realize that the concept is invalid), they taunt the evolutionist to provide one. As I shall show in every remaining chapter of the book, the fossil record of transitional forms is truly amazing, so there is no shortage of fossils that could be called missing links (however erroneous the notion). But then the creationist will play a dirty trick. To divert attention away from the successful presentation of a transitional form, they will ask, "Where is the missing link between that fossil and another?" In other words, once you provide one intermediate between two groups, they ask for the other two "links" that connect the intermediate to each relative. Instead of conceding that they are beaten, they ask for more evidence, thus moving the goalposts and dishonestly demanding more evidence even after enough evidence has already been provided. It only goes to show how badly they misunderstand the fundamental concept: there is no such thing as a chain of being or a missing link!

Shermer (1997:149) describes their tactics this way,

Creationists demand just one transitional fossil. When you give it to them, they then claim there is a gap between these two fossils and ask you to present a transitional form between these two. If you do, there are now two more gaps in the fossil record, and so on ad infinitum. Simply pointing this out refutes the argument. You can do it with cups on a table, showing how each time the gap is filled with a cup it creates two gaps, which when each is filled with a cup creates four gaps, and so on. The absurdity of the argument is visually striking.

A good analogy of how creationists abuse the evidence before them and refuse to see the obvious connections between transitional fossils is lampooned in a number of clever editorial cartoons. One cartoon displays the word EVOL_T_ON on the board. One character says, "That can't possibly spell evolution! There are too many gaps!" The other replies, "That must mean the answer to the puzzle is 'CREATION.' " Another cartoon shows the sequence of hominid fossils with a couple of question marks scattered among the individual specimens. The caption reads, "If yu cn rea ths, don' gme tht bulsit abut missng transitional forms in th evolutnry t ee!" Most people are capable of filling in the blanks and seeing patterns and connections, but creationists doggedly refuse to accept any evidence presented to them, no matter how clearcut.

Another analogy about seeing connections in transitional fossils is to imagine looking down on a bridge over a river. We can see the water flowing beneath it at one end, and the water flowing out the other side, but we can't actually see that the two bodies of water are connected if we are only looking down on the bridge. Our mind fills in that connection for us. The creationist refusal to see the connections between very similar fossils is just as illogical as not seeing the connections between the two masses of water. For them, if they can't see every bit of water flowing in full view, then there is no "transitional water" that links the water on one side to the water on the other side of the bridge.

Finally, the importance of recognizing the difference between chains/ladders and bushes/trees extends to another concept: *mosaic evolution*. Under the great chain of being metaphor, every creature up the chain or ladder is more advanced than those below it and more primitive than those above it. *But evolution is a bush, not a ladder!* Organisms evolve, but they do not always move up the ladder. As sponges and corals show, they may retain primitive features even though they have survived for 500 million years. In the case of many animals (especially many fossils), not every anatomical feature of the animal evolves at the same time. Some parts may be quite advanced, while others retain their primitive states. This is the idea of mosaic evolution. Like a mosaic, the whole organism is composed of many tiny parts, and not every part is identical or changes in the same way.

Human evolution, for example, is a classic mosaic. Some features, like our bipedal locomotion, appeared very early, while others, like our large brain size or tool use, appeared much later. Early anthropologists expected to find fossils of humans where every feature was evolving slowly and steadily toward the modern human condition, but that is not the case. Each feature can evolve at a different rate.

Likewise, the classic transitional fossil *Archaeopteryx* is a mosaic of both advanced bird-like features (asymmetric flight feathers, wishbone) and retained primitive dinosaurian features (long bony tail, long fingers with claws, long robust legs without grasping big toe, and many others). Creationists will exploit this misunderstanding of mosaic evolution to claim that because it has birdlike feathers, it is just a bird. Then they contradict themselves by misquoting Gould and Eldredge to the effect that *Archaeopteryx* is a mosaic, and thus is *not* a bird! The entire quotation is as follows:

> At the higher level of evolutionary transition between basic morphological designs, gradualism has always been in trouble, though it remains the "official" position of most Western evolutionists. Smooth intermediates between *Baupläne* are almost impossible to construct, even in thought experiments; there is certainly no evidence for them in the fossil record (curious mosaics like *Archaeopteryx* do not count). (Gould and Eldredge 1977:147)

Creationist quote miners, in their effort to mislead and confuse people, only quote the last sentence, and then claim that Gould and Eldredge (1977) do not think *Archaeopteryx* is a good transitional form. As the complete quotation shows, they are only arguing that *Archaeopteryx* is a mosaic, not a smooth transition between body plans where every feature is intermediate. Lest there be any question that they have lied about Gould doubting that *Archaeopteryx* is an intermediate form, his article "The Telltale Wishbone" (in Gould 1980:267–277) should lay that issue to rest!

The Cladistic Revolution

> When more evidence is garnered, whether through the analysis of additional charac-ters, through the discovery of new specimens, or by pointing out errors and problems with the original data sets, new trees can be calculated. If these new trees better explain the data (taking fewer evolutionary transformations), they supplant the previous trees. You might not always like what comes out, but you have to accept it. Any real systematist (or scientist in general) has to be ready to heave all that he or she has believed in, consider it crap, and move on, in the face of new evidence. That is how we differ from clerics.
> —Mark Norell, *Unearthing the Dragon*

Taxonomy and systematics may not seem like glamorous disciplines, but neither are they bor-ing, uncontroversial fields. Taxonomists are famous for getting into heated arguments with each other about how to define species, how to classify organisms, and how to draw their family trees. There are rules of taxonomy (International Codes of Zoological, Botanical, and Bacterial Nomenclature), but there is also a lot of room for interpretation as well. To a large extent, taxonomists learn their trade through sheer experience: studying enough specimens of their organisms and their close relatives, watching how other taxonomists practice their trade and solve tough problems, and doing their research in a way that the scientific commu-nity will approve and see fit to publish. For a century or more, there were some general rules about how to go about this, but nothing in the way of a truly rigorous method of deciding how to classify organisms, or how to draw their family trees. Many biologists (especially in the 1950s) regarded this state of affairs as deplorable and railed against the prevailing subjec-tivity of the "art of taxonomy." In their minds, there must be a better way to make systematics more objective and quantifiable and less arbitrary to the whims of the systematist.

The first such unorthodox attempt of the 1950s and 1960s to reform the subjectivity and fuzzy methodology of the old systematics was known as numerical taxonomy, or *phenetics*. Its proponents tried to turn taxonomy into something that could be measured and coded into a computer program, and then "objective" results would emerge. Although this movement made some progress and pointed out many problems with the older system, eventually it failed because some of its assumptions were faulty and unworkable. In addition, it turned out that phenetics was not as objective as originally claimed. Subjectivity is still involved when scientists measure and record data and decide which characters to use and even when they run different computer programs on the same data. Numerical taxonomy ultimately lost steam as a movement when it turned out that the same computer program might give different answers for the same data, which made the entire objectivity advantage disappear.

But in the late 1960s, another systematic philosophy emerged to challenge the mainstream orthodoxy. Known as phylogenetic systematics, or *cladistics*, it was originally proposed by German entomologist Willi Hennig in 1950, but not widely followed until his German text was translated into English in 1966. Unlike the phenetic movement, which fizzled out soon after it was proposed, cladistics challenged the orthodoxy and eventually became mainstream itself, primarily because of its clear and rigorous methods and also because it worked to solve many previously insoluble problems. But in the 1960s and 1970s, the introduction of cladistics met with much resistance and controversy, as the more outrageous ideas proposed were rejected by older scientists who could not imagine changing the concepts that they learned as students. By the late 1980s, however, cladistic methods prevailed in the systematics of nearly every group of organisms.

I was fortunate to witness most of the stages of this revolution in systematics. I began as a graduate student at the American Museum of Natural History in New York in 1976, just as the cladistic revolution was in full swing, so I got to see all the major debates and got to know the key players. At the time, only the American Museum and a few other places accepted these ideas; the rest of the country looked at the "New York cladists" in horror as if we had a contagious disease. I gave my first professional talk at the Society of Vertebrate Paleontology meeting in 1978 on cladistics of Jurassic mammals, and I was one of the few people to mention cladistics at the entire meeting. Less than a decade later, all of the systematic talks at this same meeting were entirely cladistic, and those who stubbornly tried to do things the old way struck us young Turks as musty old dogs who couldn't learn new tricks.

New and outrageous ideas and challenges to the orthodoxy are always being proposed in science, but most unorthodox ideas don't go very far. It's not because science is inherently reactionary or conservative. On the contrary, there are always incentives for ambitious young scientists to make a name for themselves by challenging orthodoxy. But all new ideas must meet the test of peer review and scientific scrutiny and survive in the furnace of trial and error. Most such ideas fail because their limitations eventually become apparent. Cladistics became mainstream because it cleared up a lot of clutter in thinking and in practice and *because it works*.

What is cladistics, and why is it different from the older methods of classification? Hennig's main insight was that the anatomical features, or *characters*, that we use to name and describe organisms, are not all the same. Every organism is a mosaic of *advanced* (or *derived*) features inherited from a very recent common ancestor and *primitive* features inherited from distant ancestors. For example, we humans have advanced features such as our large brain and bipedal posture, but we inherited our absence of tails from our ape ancestors (all apes lack tails) and our grasping hands and stereovision from our earliest primate ancestors (nearly all primates have stereovision and grasping hands with opposable thumbs). Likewise, we inherited our hair and mammary glands from our distant mammalian ancestors (all mammals have them) and our four-legged bodies and lungs from our distant tetrapod ancestors (all tetrapods, including amphibians, reptiles, birds, and mammals, have them). If we wanted to define what makes us human, it would involve features related to our large brains and bipedality, our most recently developed evolutionary novelties, not our primitive features, such as a lack of tail, stereovision, grasping hands, hair, mammary glands, four legs, or lungs. The older classification schemes often mixed together primitive features and advanced features in their definitions, but Hennig pointed out that only the *shared derived*, or shared advanced, *characters* are really valid in defining natural groups.

We can define relationships based on these shared derived characters. Humans and monkeys (fig. 5.3) are more closely related to each other than they are to any other organisms in this diagram because they have many shared evolutionary novelties not found elsewhere in the animal kingdom, including opposable thumbs, stereovision, and many other features that define primates. A group including humans, monkeys, and cows could be defined by shared derived characters such as hair and mammary glands, unique features that define the class Mammalia. A group including mammals plus frogs could be defined on the shared presence of four legs and lungs, features that are found in all tetrapods. Likewise, sharks are more closely related to frogs and mammals than they are to lampreys because they have the advanced features of jaws and true vertebrae in their backbones. Notice that we did not use characters like jaws to define a group such as the mammals because jaws are primitive for mammals but derived at a much deeper level, at the level of the earliest jawed vertebrates (the gnathostomes). Characters are primitive or derived relative to the level at which they are being used, and according to Hennig, we must use them only at the level at which they first appear as evolutionary novelties.

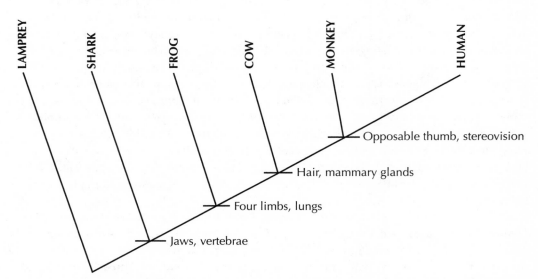

FIGURE 5.3. Evolutionary relationships of an assortment of vertebrates, showing the shared specializations (evolutionary novelties) that support each branching point (node) on this cladogram.

Thus, by using only the shared derived characters, we can construct a branching diagram of relationships of any three or more organisms. This kind of diagram is known as a *cladogram* (fig. 5.3). Cladograms only make statements about who is related to whom and show the evidence of that relationship by listing the derived characters at the branching points, or *nodes*. The power of cladograms is that they make minimal assumptions about how, why, and when these changes evolved; they only show the pattern of relationships and not much more. More importantly, a cladogram is instantly testable. All of the characters are out on display on the nodes, naked and exposed for scrutiny. Anyone who wishes to do a better job can immediately look at all the evidence and try to come up with a better hypothesis by falsifying the existing cladogram. By contrast, the family trees drawn by the old school of taxonomy prior to cladistics were completely untestable. Taxonomists would draw up a vague branching diagram of relationships, but there was no way to see on what data they based their family tree and no way to test it easily without redoing all of their work.

For the uninitiated, this simplistic explanation of cladistic methods seems obvious. What was all the fuss about? Initially, it was because it was a shocking new method with unorthodox ideas and assumptions, a lot of new alien terminology (most of which I skipped for simplicity's sake), and it was pushed by scientists who were not trying to make friends but win scientific battles. For a decade, the controversy was loud and bitter, and scientists often ended up calling each other names and insulting each other, both at meetings and in print. Most of the early, more controversial ideas either have been accepted by the mainstream or some of the more outrageous ideas have been quietly forgotten.

But a lot of the controversy involves things that are genuinely novel and hard for many who were accustomed to the older system to accept. Cladistics is called *phylogenetic systematics* because the cladogram is a kind of phylogeny or family tree of life, and it gives us a natural branching scheme for classification, the ultimate result of Linnaeus's work. Cladists assert that *classification should be a strict reflection of this phylogeny and nothing more*. Any mixing of other factors (such as ecology) mixes shared primitive and derived characters and leads to confusion. Thus, many of the groups in the classification schemes we have learned are natural and phylogenetic, but others are not.

For example, we can draw a cladogram of the tetrapods (fig. 5.4) that shows their relationships based on hundreds of shared derived characters. This part of the process is not controversial and is accepted by all scientists. But a traditional scientist might want to cluster turtles, lizards, snakes, and crocodilians in the Reptilia but not want to place the birds within the Reptilia. Instead, they usually place birds in their own class Aves, parallel to and equal in rank to class Reptilia. A cladist, however, would say that defining Reptilia without including all of their descendants (including birds) mixes ecology with phylogeny and is unacceptable. For Reptilia to be a natural group, it must include birds as a subgroup because they are descended from reptiles. To a traditionalist, it seems natural (and comforting and familiar) to cluster with "reptiles" all of the four-legged land vertebrates that are sluggish and scaly and cold-blooded, and to elevate birds to their own class because they have changed so much from their reptilian ancestors. But this mixes ecology with phylogeny because scales and cold-bloodedness and the other features that turtles and crocodilians have (but birds don't) are primitive for the entire group. To a cladist, those primitive characters are irrelevant to phylogeny and classification. Birds are a subgroup of reptiles in their classifications, and crocodiles are closer to birds than they are to snakes, turtles, and lizards. The only *natural*

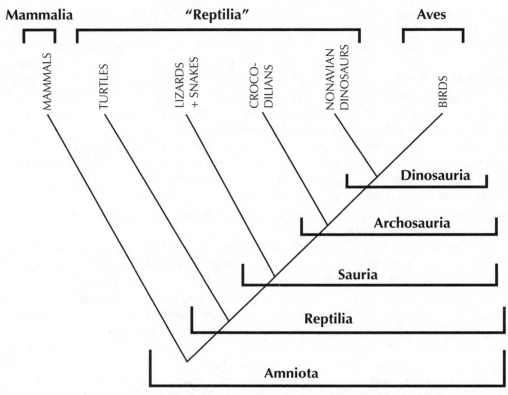

FIGURE 5.4. Different ways of classifying the same groups of organisms. Traditional classifications (top brackets) emphasize overall similarity and prefer to focus on the great evolutionary radiations of birds and mammals by placing them in their own classes, equal in rank to class Reptilia. A cladistic classification (lower brackets) only recognizes monophyletic groups, that is, groups that include all descendants of a common ancestor. "Reptilia," as traditionally defined, is not monophyletic, because it does not include one descendant group, the birds. Instead, the natural monophyletic groups include amniotes (all land vertebrates), Reptilia (only if it includes birds), Sauria (the nonturtle reptiles), Archosauria (the group including crocodilians, dinosaurs, and birds), and Dinosauria (birds and the nonavian dinosaurs).

(or *monophyletic*) *groups* are those that are *defined on shared derived characters and include all descendants of a common ancestor.*

    Let us look at another example closer to home. In older classification schemes, it was common to lump all of the great apes in the family Pongidae and place ourselves in the separate family Hominidae. In our anthropocentric arrogance, we always considered ourselves in a special group and exclude humans from the rest of the animal kingdom. But the evolutionary relationships of apes and humans are well established (fig. 5.5), and humans are just one branch among the great apes. Chimps and gorillas are much more closely related to us than they are to orangutans or gibbons, even though they all share primitive features of long, strong limbs and long hair and smaller brains and long snouts, and we have diverged the most in our bipedalism and nearly hairless skin and large brains with small faces. Cladistically speaking, it is invalid to place all the rest of the apes in a group without humans because that is a "wastebasket" group (like Reptilia without birds)

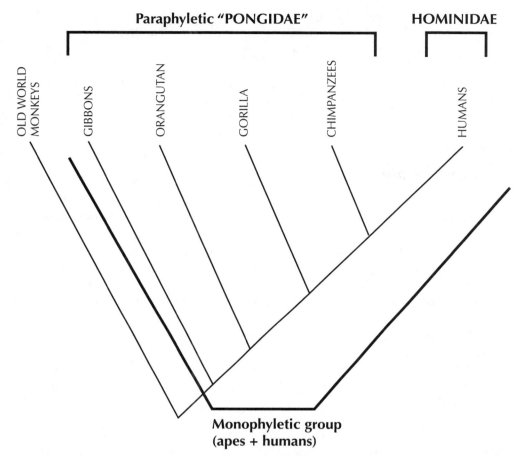

FIGURE 5.5. Traditional classifications emphasize the shared primitive similarity of the "great apes" and place them in their own family Pongidae, while acknowledging the big differences between apes and humans by placing humans in their own family, Hominidae. However, to a cladist the family Pongidae is paraphyletic because it does not include all descendants (i.e., humans) of the common ancestor of the human-ape stem. In this framework, Hominidae must be expanded to include all the great apes (as most anthropologists now agree).

that does not include all descendants of a common ancestor. To a cladist, the wastebasket (*paraphyletic*) family Pongidae is invalid, just as Reptilia without birds is paraphyletic and invalid. A natural (monophyletic) group would be to place us within the ape family Pongidae or to expand the Hominidae to include most or all of our close ape relatives. In fact, that is what has happened: most of the great apes are now in the family Hominidae. This makes a lot of sense logically, but it is tough for many to accept when they were trained to believe in the Pongidae.

This is the brave new world of cladistic classification. Most classifications of animals and plants are partly monophyletic (e.g., birds and mammals are natural monophyletic groups) but mix in a lot of paraphyletic wastebaskets as well (including reptiles and amphibians, as traditionally defined). Some groups, like the invertebrates, are unnatural by definition, because invertebrates are defined by the shared primitive *lack* of a specialized feature, the vertebrae of the backbone. Bit by bit, however, more biologists are coming to terms with the cladistic revolution in systematics, accepting the results, and learning to use less familiar but natural groups such as "amniotes" and "tetrapods" rather than "reptiles" and "amphibians" as traditionally used. I am proud to say that my historical geology textbook (Dott and Prothero 1994) was the first to introduce cladistics to its textbook market and to avoid paraphyletic groups, and many other textbooks are beginning to catch on to what professional systematists had accepted more than 30 years ago.

Ancestor Worship

That a known fossil or recent species, or higher taxonomic group, however primitive it might appear, is an actual ancestor of some other species or group, is an assumption scientifically unjustifiable, for science never can simply assume that which it has the responsibility to demonstrate. . . . It is the burden of each of us to demonstrate the reasonableness of any hypothesis we might care to erect about ancestral conditions, keeping in mind that we have no ancestors alive today, that in all probability such ancestors have been dead for many tens of millions of years, and that even in the fossil record they are not accessible to us.
—Gary Nelson, "Origin and Diversification of Teleostean Fishes"

Fossils may tell us many things, but one thing they can never disclose is whether they were ancestors of anything else.
—Colin Patterson, *Evolution*

Some aspects of cladistic theory have proven more difficult for many scientists to accept. For example, a cladogram is simply a branching diagram of relationships among three or more taxa. It does not specify whether one taxon is ancestral to another; it only shows the topology of their relationships as established by shared derived characters. In its simplicity and lack of additional assumptions, it is beautifully testable and falsifiable, so it meets Popper's criterion for a valid scientific hypothesis. The nodes are simply branching points supported by shared derived characters, which presumably represent the most recent hypothetical common ancestor of the taxa that branch from that node. But strictly speaking, cladograms never put real taxa at any nodes, but only at the tips of the branches.

Many scientists, however, would like to say more than just "taxon A is more closely related to taxon B than it is to taxon C." Instead, they would draw relationships with one taxon being suggested as ancestral to another. This is the more traditional *family tree* type of phylogeny, which not only suggests relationships, but shows a pattern of ancestry and descent as well. But as Tattersall and Eldredge (1977) point out, a family tree makes far more assumptions than does a cladogram. Some people are happy to make those assumptions, but the strict cladists are not so comfortable with them.

The biggest sticking point is the concept of ancestry. We tend to use the term "ancestor" to describe certain fossils, but we must be careful when making that statement. If we want to be rigorous and stick to testable hypotheses, it is hard to support the statement that "this particular fossil is the ancestor of all later fossils of its group," because we usually can't test that hypothesis. Because the fossil record is so incomplete, it is highly unlikely that any particular fossil in our collections is the remains of the actual ancestor of another taxon (Schaeffer et al. 1972; Engelmann and Wiley 1977).

But there's another reason why cladists avoid the concept of ancestry. To be a true ancestor, the fossil must have nothing but shared primitive characters compared to its descendants. If it has any derived feature not found in a descendant, it cannot be an ancestor. Consequently, for decades, traditional taxonomists looked only at shared primitive characters so they could construct ancestor-descendant trees, thereby missing all the derived characters that showed they were on the wrong track. One of the great advantages of cladistics is that it has solved many previously insoluble problems by getting away from paraphyletic wastebasket groups and "ancestor worship" and focusing on derived characters only. For these reasons, hard-core cladists like Gary Nelson (quoted earlier) refuse to recognize the concept of ancestor at all, except in the hypothetical sense of the taxa at the nodes of the cladogram. Instead of ancestor and descendant, cladists prefer to talk about two taxa at the tips of the branches as being *sister groups*. Neither is ancestral to the other, but they are each other's closest relatives.

But there *are* circumstances where the fossil record is so complete that it is possible to say that "the fossils in this population represent the ancestors of this later population." My friend and fellow former graduate student Dave Lazarus (now a curator at the Museum für Naturkunde in Berlin) and I (Prothero and Lazarus 1980) provided just such an example from the extraordinary fossil record of planktonic microfossils. In these unusual circumstances, we have deep-sea cores covering all of geologic time since the Jurassic for most of the world's oceans and every centimeter of sediment in most of those cores is filled with thousands of microfossils. With an extraordinarily dense and continuous record such as this, we really can say that we have sampled all the fossil populations that lived in the world's oceans and can establish which samples are most likely the ancestors of later populations. Since our paper, a number of studies have been done to establish how complete the fossil record needs to be to determine the probability that one population is ancestral to another (Fortey and Jefferies 1982; Lazarus and Prothero 1984; Paul 1992; Huelsenbeck 1994; Fisher 1994; Smith 1994; Clyde and Fisher 1997; Hitchin and Benton 1997; Huelsenbeck and Rannata 1997). Nowadays, paleontologists are a lot more relaxed about the concept of ancestry than they were during the early, bitterly polarized debates over cladistics in the 1970s. Most paleontologists use the word *ancestor* (as I will throughout this book) very loosely to describe a fossil that has all the right anatomy and is older in time to potentially be ancestral to some later form. But we all recognize subconsciously that, in the strictest sense, telling whether a particular fossil is *actually* the ancestor of another is not a testable hypothesis. Instead, we look to fossils to show us the *transitional anatomical features of ancestors* that illustrate the path that evolution took.

Naturally, this debate, which revolves around subtle philosophical distinctions about ancestry, has been a bonanza for the quote-mining creationists. They pull dozens of quotations (like the one from Gary Nelson) completely out of the context I have just described and claim that these statements show that there are no such things as ancestors in the fossil record. (Prothero and Lazarus [1980] disproved that decisively.) The debate was all about *whether we could tell whether a particular fossil could be recognized as an ancestor and how to do phylogeny*, but even the most hard-core cladists *do not doubt that ancestors existed!* None of the debate is about whether life has evolved. After all, what would be the point of doing a cladogram, which is a phylogeny, if you didn't accept the fact of evolution?

The most outrageous case was when a creationist spy named Luther Sunderland snuck into a closed scientific meeting of the Systematics Discussion Group at the American Museum in 1981 with a hidden tape recorder. At this time in the long history of debates on cladistics, many of the most extreme advocates were calling themselves "pattern cladists." They no longer followed neo-Darwinism (as we discussed in chapter 4) and all the convoluted scenarios that had been built on top of simple phylogenetic diagrams to create complicated family trees with many additional assumptions. Instead, they argued that pure science was simply the testable hypotheses of the *patterns* of cladograms and nothing more. My friend, the distinguished paleoichthyologist Colin Patterson of the Natural History Museum in London, was talking about pattern cladism and how he had abandoned many of the assumptions about evolution that he had once held, including the recognition of ancestors in the fossil record. He was now only interested in the simplest hypotheses that were easily tested, such as cladograms. But, of course, taken out of context, it sounds as though Colin doubted that evolution had taken place, yet he said nothing of the sort! Colin was speaking in a kind of "shorthand" that makes sense to the scientists who understand the subtleties of the debate, but means something entirely different when taken out of context. I was at that meeting and was stunned to read afterward about Sunderland's account of what had happened because I remembered Colin's ideas clearly and could not imagine how they could be misinterpreted. For decades afterward, Colin had to explain over and over again what he had meant, and why he did not doubt the fact that evolution had occurred, only that he no longer accepted a lot of the other assumptions about evolution that neo-Darwinists made. Unfortunately, Colin died in 1998 while he was still in his scientific prime, unable to continue fighting these misinterpretations of his ideas that continue to be propagated by the creationists.

The Molecular Third Dimension

One no longer has the option of considering a fossil older than about eight million years as a hominid no matter what it looks like.
—Vincent Sarich, 1971

As was pointed out in chapter 4, one of the most powerful corroborations of Darwin's evidence of the branching structure of life is that we can see that branching pattern by comparing the molecules in nearly every cell in every living thing (e.g., fig. 4.7). This was evidence that not even Darwin could have anticipated and convincing proof of the fact that life has evolved. Whether one looks at the DNA sequences directly or at other nucleic acids like mitochondrial DNA, or the RNAs, or the protein sequences in any other number of biochemicals—cytochrome *c*, hemoglobin, alpha lens crystalline, and many other proteins—the answer nearly always comes out the same. To paraphrase the Bible, every one of our cells declare the handiwork of evolution! It is a simple calculation to show that these identical branching patterns in every biochemical system in an organism are not random and could not occur unless that branching pattern were due to common ancestry.

Molecular phylogeny emerged in the 1960s with very crude methods, such as hybridizing DNA strands from different animals to see how similar their DNAs were (and therefore their evolutionary distance). Other methods compared the strength of the immune response (more closely related organisms have stronger immune responses than distantly related organisms, because the former have more genes in common than the latter). But since the 1990s, especially with the ability of the PCR (polymerase chain reaction) to produce lots of copies of DNA, allowing us to read the DNA rapidly, we now have the full DNA sequence of a number of organisms, including fruit flies, lab rats, mice, and rabbits, several domesticated animals, the nematode worm *Caenorhabditis elegans*, and most of our ape relatives. The mitochondrial DNAs of many apes were sequenced as early as 1982, but the entire chimpanzee nuclear DNA sequence was not completed until August 2005. Human nuclear DNA was sequenced in 2001 by the Human Genome Project and also by Craig Venter's lab. From all these studies, we now have a powerful tool to compare the genetic codes of a wide range of organisms. These data not only show us how we differ from other organisms but also (especially in the case of human DNA) allow us to find out what genes code for what parts of the body and where in the DNA the genes for inherited diseases occur. Many scientists hail the decoding of human DNA as one of our greatest scientific achievements ever because of its potential not only to answer scientific questions but also to cure many diseases.

Molecular approaches have been particularly useful where there isn't much evidence from the anatomy or the fossil record of organisms to determine their evolutionary relationships. For example, the external form of most bacteria is pretty stereotyped, and most early bacteriologists underestimated their diversity. With the advent of genetic analysis, however, scientists such as Carl Woese have shown that there are several different kingdoms of bacteria, including the most primitive organisms of all, the Archaebacteria, which mostly live in extreme environments such as hot springs and anoxic conditions. Scientists have debated for years about how animals, plants, fungi, and bacteria might be related, but molecular phylogeny has provided an answer that could never have been solved by traditional methods (fig. 5.6). The relationships of the major groups of multicellular animals were also hotly debated for over a century, but molecular techniques, in combination with newer ideas about embryology and anatomy, have provided an answer that is no longer disputed (fig. 5.7). Thus the molecular evidence provides an independent way of discovering the family tree of life and, in many instances, has given us answers that could not have been obtained by any other method.

This is not to say that every molecular study works perfectly or that molecular phylogenies are always superior to other methods. Like any other characters used in a phylogenetic analysis, molecular changes can be viewed as primitive and derived. But unlike anatomical characters, most molecular changes are limited to the four nucleotides (adenine, thymine, guanine, and cytosine), so there are only a limited number of possible changes. Consequently, if a gene is going to change, it can very easily return to the primitive condition, since that is the only alternative. This generates some "noise" in the molecular signal. In recent years, sophisticated methods have been used to detect this noise and filter it out. Likewise, the "clocklike" mutation rate hypothesis sometimes fails, because some organisms or taxonomic groups seem to have higher or lower than average rates of mutation, so there are still debates when molecular clock estimates give ages of branching points in evolution that are much too old to match the fossil record.

Still, there have been some remarkable successes. For years, paleoanthropologists such as Elwyn Simons and David Pilbeam argued that *Ramapithecus*, a fossil from beds in Pakistan that are 12 million years old, was the earliest member of our family Hominidae. If that were true, then the ape-human split would be older than 12 million years ago. But molecular biologists Vince Sarich and Allan Wilson of Berkeley looked at many different

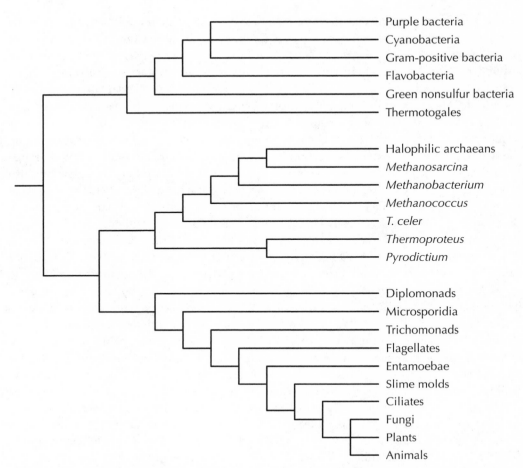

FIGURE 5.6. The fundamental tree of life derived from molecular data, showing the major kingdoms of pro-karyotes (Eubacteria, Archaebacteria, and many other microbes), and the small side branch of eukaryotes (plants, animals, and fungi).

biochemical systems (starting with the simple immunological distance method) and always concluded that the divergence point between apes and humans was 5–7 million years ago. During the 1970s and early 1980s, there were many bitter debates over the point, so that Sarich was even quoted as saying that a fossil older than 8 million years old cannot be a hominid, no matter what it looks like! But in the late 1980s, additional, more complete fossils were found in Pakistan that showed that *Ramapithecus* was more like an orangutan and was actually a member of the genus *Sivapithecus*. In this case, the molecular biologists were right, and the paleontologists who staked their careers on *Ramapithecus* had to lick their wounds.

Likewise, Vincent Sarich was one of the first (based on molecular evidence) to say that whales were descended from even-toed hoofed mammals (the Artiodactyla) and particularly closely related to the hippopotamus. That radical idea was backed up by many more molecu-lar studies and was finally corroborated in 2001 when two different groups of paleontologists (Gingerich et al. 2001; Thewissen et al. 2001) found the ankle bones of two different kinds of primitive whales from Pakistan that showed their relationships to the artiodactyls (see

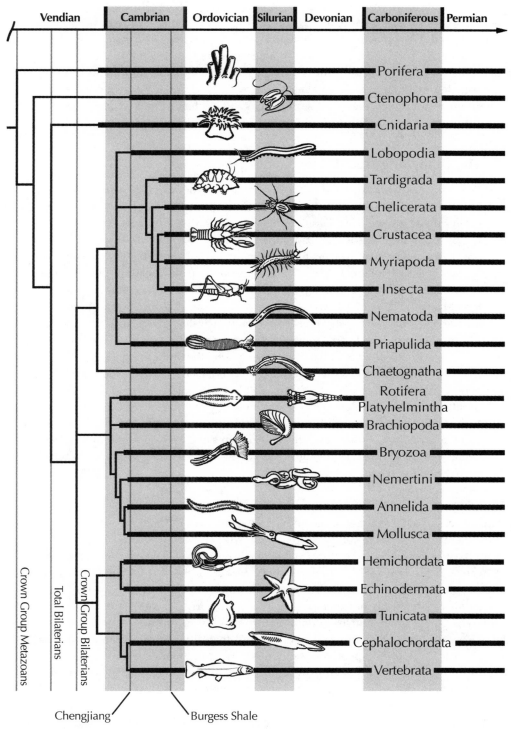

FIGURE 5.7. The branching history of the animals, based on molecular data. (Drawing by Carl Buell; modified from Briggs and Fortey 2005: fig. 2)

chapter 14). But for the many successes of molecular biologists, there have been some embarrassments. One study (Graur et al. 1991) concluded that guinea pigs were not rodents! This was quickly shot down by numerous other molecular labs (e.g., Cao et al. 1994) that showed the flaws in the analysis. No method in science is perfect, but molecular methods have proven very powerful, and there are so many checks and balances in the peer-review process that if one lab makes a mistake, other labs correct it. But if many labs get the same results from different molecules, then that is probably good evidence that they are onto something.

The Branching Tree of Life

From the most remote period in the history of the world organic beings have been found to resemble each other in descending degrees, so that they can be classed in groups under groups. This classification is not arbitrary like the grouping of the stars in constellations. The existence of groups would have been of simpler significance, if one group had been exclusively fitted to inhabit the land and another the water; one to feed on flesh, another on vegetable matter, and so on; but the case is widely different, for it is notorious how commonly members of even the same subgroup have different habits. . . . Naturalists, as we have seen, try to arrange the species, genera, and families in each class, on what is called the Natural System. But what is meant by this system? Some authors look at it merely as a scheme for arranging together those living objects which are most alike, and for separating those which are most unlike. . . . But many naturalists think that something more is meant by the Natural System; they believe that it reveals the plan of the Creator; but unless it be specified whether order in time or space, or both, or what else is meant by the plan of the Creator, it seems to me that nothing is thus added to our knowledge. . . . I believe that this is the case, and that community of descent—the one known cause of close similarity in organic beings—is the bond, which though observed by various degrees of modification, is partially revealed to us by our classifications.
—Charles Darwin, *On the Origin of Species*

As this discussion of systematics has shown, there are many different methods for deciphering the history of life. From comparison of their external and internal anatomical features, to the similarities in embryonic history, to the details of the molecules in every cell, the branching history of life is revealed in nearly every aspect of organisms. The fact that so many systems give the same answer makes it very robust. If we can't resolve the phylogeny from the anatomy, perhaps the molecules will help out. If the fossil record is poor in one particular group, we look to other sources of data. But if the fossil record or anatomical data is excellent, we are cautious about molecular conclusions that differ widely from our paleontological estimates. Thus, although no one method is perfect all the time, each method has its own strengths and weaknesses that allow us to decipher the problem, one way or another.

In addition, the advent of rigorous testable methods of phylogenetic reconstruction through cladistic analysis of anatomical details and of molecular sequences has made our efforts at determining the true history of life more successful than ever before. When I was a graduate student, many areas of evolutionary history were controversial and poorly resolved, based on the evidence that was available in the 1970s. But one after another, the

Gordian knots have been cut. The phylogeny of the placental mammals was an unsolved mystery that was already two centuries old when scientists at the American Museum, such as Malcolm McKenna, Mike Novacek, Earl Manning, and I, tackled the problem. By the late 1980s, we had solved many aspects of the problem using cladistic analysis (see the papers in the volumes edited by Benton [1988a, 1988b] and Szalay et al. [1993]), and most studies published since then have corroborated our original topology (although there are some slightly different answers from the molecular world, and the differences are still to be resolved). Likewise, the deciphering of the phylogeny of life (fig. 5.6) and of the major groups of animals (fig. 5.7) are great achievements that have resolved over a century of controversy. Today, hundreds of systematists are at work solving these important problems, trying to test older hypotheses with new data, and working to resolve the issues even more definitively.

From the vaguely formulated hypotheses of ancestry published only a generation ago, we are now in a new age of understanding of life's history (see Dawkins 2004). Where we once knew very little about its true pattern, we now have multiple lines of evidence that converge on a common answer and give a robust solution corroborated and tested many different ways that is almost certainly "the truth" (as much as we can use that term in science). In the chapters that follow, we may find that we are missing evidence from fossils at a certain interval or from anatomy at another. Contrary to creationists' claims, there are always additional lines of evidence that allow us to decipher that phylogenetic problem. We are always continuing to move forward and find new answers, but we have already learned more in the past two decades than we did in all previous centuries. We no longer have to use the guesswork that creationists criticize because the tree of life is now as well known as almost any other fact of nature.

The great molecular biologist Emile Zuckerkandl and the Nobel-Prize-winning chemist Linus Pauling said it best over 50 years ago:

> It will be determined to what extent the phylogenetic tree, as derived from molecular data in complete independence from the results of organismal biology, coincides with the phylogenetic tree constructed on the basis of organismal biology. If the two phylogenetic trees are mostly in agreement with respect to the topology of branching, the best available single proof of the reality of macro-evolution would be furnished. Indeed, only the theory of evolution, combined with the realization that events at any supramolecular level are consistent with molecular events, could reasonably account for such a congruence between lines of evidence obtained independently, namely amino acid sequences of homologous polypeptide chains on the one hand, and the finds of organismal taxonomy and paleontology on the other hand. Besides offering an intellectual satisfaction to some, the advertising of such evidence would of course amount to beating a dead horse. Some beating of dead horses may be ethical, when here and there they display unexpected twitches that look like life. (Zuckerkandl and Pauling 1965:101)

For Further Reading

Adoutte, A., G. Balavoine, N. Lartillot, O. Lespinet, B. Prudhomme, and R. de Rosa. 2000. The new animal phylogeny: Reliability and implications. *Proceedings of the National Academy of Sciences USA* 97:4453–4456.

Arthur, W. 1997. *The Origin of Animal Body Plans: A Study in Evolutionary Developmental Biology*. New York: Cambridge University Press.

Dawkins, R. 2004. *The Ancestor's Tale: A Pilgrimage to the Dawn of Evolution*. Boston: Houghton Mifflin.

Foote, M. 1996. On the probability of ancestors in the fossil record. *Paleobiology* 22:141–151.

Hennig, W. 1966. *Phylogenetic Systematics*. Urbana: University of Illinois Press.

Hillis, D. M., and C. Moritz, eds. 1990. *Molecular Systematics*. Sunderland, Mass.: Sinauer.

Lazarus, D. B., and D. R. Prothero. 1984. The role of stratigraphic and morphologic data in phylogeny reconstruction. *Journal of Paleontology* 58:163–172.

Nielsen, C. 2001. *Animal Evolution: Interrelationships of the Living Phyla*. 2nd ed. New York: Oxford University Press.

Patterson, C. 1981. Significance of fossils in determining evolutionary relationships. *Annual Review of Ecology and Systematics* 12:195–223.

Patterson, C., ed. 1987. *Molecules or Morphology in Evolution: Conflict or Compromise?* New York: Cambridge University Press.

Prothero, D. R. 2004. *Bringing Fossils to Life: An Introduction to Paleobiology*. 2nd ed. New York: McGraw-Hill.

Prothero, D. R., and D. B. Lazarus, 1980. Planktonic microfossils and the recognition of ancestors. *Systematic Zoology* 29:119–129.

Runnegar, B., and J. W. Schopf, eds. 1988. *Molecular Evolution in the Fossil Record*. Lancaster, Pa.: Paleontological Society Short Course Notes 1.

Schaeffer, B., M. K. Hecht, and N. Eldredge. 1972. Phylogeny and paleontology. *Evolutionary Biology* 6:31–46.

Schoch, R. M. 1986. *Phylogeny Reconstruction in Paleontology*. New York: Van Nostrand Reinhold.

Smith, A. B., and K. J. Peterson. 2002. Dating the time of origin of major clades: Molecular clocks and the fossil record. *Annual Reviews of Earth and Planetary Sciences* 30:65–88.

Tudge, C. 2000. *The Variety of Life: A Survey and a Celebration of All the Creatures That Have Ever Lived*. Oxford: Oxford University Press.

Wiley, E. O. 1981. *Phylogenetics: The Theory and Practice of Phylogenetic Systematics*. New York: Wiley Interscience.

Woese, C. R., and G. E. Fox. 1977. Phylogenetic structure of the prokaryotic domain: The primary kingdoms. *Proceedings of the National Academy of Sciences USA* 74:5088–5090.

Part II

EVOLUTION? THE FOSSILS SAY YES!

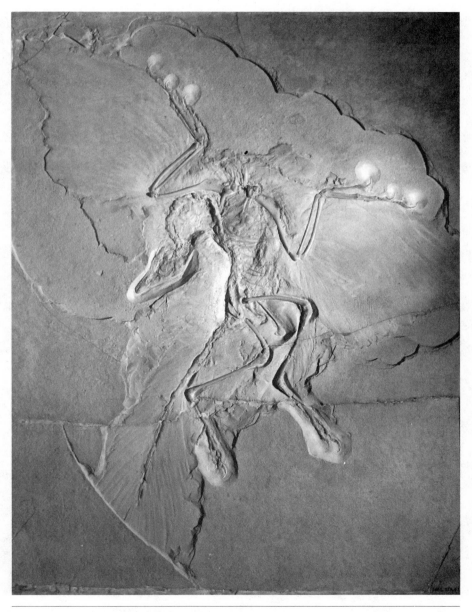

The beautifully complete specimen of the transitional fossil between dinosaurs and birds, known as *Archaeopteryx*. (Photo by H. Raab, courtesy Wikimedia Commons)

FIGURE 6.1. The concentric layers of the cabbage-like domed cyanobacterial mats known as stromatolites, here sliced off at the top by a glacier, now found in Lester Park, New York. (Photo by the author)

LIFE'S ORIGINS

6

To the Blazing Hot North Pole

On a sunny day in late July, I am off to North Pole to examine these vestiges [of ancient life]. Heat and dust permeate the cab as our Land Rover rattles over the rutted dirt track. There are flies everywhere. This North Pole, you see, lies in northwestern Australia—its name, with characteristic Aussie humor, marking one of the hottest places on earth.

—Andrew Knoll, *Life on a Young Planet*

How did life on earth begin? This is one of the most fascinating and controversial topics in science. Unlike many of the other areas of life's history that we discuss in the remaining chapters, we have only a few fossils to guide us. The oldest fossils that are clearly formed by living things are microscopic fossils of *cyanobacteria* (formerly mislabeled "blue-green algae," but they are not true algae) from 3.5-billion-year-old rocks of the Warrawoona Group near North Pole in Western Australia. In addition to these microfossils, there are layered domed structures known as stromatolites (fig. 6.1) that are produced by cyanobacterial mats (which are still producing such structures today) from the Warrawoona Group (fig. 6.2). Just slightly younger are numerous even better microfossils and stromatolites from rocks dated at 3.4 billion years from the Fig Tree Group of South Africa. As this book went to press, possible stromatolites dated at 3.8 billion years old were reported from Greenland, which would be the oldest fossils known if they are indeed stromatolites.

All of these microfossils are simple filaments with a number of cells in a row, many with distinctive structures that are virtually indistinguishable from their modern counterparts. So we know for sure that bacteria and other simple *prokaryotic* (lacking a discrete nucleus) cells were present 3.4–3.5 billion years ago, but so far we have not found any older fossils. This is not surprising, given that there are very few places on earth with rocks any older than 3.5 billion years, and most rock that old has already been heavily deformed, cooked by metamorphism, or otherwise so altered that there is little chance of any fossils surviving. There are rocks from the Isua Supracrustals in west Greenland that have distinctive organic molecules in them that are unique to living systems, suggesting that life was present 3.8 billion years ago (Schidlowski et al. 1979; Mojzsis et al. 1996), and these are the rocks that also yield the possible stromatolites as well (Nutman et al. 2016).

Certainly, it is likely that life was already established by 3.8 billion years ago. Whether it was possible for life to get a foothold on earth before then is still debated. Before 3.9 billion years ago, the earth was still heavily bombarded by leftover debris from the formation of the solar system, which probably vaporized the oceans of the earth many times. (We know this because most of the impact craters on the moon, which must have suffered through the same bombardment, date from 3.9 billion years and older.) Scientists call this time the period of "impact frustration." Most geologists did not think the earth could have cooled enough for

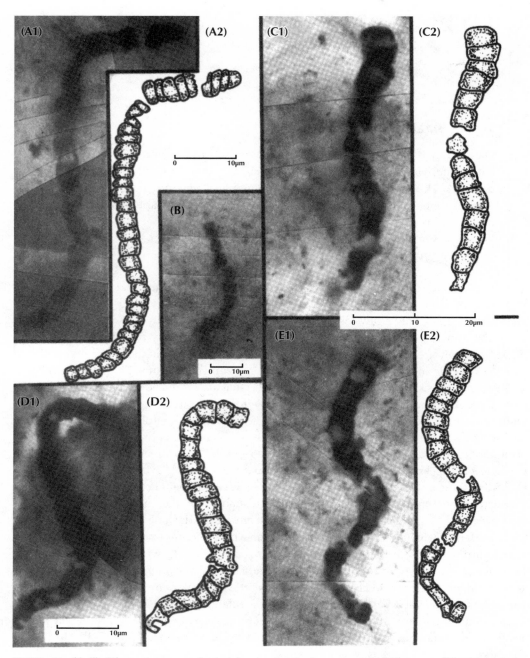

FIGURE 6.2. (A–E) Filamentous cyanobacterial microfossils from the 3.46-billion-year-old rocks of the Warrawoona Group, North Pole, Western Australia, showing the distinctive chains of cells. These are virtually identical to (F) (shown on next page), the modern filamentous cyanobacterium, *Lyngbya*, in every detail of the size, shape, and arrangement of the cells. (Photos courtesy J. W. Schopf)

(F)

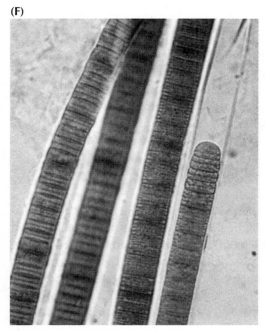

FIGURE 6.2. (*Continued*)

liquid water to condense and form oceans much before 4.0 billion years ago, but tiny zircon grains from Australia were recently discovered that seem to have a distinctive chemistry indicating oceans as early as 4.3–4.4 billion years ago (Wilde et al. 2001; Valley et al. 2002). If so, then the earth's surface cooled down to below 100°C (the boiling point of water) in only 250,000 years after its formation 4.65 billion years ago. There is nearly a billion years of time between when the first oceans form and the first clear fossils are known, plenty of time for life to evolve (more than once, if necessary), given these constraints.

Because these earliest fossils are tiny carbonized films preserved in cherts and flints, they provide little evidence for the chemical processes that formed life. All the organic chemicals in them have long been transformed by heat and pressure into other carbon compounds (usually pure carbon in the form of graphite, the "lead" in your pencil). We have only their shapes (fig. 6.2) to demonstrate that they were once living things. For earlier stages in this process, we cannot use the fossil record. Instead, we must try an experimental approach, using the constraints imposed by our knowledge of earth's history and the nature of organic chemistry to find possible solutions. Naturally, some people (especially creationists) label this as "guesswork" and refuse to accept any of the experimental evidence to understanding life's origins. They prefer to just say "God did it" and stop there. They are welcome to their opinions, of course, but as we explained in chapter 1, no true scientist falls back on this "god of the gaps" approach. The supernatural hypothesis is simply untestable and leads nowhere. Scientists may not have all the answers to this very complex and difficult problem, but they are not sitting on their hands or surrendering to the unscientific supernatural approach. They are constantly trying new experimental approaches and (as we shall soon see) have made a huge amount of progress—much more than creationists realize.

Recipe for Primordial Soup

It is often said that all the conditions for the first production of a living organism are now present, which could ever have been present. But if (and oh! what a big if!) we could conceive in some warm little pond, with all sorts of ammonia and phosphoric salts, light, heat, electricity, &c., present, that a proteine [*sic*] compound was chemically formed ready to undergo still more complex changes, at the present day such matter would be instantly absorbed, which would not have been the case before living creatures were found.

—Charles Darwin, letter to Joseph Hooker

The first scientific suggestions about the origin of life were made by several people, including Darwin himself in this 1871 letter to his friend, botanist Joseph Hooker. Darwin speculated that a "warm little pond" with the right combination of chemical compounds (usually called the "primordial soup") and the right sources of energy could produce proteins. But organic chemistry was still in its infancy back then, so little could be done to follow this suggestion. In the 1920s, the Russian biochemist A. I. Oparin and the British geneticist J. B. S. Haldane (also one of the fathers of neo-Darwinism mentioned in chapter 4) independently suggested that the earth with a reducing atmosphere of nitrogen, carbon dioxide, ammonia (NH_3), and methane or "natural gas" (CH_4) would be the ideal primordial soup for producing simple organic compounds.

The most important breakthrough occurred in 1953, when a young graduate student at the University of Chicago named Stanley Miller heard about Oparin's hypothesis from his advisor, Nobel Prize–winning chemist Harold Urey. They decided to try an experiment along the lines suggested by Oparin and Haldane to see whether such a primordial soup could generate basic biochemicals. Miller built a simple apparatus (fig. 6.3) out of sealed tubes that formed a continuous loop, with all the air removed by vacuum. A new atmosphere rich in carbon dioxide, nitrogen, methane, ammonia, and water (but *no* free oxygen) was placed in the evacuated tubes. Miller then put a source of heat below the "ocean" flask at the base to start the steam circulating, and in another flask he used electrodes that created sparks to simulate "lightning" as an energy source (fig. 6.3). Below the "lightning" chamber, a condenser returned the gases to liquid state, where they recirculated back to the "ocean." This graduate student experiment (not even his original thesis topic) yielded the most startling results. Within days, the ocean became brown with new chemicals, and within a week, it was an organic-rich gunk. When Miller analyzed it, he had already produced 4 of the 20 amino acids that life uses to make proteins, plus many other organic molecules, such as cyanide (HCN) and formaldehyde (H_2CO). As Knoll (2003:74) writes, "In one remarkable experiment, Miller jump-started research on life's origins. Powered by the energy of nature, simple gas mixtures could give rise to molecules of biological relevance and complexity." Even though amino acids are much more complex than the chemicals he started with, Miller showed they were remarkably easy to produce. Later experiments produced 12 of the 20 amino acids found in life. Another experiment with a dilute cyanide mixture produced seven amino acids. No matter how you cut it, it does not require divine intervention or even more than a few days in the lab to make the basic building blocks of life. Since Miller's experiments, other scientists have found 74 different amino acids in meteorites (including all 20 found in living systems), so apparently organic compounds

Stopcock for withdrawing samples during run

Tungsten electrode   5-liter flask   Tungsten electrode

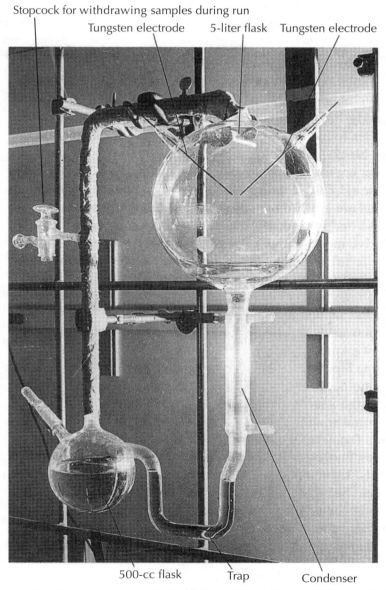

500-cc flask   Trap   Condenser

FIGURE 6.3. An apparatus like this was used by Stanley Miller and Harold Urey in 1953 to simulate the synthesis of complex organic compounds on the early earth. The system was evacuated of air, then the large flask held an "atmosphere" rich in carbon dioxide, water, nitrogen, ammonia, and methane (but no oxygen). Sparks from electrodes simulated lightning. The product of this reaction then flowed through the condenser and accumulated in the flask, which became a brew of "primordial soup." After about a week, the clear solution had turned into a thick murky brown sludge full of newly synthesized organic compounds, including many of the amino acids necessary to build life. (Photo courtesy S. Miller)

have been produced in many other places in the universe. Some scientists even speculate that the earth was "seeded" with organic compounds from space and that sparked the origin of life, although given how easily they are made here on earth, we don't need this more complex hypothesis.

In the 1980s and 1990s, scientists thought that ammonia and methane were not common in the early atmosphere. This does not invalidate the Urey-Miller experiment. It only makes the task of creating organic materials a bit harder. Many later experiments have been tried with atmospheric compositions lower in methane and ammonia, and they still produced the same results as the original experiment. Since that time, scientists have returned to thinking that the early earth's atmosphere was indeed rich in methane, and probably ammonia, too, so this criticism is no longer valid (Fegley and Schaefer 2005). In fact, the presence of amino acids in many meteorites tells us that a wide variety of atmospheric conditions throughout the universe are capable of producing them, not just the original conditions of the Urey-Miller experiment. But the creationists like Wells only want to focus on old outdated objections to experiments that may or may not be representative of current thinking. As we shall see throughout this book, creationists ignore more recent results if they can criticize outdated results and appear to invalidate a whole field of study.

So the initial building blocks are incredibly easy to produce, and it's a fair assumption that the earth's oceans had plenty of amino acids and other simple organic molecules floating around. The next step is a bit more difficult: assemble the simple building blocks of life into longer-chain molecules, or *polymers*. Amino acids link up to form longer polymers we know as *proteins*, which are the fundamental components of most living systems (fig. 6.4A). Simple fatty acids plus alcohols link up to form *lipids*, the "oils" and "fats" so common on earth. Simple sugars like glucose and sucrose link together to form complex *carbohydrates* and starches (fig. 6.4B). Finally, the nucleotide bases (plus phosphates and sugars) link up to form *nucleic acids*, the genetic code of organisms, known as RNA and DNA (fig. 6.4C).

There are lots of ways of approaching this complex problem of linking simple molecules into polymers like proteins, lipids, starches, and nucleic acids. Using the primordial soup approach has produced some successes. In the 1950s, Sidney Fox showed that splashing amino acids on hot dry volcanic rocks produced most of the proteins found in life instantly. In the presence of formaldehyde, certain sugars readily form complex carbohydrates. Some of Stanley Miller's early experiments produced the components of nucleic acids, such as the nucleotide base adenine (by heating aqueous solutions of cyanide) and adenine and guanine (by bombarding dilute hydrogen cyanide with ultraviolet radiation).

It is even easier to polymerize lipids. We all know that "oil and water don't mix," but most people don't know *why* this is so. Fatty acids are polar molecules, with a "head" end that is naturally attracted to water and a "tail" end that is repelled by water (fig. 6.5). As soon as fatty acids are mixed with water, the individual molecules naturally line up with their heads facing the water and their tails pointed away. These then clump together to form an oil droplet (lipid in excess water) or a water droplet in oil. Once these droplets form, they have an automatic outer membrane of fatty acids that link together to form a lipid. In fact, the cell wall of most simple cells is composed of the same kind of lipid bilayer. When these lipid droplets are dried and then rehydrated, they form spherical balls that also concentrate any DNA present up to 100 times. Thus, little lipid bilayer droplets with nucleic acids trapped inside have all the properties of "protolife." In fact, Sidney Fox produced just such structures that he called *proteinoids*, and Oparin produced droplets he called *coacervates*. These structures behave much like living cells, holding together when conditions change, growing, and budding spontaneously into daughter droplets. They selectively absorb and release certain compounds in a process similar to bacterial feeding and excretion of waste products. Some even metabolize starch! Even though they are not living, they have most of the properties of living cells—all without much more than simple chemical reactions plus heat.

(A)

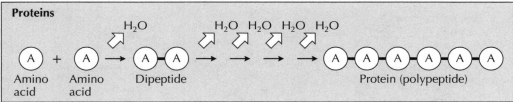

(B)

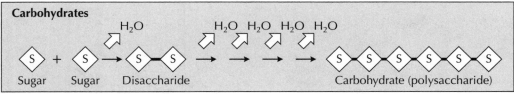

(C)

FIGURE 6.4. The next step in the origin of life is arranging the smaller building blocks into longer, more complex chains (polymerization). The common reactions include linking together a number of amino acids to form proteins, the basic building blocks of life; polymerizing simple sugars into complex carbohydrates, the basic component of cell walls and also a critical energy source in metabolism; and linking together sugars, phosphates, and nucleosides to make nucleic acids (DNA and RNA), the basic genetic code of all life. (Modified from Schopf 1999: fig. 4.12)

Mosh Pits, Fool's Gold, Kitty Litter, Black Smokers, and Mud

Their bacterial plight was pathetic
It's hard to be unsympathetic
Volcanic heat diminished
Organic soup finished
Their solution was photosynthetic
—Richard Cowen, *History of Life*

For origin of life research, the biggest challenge is how to assemble longer and more complex polymers, especially the long proteins that are so important for life. Most of the primordial soup chemical experiments have produced only shorter proteins. But a number of scientists have suggested that we've been going about it in the wrong way. Mixing chemicals randomly in a beaker will only link things together so much. For a longer, more complex polymer, you need a "scaffold" or "template," some other material that will attract and line up all the organic molecules in the same direction until they are closely packed like dancers

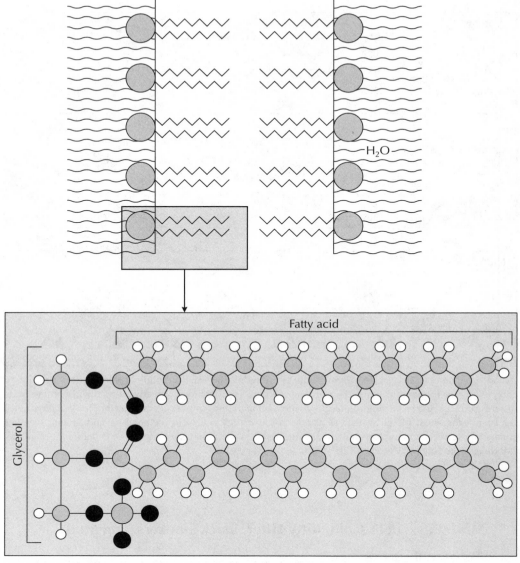

FIGURE 6.5. Some organic chemicals have properties that enable cells to form naturally without complex organic reactions. Lipids, the building blocks of fats and oils, have an end that repels water and an end that bonds to water. The properties that repel or attract lipids to water naturally line them up and then combine them to form membranes. Whenever oil mixes with water, it forms a natural membrane that encloses a droplet, comparable to the lipid bilayer membrane that surrounds all cells.

in the mosh pit in a hot nightclub. Once they are all lined up facing the same way (toward the stage) and closely packed, it is easy to link them up side by side and make complex organic polymers—just like the dense mass of people on the dance floor can easily link up and form a solid carpet of arms who can pass bodies above them.

There are many possibilities for natural substances that could produce just such a template for organic molecules. The best-known candidates are the minerals known as *zeolites*, which are complex silicate minerals typically formed by the breakdown of volcanic glass in hot gas bubbles left in lava. Zeolites have a complex, repetitious mineral structure that helps them catalyze organic reactions and make these reactions go much faster. In fact, they are heavily used in industrial settings for just that purpose, especially in petroleum refining, in filtration, and in absorbing chemicals (hence their use in kitty litter). All it would take is a few zeolites in a primordial soup, and the amino acids could be lined up into much more complex proteins.

An even more daring idea was posed by Alexander Graham Cairns-Smith, who suggested a template more humble than zeolites: common clay. Just as God supposedly created Adam and Eve out of clay, Cairns-Smith argues that the complex open-layered structures of clay minerals (which are sheet silicates, like micas) are ideal for absorbing organic molecules and lining them up along the clay mineral structure. The basic sheet structure of clay minerals is repeated again and again, with small imperfections in the crystals, comparable to mutations in the genetic code. In fact, Cairns-Smith takes the idea one step further. He argues that life began as clay minerals that copied themselves over and over again (with the mutations) during the crystallization process. Clays can grow, modify their environment, and replicate in a very low-tech version of life. Cairns-Smith then argues that the high-tech nucleic acids that had been lined up along these replicating clay "life forms" then underwent a *genetic takeover event*, and organic "replicators" replaced silicate-based replicators.

Naturally, these highly speculative ideas are very controversial but not impossible given what we know about the chemistry of clays and organic molecules. But there is one more possible template to consider: the mineral pyrite, or iron sulfide, FeS_2, better known as "fool's gold." Gunter Wächtershäuser has shown that pyrite crystals have a positive charge on the surface that could attract the negatively charged ends of many organic molecules. Once they are attracted, lined up, and packed close to one another, the organic molecules could easily link together to form complex polymers. Once linked together, they could unzip from the pyrite template and float as free biochemicals.

The strength of this suggestion is that it fits another astonishing scientific discovery: the "black smokers" at the bottom of the sea (fig. 6.6). In 1977, scientists using small submarines floated over the midocean ridges on the deep seafloor, where seafloor spreading takes place and new oceanic crust is generated. They were astonished to find places where the volcanic heat was superheating the seawater and producing submarine hot springs, with jets of near-boiling water shooting upward through chimneys (known as "black smokers") composed of pyrite, calcium sulfate, lead sulfide, and zinc sulfide. Living in this hot, dark environment is a dense population of sulfide-reducing bacteria, some of the most primitive life forms on earth, which take hydrogen sulfide (H_2S, which produces the "rotten egg" smell) and metabolize it, using it instead of water as the source of their hydrogen. These bacteria thus use *chemosynthesis* to power life because there is no light and therefore no possibility of photosynthesis or plant life (in contrast to most other environments on earth). Feeding on this bacterial population was a huge community of bizarre animals never seen before, including gigantic clams, huge tube worms, weird crabs, and some animals that were entirely new to science. I vividly remember working in the lab as a graduate student at Woods Hole Oceanographic Institution in Cape Cod in the summer of 1978 and attending

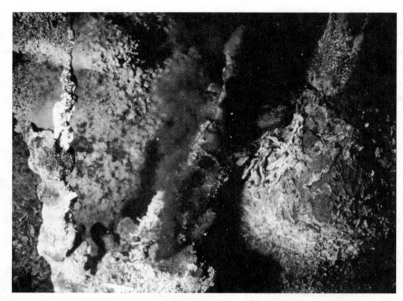

FIGURE 6.6. In the deep volcanic rift valleys of midocean ridges, fresh lava erupts as the oceanic crust pulls apart. The hot magma heats the seawater percolating through it to superheated temperatures, forming plumes of boiling water and dissolved minerals known as "black smokers." The main precipitate of this reaction is pyrite (iron sulfide, or "fool's gold"), which is also a good template for bonding together complex organic materials. Consistent with the hypothesis that life originated in deep-sea vents, biologists have found that the genetically simplest forms of life, the Archaebacteria, are common in the black smokers. These are the base of a food chain that includes a huge community of giant clams, tube worms, crabs, and many other unique creatures found only in these dark submarine communities. Since there is no light at this depth, this entire system relies not on photosynthesis, but on chemosynthesis, with sulfur-reducing bacteria (rather than plants) at the base of the food chain. (Photo courtesy NOAA)

a presentation of these remarkable discoveries given by the very scientists who had just left the submarine. This amazing chemosynthetic community is totally different from the plant-based photosynthetic ecologies elsewhere on earth. The most important clue is that it is inhabited by the most primitive life forms known on earth, the sulfide-reducing Archaebacteria (fig. 5.6). To many scientists, this suggests that the simplest forms of life arose not on the surface in Darwin's "warm little pond," but in a deep-sea hot spring, where they would have been protected if impacts vaporized the shallow oceans.

But there is one other issue that has been widely argued among scientists working on the origins of life: What was the first genetic material? Today, the information for reproduction and making more copies of living organisms is encoded in the nucleic acids, RNA or DNA, of each cell. The nucleic acids then code for certain strings of proteins, which are the stuff of life. But nucleic acids are far more complex and difficult to produce than are proteins, which we saw are among the easiest long-chain biomolecules to generate. Protein biochemists like Sidney Fox have long advocated that it would be easier for the first self-replicating organism to make its genetic code out of readily available protein chains (which still execute the commands of the nucleic acids today). At some later point in time, more complex nucleic acids were produced that eventually hijacked the system of replication from one protein to its descendants.

On the other hand, many scientists have suggested that this is a highly unparsimonious and implausible hypothesis: build a genetic code of proteins first, then replace it with another more complex one. Instead, they argue, it makes more sense to evolve the genetic code in nucleic acids from the very beginning, even if nucleic acids are harder to produce in chemical reactions than are proteins. Thus, we have a classic "chicken or the egg" problem. Which came first: the protein replication system or the nucleic acid replication system?

Fortunately, there is a way to resolve this conundrum. In 1968, DNA co-discoverer Francis Crick first suggested that the earliest protocell was a strand of RNA. In the early 1980s Tom Cech and other scientists discovered certain types of RNA, known as *ribozymes*, that perform multiple functions. It was such a momentous discovery that they won the 1989 Nobel Prize in Chemistry for it. These molecules act not only as a genetic code, but also catalyze reactions and bond together proteins. In fact, the functional part of the ribosome in the cell, which translates the RNA into proteins, *is* a ribozyme. Thus, ribozymes perform not only their familiar role as replicators, but also the role that proteins play. Further research led to the idea that the simplest scenario for the origin of living, self-replicating systems would be an "RNA world" (a term first proposed by Walter Gilbert in 1986, but Francis Crick, Leslie Orgel, Carl Woese, and others first argued that it was plausible back in the late 1960s). The very first self-replicating form of life would be a single-stranded RNA, perhaps enclosed in a lipid bilayer membrane, and perhaps using simple carbohydrates for food storage. Using both its replication powers and enzymatic powers, it would make more copies of itself and perform the role of the proteins as well until later more complex reactions involving many different proteins could evolve.

Every year, more discoveries are made that add details to our understanding of the origin of life and the RNA world. For example, coding sequences of amino acids are easily built on small RNA templates in normal prebiotic conditions (Lehmann et al. 2009). Experiments show that the first ribozymes in RNA world were much longer and more stable (Santos et al. 2004; Kun et al. 2005). Other experiments have shown that nucleotides easily merge in water to form RNA over 100 nucleotides long (Costanza et al. 2009). Pino et al. (2008) demonstrated that RNA molecules link up into long chains easily under normal earth conditions. And finally, a range of experiments have shown that new genes have been produced repeatedly by evolution (Long 2001; Long et al. 2003; Patthy 2003).

The "RNA world" hypothesis is now accepted as the most likely scenario for the origin of the first self-replicating system that can be truly called "life," although there are still additional conundrums that are being worked on: How did the RNA world get replaced by the DNA world of today? And what preceded the RNA world? Could it have been (as some suggest) a PNA world (peptide-nucleic acid) system that had amino acids in the nucleic acid chains instead of the sugar ribose? Or something else? Like any good scientific problem, the solution of one mystery then leads to additional new and more interesting problems to solve. This is how science *should* operate.

What scientists *don't* do is point to a complex system, say they can't imagine how it could have arisen by natural causes, and throw up their hands in surrender as creationists do. Instead of claiming the origin of life is impossible to solve, and falling back on untestable, unscientific, god of the gaps arguments, scientists have made enormous progress showing how life must have arisen. We may never watch life evolve from non-life in a test tube (although we are coming close), but we certainly have good experimental evidence about how nearly all the steps took place, so the problem does not require any supernatural intervention or other cop-out to solve.

Communal Living Builds Complex Cells

We are symbionts on a symbiotic planet, and if we care to, we can find symbiosis everywhere. Physical contact is a nonnegotiable requisite for many differing forms of life.

—Lynn Margulis, *Symbiotic Planet*

Creationists love to point to the complexity of the eukaryotic cell (fig. 6.7), with all its diverse organelles (such as mitochondria, chloroplasts, flagella, and so on), and try to persuade their nonscientific audiences that evolution could never construct such an amazing arrangement. What they don't mention is that the solution to how to make a complex eukaryotic cell has been known for decades and that it doesn't require anything more complex than living together in peace and harmony. If we were to try to take a simple prokaryote like a bacterium and develop all the organelles from scratch within it, such a task would seem improbably difficult. But in 1967, Lynn Margulis proposed a radical idea that solved the problem (independently and unknowingly reviving an obscure older idea suggested by K. S. Merezhkovsky in 1905) and gave us a much simpler solution: *endosymbiosis*. Instead of "inventing" mitochondria and chloroplasts and the rest from scratch, Margulis argued that these organelles were originally independent prokaryotic cells that came to live within the walls of a larger cell in exchange for food or protection (fig. 6.8). Chloroplasts apparently started out as cyanobacteria, which are photosynthetic even though they are prokaryotes without organelles. Purple nonsulfur bacteria have much the same structure and function as mitochondria, and apparently that's where these organelles came from. The flagellum

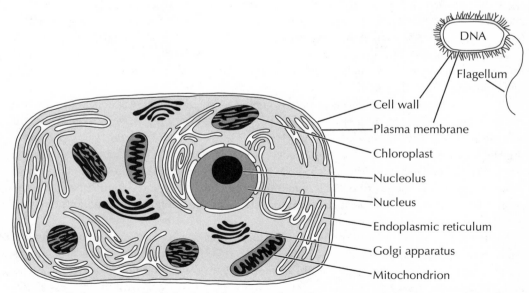

FIGURE 6.7. Prokaryotes, such as the Archaebacteria and true bacteria, are small cells only a few microns in diameter. Their genetic material (DNA) is not enclosed within a nucleus but floats within the cell, and they lack organelles. Eukaryotes (all other living organisms) have larger, more complex cells, with discrete nuclei containing their DNA. They also may have a number of other organelles, including mitochondria, chloroplasts, Golgi apparatus, endoplasmic reticulum, cilia, flagella, and other subcellular structures.

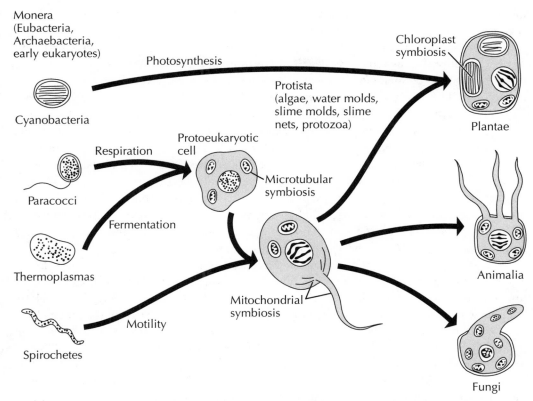

Monera
(Eubacteria,
Archaebacteria,
early eukaryotes)

Cyanobacteria

Photosynthesis

Chloroplast
symbiosis

Protista
(algae, water molds,
slime molds, slime
nets, protozoa)

Plantae

Respiration

Protoeukaryotic
cell

Paracocci

Microtubular
symbiosis

Fermentation

Thermoplasmas

Mitochondrial
symbiosis

Animalia

Motility

Spirochetes

Fungi

FIGURE 6.8. According to Lynn Margulis, complex eukaryotic cells arose from two or more prokaryotic cells that combined to live symbiotically. Cyanobacteria are apparently the precursors of the photosynthetic chloroplasts, which provide photosynthesis in plant cells. Purple nonsulfur bacteria have the same structure and genetic code of the mitochondria, which provide energy in the cell. And the flagellum has the same structure as the prokaryotes known as spirochetes, which are also responsible for causing syphilis.

has the identical 9 + 2 fiber structure (nine sets of microtubule doublets surrounding a pair of single microtubules in the center) as the prokaryotes known as spirochetes, which cause syphilis. As each of these smaller prokaryotes came to live within a larger cell, they sublimated their functions to that of their host, so that the cyanobacteria became chloroplasts that are now homes for photosynthesis, and the purple nonsulfur bacteria became mitochondria and performed the role of the energy converter for the cell.

In addition to the detailed similarities of these prokaryotes to the organelles, Margulis pointed to many other suggestive lines of evidence. Organelles are not usually enclosed within the eukaryotic cell membrane but separated from the rest of the cell by their own membranes, strongly suggesting that they are foreign bodies that have been partially incorporated within a larger cell. Mitochondria and chloroplasts also make proteins with their own set of biochemical pathways, which are different from those used by the rest of the cell. Chloroplasts and mitochondria are also susceptible to antibiotics like streptomycin and tetracycline, which are good at killing bacteria and other prokaryotes, but the antibiotics have no effect on the rest of the cell. Even more surprising, mitochondria and chloroplasts can multiply only by dividing into daughter cells like prokaryotes and thus have their own independent reproductive mechanisms; they are not made by the cytoplasm of the cell. If a cell loses its mitochondria or chloroplasts, it cannot make more.

When Margulis's startling ideas were first proposed 50 years ago, they were met with much resistance. But as biologists began to see more and more examples of symbiosis in nature, the notion became more plausible. We humans have many symbiotic bacteria living in and on us. Our intestines are full of the bacterium *Escherischia coli* (*E. coli* for short), familiar from petri dishes and those news alerts about sewage spills or contaminated kitchens. These bacteria actually do most of our digestion for us, breaking down food into nutrients in exchange for a home in our guts. Most of our fecal matter is actually made of the dead bacterial tissues after digestion, plus indigestible fiber and other material that we cannot metabolize. There are many other examples of endosymbiosis in nature. Termites, sea turtles, cattle, goats, and many other organisms have specialized gut bacteria that help break down indigestible cellulose so these animals can eat plant matter wholesale. Tropical corals, large foraminifera, and giant clams all house algae in their tissues that produce oxygen and help secrete the minerals for their large skeletons.

The strongest evidence came when people started studying the organelles more closely and found that not only did they have the right structure to have once been independent prokaryotic cells, but they also *have their own genetic code!* Mitochondria and chloroplasts both have their own DNA, which has a different sequence than the DNA in the cell nucleus. This would make absolutely no sense unless mitochondria and chloroplasts had once been free-living prokaryotes that reproduced independently. In fact, the mitochondrial DNA is different enough and evolves at a different rate from nuclear DNA, so it can be used to solve problems of evolution that the nuclear DNA cannot. This evidence would make no sense if the eukaryotic cell had tried to generate the organelles from scratch (they would not have a genetic code if that were true), and certainly it makes no sense in a creationist explanation of the cell as having been created the way we see it. If so, then why did God give the organelles their own DNA as if they had once been free-living prokaryotes? The creationists may fall back on Gosse's *Omphalos* idea again, but that's not science.

The final clincher is that we have many *living transitional forms* that show that this process is occurring right now! The simpler eukaryotes, such as the freshwater amoebae *Pelomyxa* and *Giardia* (famous for causing dysentery in hikers who drink contaminated water), lack mitochondria but contain symbiotic bacteria that perform the same respiratory function. In the laboratory, scientists have observed amoebae that have incorporated certain bacteria in their tissues as endosymbionts. The parabasalids, which live in the guts of termites, use spirochetes for a motility organ instead of a flagellum. Thus, from the wild speculation of 1967, Margulis's idea is now accepted as the best possible explanation of the origin of eukaryotes and organelles. Lynn Margulis has even received the National Medal of Science for her groundbreaking and daringly original ideas.

One final point: Margulis showed that the eukaryotic flagellum was derived from the syphilis-causing spirochete prokaryotes. It so happens that the flagellum is one of the ID creationists' (e.g., Behe 1996) favorite examples of "irreducible complexity." Behe argues that the structure of the flagellum is too complex to explain by evolution. Apparently, he is completely unaware that this distinctive 9 + 2 structure of the flagellum already exists in nature in the structure of the prokaryotic spirochete, a much simpler life form than the eukaryotic cell. As Miller (2004) has shown, even the biochemical processes are the same. The basal body of the flagellum has been found to be similar to the type III secretion system, which many bacteria use to secrete toxins. This example of co-option is regarded as strong evidence against Behe's example of "irreducible complexity."

Probability and the Origin of Life

In what manner the mental powers were first developed in the lowest organisms, is as hopeless as how life itself first originated. These are problems for the distant future, if they are ever to be solved by man.
—Charles Darwin, *The Descent of Man*

As this quote shows, even Darwin was reluctant to speculate in print about the origin of life (although he did so privately in his letter to Hooker). The evidence of evolution from the fossil record since life originated is very clear (as the next few chapters shall show), but scientists must speculate and use chemical and physical experiments to try to reconstruct the origin of life. Nonetheless, we have seen that scientists have made enormous strides in the past seven decades, from the first Urey-Miller experiment to the many abiotic syntheses of amino acids; to the many mechanisms that allow us to assemble complex polymers of proteins, lipids, carbohydrates, and nucleic acids from simpler components using templates like zeolites, clay, or pyrite; to Margulis's endosymbiotic origins of the eukaryotic cell. Not every problem has been solved or every answer revealed, but the research on the origins of life is a relatively young, healthy field of science with much more to learn and much more to do. Given the progress that has been made so far, we seem to be close to having many of the steps in the origin of life nailed down scientifically.

Yet you would never know this from the creationist literature. The creationists view the origin of life as a weak spot in evolutionary theory and love to attack it because it is complicated and difficult to discuss or defend in a debate format. They know that most of their readers have no science background, are impressed and baffled by all that talk of cells and biochemistry, and are easily persuaded to believe simplistic arguments about stuff they don't intuitively understand. Typically, creationists wow the audience with a presentation on the complexity of the cell and its many biochemical mechanisms and challenge the evolutionist to assemble this fantastically complex system by chance.

As many others have shown, there are many simple clear-cut answers to this challenge. First of all, we have just run through the steps that show us how to gradually build life from the simplest chemicals to amino acids to proteins and other polymers to prokaryotic cells to eukaryotic cells, all using relatively small steps that could be driven by natural selection. None of the steps require extraordinary conditions, and none are outside the realm of plausibility. In most cases, each step can either be simulated in the lab or seen in examples of the process (such as endosymbiosis) still working in nature today. Second, no evolutionary biologist says that this all arose by chance. As we discussed in chapter 2, chance may supply the raw material of variation on which selection acts, but selection is definitely a nonrandom agent (the "monkey with the word processor" analogy we used earlier). Creationists will point to some complex biochemical pathway and claim that it is "irreducibly complex" and cannot be built by natural selection. But many biochemists have torn these arguments apart because nearly every biochemical process or pathway exists in multiple forms from simple to complex, and it is easy to show that just by adding on a few steps here and there, you can start with a simple pathway (which is still adaptive) and improve it until it is as complex as the Krebs cycle. Finally, creationists invoke the fallacy of pointing to a finished product and arguing after the fact that the odds needed to produce this structure are astronomical. As we already discussed in chapter 2, you can *never* argue the probability of something after the fact because almost every event is extremely improbable if we start with the initial conditions and build forward—and yet these "extremely improbable" events have happened!

If the reader still feels uncomfortable with the speculative nature of the research into the origins of life, we can put the whole issue aside for now. Whether or not you agree that we can explain life's origins by naturalistic methods, the fact that life has evolved since its origins is not subject to dispute but proven beyond a reasonable doubt by an amazing convergence of evidence from the fossil record, molecules, and the embryology and anatomy of organisms. We will focus on that evidence in the remaining chapters.

For Further Reading

Cairns-Smith, A. G. 1985. *Seven Clues to the Origin of Life.* New York: Cambridge University Press.

Cone, J. 1991. *Fire Under the Sea: The Discovery of the Most Extraordinary Environment on Earth—Volcanic Hot Springs on the Ocean Floor.* New York: Morrow.

Costanza, G., S. Pino, F. Ciciriello, and E. Di Mauro. 2009. Generation of long RNA chains in water. *Journal of Biological Chemistry* 284:33206–33216.

Fry, I. 2000. *The Emergence of Life on Earth: A Historical and Scientific Overview.* Piscataway, N.J.: Rutgers University Press.

Hazen, R. M. 2005. *Gen-e-sis: The Scientific Quest for Life's Origins.* Washington, D.C.: Joseph Henry.

Knoll, A. H. 2003. *Life on a Young Planet: The First Three Billion Years of Evolution on Earth.* Princeton, N.J.: Princeton University Press.

Kun, A., M. Santos, and E. Szathmary, E. 2005. Real ribozymes suggest a relaxed error threshold. *Nature Genetics* 37:1008–1011.

Lehmann, J., M. Cibils, and A. Libchaber. 2009. Emergence of a code in the polymerization of amino acids along RNA templates. *PLoS ONE* 4:e5773.

Long, M. 2001. Evolution of novel genes. *Current Opinions in Genetics and Development.* 11:673–680.

Long, M., E. Betran, K. Thornton, and W. Wang. 2003. The origin of new genes: glimpses from the young and old. *Nature Review of Genetics* 4:865–875.

Margulis, L. 1981. *Symbiosis in Cell Evolution.* San Francisco: Freeman.

Margulis, L. 1982. Early animal evolution: emerging view from comparative biology and geology. *Science* 284:2129–2137.

Margulis, L. 2000. *Symbiotic Planet: A New Look at Evolution.* New York: Basic.

Miller, K. 2004. The flagellum unspun: the collapse of "irreducible complexity." In *Debating Design: From Darwin to DNA*, ed. M. Ruse and W. Dembski. New York: Cambridge University Press, 81–97.

Miller, Stanley L. 1953. A production of amino acids under possible primitive earth conditions. *Science* 117:528–529.

Nutman, A. P., V. C. Bennett, C. R. L. Friend, M. J. van Kranendonk, and A. R. Chivas. 2016. Rapid emergence of life shown by 3700-million-year-old microbial structures. *Nature* 537:535–538.

Patthy, L. 2003. Modular assembly of genes and the evolution of new functions. *Genetica* 118:217–231.

Pino, S., F. Ciciriello, G. Costanzo, and E. Di Mauro, E. 2008. Nonenzymatic RNA ligation in water. *Journal of Biological Chemistry* 283:36494–36503.

Schidlowski, M., P. W. U. Appel, R. Eichmann and C. E. Junge. 1979. Carbon isotope geochemistry of the 3.7 × 109 yr old Isua sediments, West Greenland; implications for the Archaean carbon and oxygen cycles. *Geochimica Cosmochimica Acta* 43:189–200.

Schopf, J. W. 1999. *Cradle of Life*. Princeton, N.J.: Princeton University Press.

Schopf, J. W. 2002. *Life's Origin: The Beginnings of Biological Evolution*. Berkeley: University of California Press.

Shapiro, R. 1986. *Origins, A Skeptic's Guide to the Creation of Life on Earth*. New York: Summit.

Wächtershäuser, G. 2006. From volcanic origins of chemoautotrophic life to Bacteria, Archaea, and Eukarya. *Philosophical Transactions of the Royal Society of London B* 361:1787–1806.

Wächtershäuser, G. 2008. Origin of life: life as we don't know it. *Science* 289:1307–1308.

Wills, C., and J. Bada. 2000. *The Spark of Life: Darwin and the Primeval Soup*. New York: Perseus.

FIGURE 7.1. A diorama of how life might have looked on this planet for almost the first 80 percent (3 billion years) of its history. The most complex form of life was the cyanobacterial mats forming domed stromatolites. If an alien had visited earth through most of its first 3.5 billion years, it would have seen nothing more complex than a mat of scum on the surf zone and probably left unimpressed with life on this planet. (Drawing by Carl Buell)

CAMBRIAN "EXPLOSION"— OR "SLOW FUSE"?

7

The Creationists' Favorite Myth

To the question why we do not find rich fossiliferous deposits belonging to these assumed earliest periods prior to the Cambrian system, I can give no satisfactory answer. . . . Nevertheless, the difficulty of assigning any good reason for the absence of vast piles of strata rich in fossils beneath the Cambrian system is very great, and may be truly urged as a valid argument against the views here entertained.
—Charles Darwin, *On the Origin of Species*

Of all the distortions of the fossil record that the creationists promote, the worst is their version of the "Cambrian explosion." The idea that most invertebrate fossils might first appear "suddenly" at the beginning of the Cambrian period with no fossils preceding them seems to suggest special creation. Creationists love to quote a variety of legitimate scientists about the "mystery" of the Cambrian explosion, although most of their quotes are grossly out of date, and many are out of context and say just the exact opposite when the full quote is read carefully. The quote by Darwin above is representative—but it was written over 150 years ago, when we knew very little about the Cambrian or Precambrian. This is true not only of the usual suspects, such as Gish, Morris, and Sarfati, but especially the intelligent design creationists, such as Meyer, Davis and Kenyon. Even the ID creationist biochemist Michael Behe (1996), who generally avoids talking about fossils because they are way outside his range of expertise, brings up the Cambrian explosion.

The problem with the creationists' fascination with the Cambrian explosion is that *it's all wrong!* The major groups of invertebrate fossils *do not all appear suddenly at the base of the Cambrian* but are spaced out over strata spanning 80 million years—hardly an instantaneous "explosion"! Some groups appear *tens of millions of years* earlier than others. And preceding the "Cambrian explosion" was a long slow buildup to the first appearance of typical Cambrian shelled invertebrates.

In this chapter, we'll go step by step through many of the recent developments that turned the Cambrian explosion into the "Cambrian slow fuse." Let us hope that even if creationists can't keep up with the recent discoveries in science, their audiences will know when they hear such baloney.

Step 1: Planet of the Scum

For four-fifths of our history, our planet was populated by pond scum.
—J. W. Schopf, *Cradle of Life*

The solution to Darwin's dilemma about the lack of apparent fossils in Precambrian rocks is that we were looking for them in the wrong way. The fossils were there all along, but they are nearly always microscopic. It wasn't until the 1940s and 1950s that Stanley Tyler and Elso Barghoorn found cherts and flints like the 2-billion-year-old Gunflint Chert in Canada that preserve these delicate microfossils and made it possible for us to study them. So the answer to the first creationist misconception about the Precambrian fossil record is yes, there *are* many fossils before the trilobites—but you need a microscope to see them, and they're only preserved in certain circumstances.

We have already discussed in chapter 6 the evidence for the earliest fossils (fig. 6.2) from rocks 3.5 billion years old in Australia and 3.4 billion years old in South Africa, as well as organic carbon and possible stromatolites from rocks 3.8 billion years in age. These fossils are all of simple prokaryotic bacteria and cyanobacteria (formerly but incorrectly known as "blue-green algae"). Their equivalents among the modern cyanobacteria are virtually indistinguishable from their fossil counterparts, showing that they have evolved very little (at least in the external anatomical sense) for the last 3.5 billion years. The earliest forms of life made simple microbial mats on the seafloor, and that way of living was so successful that they saw no reason to change it since.

As it was in the beginning (3.5 billion years ago), so it was for almost another 2 billion years. There are hundreds of microfossil localities around the world (see Schopf [1983] and Schopf and Klein [1992] for documentation of these fossils) in rocks dated between 3.5 and 1.75 billion years ago, and they yield plenty of good examples of prokaryotes (and occasionally their macroscopic sedimentary structures, the layered fossilized bacterial mats known as stromatolites). Schopf (1999) calls this extraordinarily slow rate of evolution *hypobradytely*, after George Gaylord Simpson's (1944) term for slow rates of evolution (*bradytely*) with the additional prefix "hypo-" (meaning "below") to indicate that cyanobacteria evolve slower than anything we know. Indeed, they show almost no visible change in 3.5 billion years. Everywhere we look in rocks between 3.5 billion years old and about 1.75 billion years old, we see nothing more complicated than prokaryotes and stromatolites. The first fossil cells that are large enough to have been eukaryotes do not appear until 1.75 billion years ago, and multicellular life does not appear until 600 million years ago. For almost 2 billion years, or about 60 percent of life's history, there was nothing on the planet more complicated than a bacterium or a microbial mat, and for almost 3 billion years, or 85 percent of Earth's history, there was nothing more complicated than single-celled organisms. It was truly the "planet of the scum." If aliens existed and had visited the planet long ago, odds are they would have come at a time when there was nothing more interesting to see (fig. 7.1) than mats of cyanobacteria—and they would have probably blasted off immediately because this planet was so boring (unless they studied cyanobacteria, in which case it would be exciting).

Humans like to think of themselves as special and the center of creation, but that anthropocentric view of the universe has been shocked again and again by the discovery (starting with Copernicus) that the earth is a minor planet in a small solar system on the fringe of an immense universe, and the discovery (starting with Hutton) that geologic time is immensely long and humans appeared in only the very last part of the age of the earth. Add to that the fact that most of life's history is characterized by nothing more complicated than pond scum and that humans appeared in a tiny fraction of the final 1 percent of life's history, and the blow to our cosmic arrogance is complete. Mark Twain said it best, "If the Eiffel Tower were now representing the world's age, the skin of paint on the pinnacle-knob at its summit would represent man's share of that age, and anybody would perceive that that skin was what the tower was built for. I reckon they would, I dunno."

Step 2: The Garden of Ediacara

Aspiring paleontologists are typically attracted to the large, flashy specimens such as carnivorous dinosaurs and Pleistocene mammals. But to find the real monsters, the weird wonders of lost worlds, one must turn to invertebrate paleontology. Without question the strangest of all fossilized bodies are to be found among the Ediacarans.

—Mark McMenamin, *The Garden of Ediacara*

The next step in our progression from single-celled life to Cambrian trilobites is the appearance of fossils of multicellular life. Contrary to the creationists' myths, we have abundant fossils in rocks older than the Early Cambrian (prior to 545 million years ago). Some of these date to 600 million years ago, and they are known as the *Ediacara fauna* (pronounced "Ee-dee-AK-ara"). This period of time from 600 million to the beginning of the Cambrian 545 million years ago is known as the Ediacaran Period of the Proterozoic Era. First discovered in the Rawnsley Quartzite in the Ediacara Hills of Australia by Reg Sprigg in 1946, the Ediacara fauna is now known from a wide variety of localities around the world, including many spectacular localities in China, Russia, Siberia, Namibia, England, Scandinavia, the Yukon, and Newfoundland. Most of these fossils (fig. 7.2A) are the impressions of soft-bodied organisms without skeletons, so there are no hard parts that make up the bulk of the later fossil record. Instead, these impressions have reminded some paleontologists (such as Martin Glaessner, who studied the classic Australian Ediacara fauna) of the impressions made by sea jellies, worms, soft corals, and other simple nonskeletonized organisms. Over 2,000 specimens are known, usually placed in about 30–40 genera and about 50–70 species, so they were relatively diverse.

Although the Ediacara fauna clearly represents fossils of multicellular organisms (some reach almost a meter in length), paleontologists have a wide spectrum of opinions about what made these impressions. The more conventional interpretation (fig. 7.2D) is that they are like fossils of groups we know today: sea jellies, sea pens, and worms of various sorts. Some do look like sea jellies, but if so, they have symmetry unlike any living sea jellies. Others vaguely resemble some of the known marine worms, although their symmetry and segmentation do not match any groups of worms alive in the ocean today. Nor do the "worms" have evidence of eyes, mouth, anus, locomotory appendages, or even a digestive tract.

For this reason, other paleontologists have suggested that the Ediacara fauna was composed of organisms unlike any that are alive today. They point to the lack of modern patterns of symmetry and the apparent large size of many of the fossils and argue that they are an early failed experiment in multicellularity. Adolf Seilacher (1989), for example, calls them the "Vendozoa" and suggests that they were constructed in a quilted or "water-filled air mattress" fashion that maximizes surface area. Instead of using internal digestive and circulatory systems to solve the problem of large multicellular bodies, Seilacher suggests that these simple organisms had no internal organs but instead received all their nutrients and oxygen and got rid of waste through the huge surface area of their outer membranes. Mark McMenamin (1998) suggested that they housed symbiotic algae (as do many living large invertebrates, such as reef corals and giant clams). In his "garden of Ediacara" hypothesis, McMenamin suggests that the large surface area of the Ediacarans maximizes the area of exposure of sunlight for these internal algae, which then help such large organisms metabolize.

FIGURE 7.2. The Ediacara fauna consists of the impressions of soft-bodied fossils without skeletons, whose biological affinities are still controversial. (A) The wormlike segmented creatures *Dickinsonia*, which reached almost a meter in length. (B) The elongated segmented creature *Spriggina*. (C) The odd form *Parvancorina*, which has been linked to arthropods. (D) A reconstruction of the Ediacaran community, with most of the fossils assumed to be related to jellyfish, sea pens, and worms. (Photos courtesy of the Smithsonian Institution)

Still other hypotheses (such as the ideas that they are lichens, proposed by Greg Retallack) have been put forth. Unfortunately, because the Ediacarans are known entirely from the impressions on the soft sea bottom and not from any body fossils with internal organs or other important features, it is very difficult to resolve this controversy. Whatever the biological affinities of the Ediacara fauna, it is very clear that they are multicellular organisms, whether animals, plants, fungi, or some early experimental kingdom not in any living group.

Even more intriguing is the fact that some molecular clock estimates of the divergence times of the major invertebrate groups (fig. 5.7) place the branching points of the major invertebrates as old as 800 to 900 million years ago (Runnegar 1992; Wray et al. 1996; Ayala et al. 1998). We still do not have any fossils nor even any undisputed burrows or other evidence to support this prediction, but either way, it is clear that advanced multicellular life (but still soft-bodied without any trace of fossilizable skeletons) was on earth 600 million years ago (more than 50 million years before the Cambrian), and possibly as early as 900 million years ago.

Step 3: The Little Shellies

The wave of discoveries that rewrote the story of the earliest Cambrian began when the former Soviet Union mustered sizable teams of scientists to explore geological resources in Siberia after the end of World War II. There, above thick sequences of Precambrian sedimentary rocks, lie thinner formations of early Cambrian sediments undisturbed by later mountain-building events (unlike the folded Cambrian of Wales). These rocks are beautifully exposed along the Lena and Aldan rivers, as well as in other parts of that vast and sparsely populated region. A team headed by Alexi Rozanov of the Paleontological Institute in Moscow discovered that the oldest limestones of Cambrian age contained a whole assortment of small and unfamiliar skeletons and skeletal components, few bigger than 1/2 in (1 cm) long. These fossils have been wrapped in strings of Latin syllables but have been more plainly baptized in English as the "small shelly fossils" (SSFs for short).

—Jack Sepkoski, *The Book of Life*

If the soft-bodied multicellular (but nonskeletonized) Ediacara fauna represents the next logical step up from single-celled life, then the next step beyond that would be the appearance of mineralized, fossilizable skeletons. But if life took almost 3 billion years to develop the ability to mineralize shells, we expect that it would be a difficult process and would not arise fully fledged. Sure enough, the earliest stages of the Cambrian (known as the Nemakit-Daldynian and the Tommotian stages, from 520 to 545 million years ago) are dominated by tiny (only a few millimeters) fossils nicknamed the "little shellies" or the "small shelly fossils (SSFs)" in the trade (fig. 7.3). For decades, these little fossils were overlooked as people hunted the beds above them for the more spectacular trilobite fossils. But as the quotation from Sepkoski points out, the Soviets were the first to study their long and detailed sequences of Proterozoic and Cambrian sediments in detail and named the stages of the Cambrian. And when they looked closer at the beds below the trilobites and took samples back to the lab to dissolve in acid or slice into thin sections, it became apparent that these long-neglected beds were chock-full of tiny fossils.

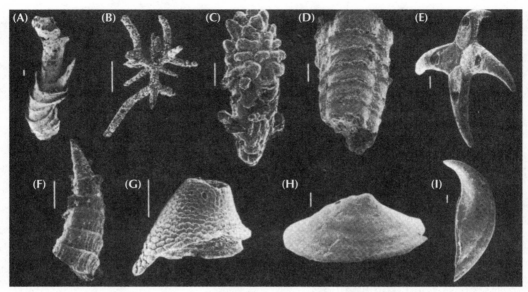

FIGURE 7.3. The earliest stages of the Cambrian (Nemakit-Daldynian and Tommotian) do not produce trilobites but are dominated by tiny phosphatic fossils nicknamed the "little shellies." Some may have been mollusk shells (E, H, and I), while others are apparently sponge spicules or pieces of the "chain-mail armor" of larger creatures, such as worms. (A) *Cloudina hartmannae*, one of the earliest known skeletal fossils, from the same beds that produce Ediacaran fossils in China. (B) A spicule of a calcareous sponge. (C) A spicule of a possible coral. (D) *Anabarites sexalox*, a tube-dwelling animal with triradial body symmetry. (E) A spicule from a possible early mollusk. (F) *Lapworthella*, a cone-shaped fossil of unknown relationships. (G) A skeletal plate of *Stoibostromus crenulatus*, another creature of unknown relationships. (H) Skeletal plate of *Mobergella*, a possible mollusk. (I) Cap-shaped shell of *Cyrtochites*, another possible mollusk. All scale bars = 1 mm. (Photos courtesy S. Bengston)

Some of the little shellies look like simple cap-shaped or coiled mollusks; others look like primitive clams (fig. 7.3). Many others are simple tubular or conical fossils whose connections to any living group are hard to establish. Many of the fossils look like miniature jacks or little spiky Christmas ornaments. These appear to have been part of the "chain-mail armor" that studded the skins of larger organisms, such as the sponge *Chancelloria*, that were soft-bodied except for these little spiky objects (much like the tiny spikes in the skin of a shark or a sea cucumber). Most of the little shellies were made of calcium phosphate, the same mineral that makes up the bones of vertebrates. Today, most marine invertebrate shells are made of calcium carbonate (the minerals calcite and aragonite). To some scientists, this suggests that some sort of environmental condition (such as low atmospheric oxygen) made it hard to secrete calcite skeletons, but phosphate skeletons were easier to produce. They suggest that the appearance of large calcified trilobites and other fossils reflects the point where atmospheric oxygen passed a critical threshold and became abundant enough to allow this chemical mineralization.

Whatever the reason, for almost 25 million years, the Cambrian explosion was burning on a slow fuse. The little shellies were abundant, but larger fossils were not. The earliest sponges had already appeared back in the late Ediacaran, but this is not surprising, considering that all lines of evidence show that sponges are the most primitive animals alive today (fig. 5.6). By the Tommotian Stage (530 million years ago), a slow trickle of other groups of larger invertebrates began to appear, including the first "lamp shells" (brachiopods), and also members of an extinct spongelike group known as archaeocyathans. Diversity in the Tommotian reached only about 50 genera, about the same as in the Vendian. In addition, the sediments of the earliest Cambrian show abundant burrowing, proof that many other types of soft-bodied worms with a true internal fluid-filled cavity (a *coelom*) must have lived at that time. Thus the earliest Cambrian shows evidence of a gradual increase in diversity from the Vendian, but no "explosion."

Step 4: In with a Whimper, Not a Bang

Was there really a Cambrian Explosion? Some have treated the issue as semantic— anything that plays out over tens of millions of years cannot be "explosive," and if the Cambrian animals didn't "explode," perhaps they did nothing at all out of the ordinary. Cambrian evolution was certainly not cartoonishly fast. . . . Do we need to posit some unique but poorly understood evolutionary process to explain the emergence of modern animals? I don't think so. The Cambrian Period contains plenty of time to accomplish what the Proterozoic didn't without invoking processes unknown to population geneticists—20 million years is a long time for organisms that produce a new generation every year or two.

—Andrew Knoll, *Life on a Young Planet*

The third stage of the Early Cambrian is known as the Atdabanian Stage (515–520 million years ago), and with this stage, we finally see a great increase in diversity: more than 600 genera are recorded (fig. 7.4). However, this number is misleading and a bit inflated. Most of the genera are trilobites, which fossilize readily and so greatly increase the volume and diversity of large shelly fossils. Most of the other animal phyla had already appeared by this time (including mollusks, sponges, corals, echinoderms) or would appear later in the Cambrian (vertebrates) or even in the Ordovician Period that followed (e.g., the "moss animals" or bryozoans).

The second misleading aspect of this apparent diversity "explosion" is that during the Atdabanian Stage we get the first good fauna of soft-bodied fossils (the Chinese Chengjiang fauna), so we get the apparent (but not real) first appearance of phyla only known from soft tissues. Then in the Middle Cambrian we have the extraordinary soft-bodied preservation of fossils from places like the Burgess Shale in Canada (fig. 7.5). As pointed out by Stephen Jay Gould in his book *Wonderful Life* (1989), the soft-bodied animals preserved in these amazing deposits allow us to see what the normal fossil record is missing. We have many bizarre wormlike and odd fossils, many of which don't fit into any living phylum. Some, like the five-eyed nozzle-nosed *Opabinia* (fig. 7.5, top left) or the soft flowerlike *Dinomischus*, are complete mysteries to zoologists. Others are apparently soft-shelled arthropods. One fossil, appropriately named *Hallucigenia* (fig. 8.17B), was a bizarre creature that seemed to have tentacles or spikes on a wormlike body, until recent better fossils from China showed it is related to the "velvet worms," phylum Onychophora (discussed in chapter 8). The largest predator (about 2 feet long) was a soft-bodied swimmer known as *Anomalocaris*, which had a strange mouth that looked like a pineapple slice and was originally found and misinterpreted as a sea jelly.

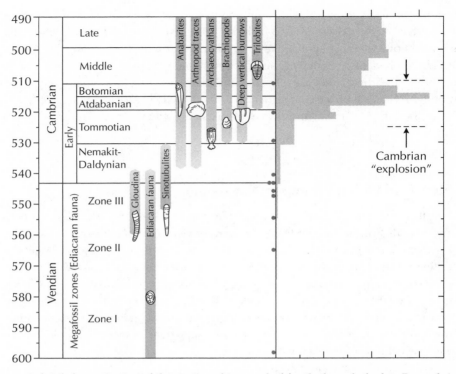

FIGURE 7.4. A detailed examination of the stratigraphic record of fossils through the late Precambrian and Cambrian shows that life did not "explode" in the Cambrian, but appeared in a number of steps spanning almost 100 million years. The large soft-bodied Ediacaran fossils (fig. 7.2) appeared first at 600 million years ago, in the late Precambrian (Vendian). Toward the end of their reign, we see the first tiny shelly fossils, including the simple conical Cloudina and *Sinotubulites*. The Nemakit-Daldynian and Tommotian stages are dominated by the "little shellies" (fig. 7.3) plus the earliest lamp-shells, or brachiopods, and the conical spongelike archaeocyathans, and many burrows showing that wormlike animals without hard skeletons were also common. Finally, in the third stage of the Cambrian (Atdabanian, around 520 million years ago), we see the radiation of trilobites and a huge diversification in total number of genera (histograms on the right side of the diagram). Thus the Cambrian explosion took over 80 million years to develop and was no "sudden" event, even by geological standards. (Modified from Dott and Prothero 2010: fig. 9.14, and from Kirschvink et al. 1997: fig. 1. Copyright © 1997 American Association for the Advancement of Science. Reprinted with permission.)

Thus we have seen that the "Cambrian explosion" is a myth. It is better described as the Cambrian slow fuse. It takes from 600 to 520 million years ago before the typical Cambrian fauna of large shelly organisms (especially trilobites) finally develops. Eighty million years is not explosive by any stretch of the imagination! Not only is the explosion a slow fuse, but it follows a series of logical stages from simple and small to larger and complex and mineralized. First, of course, we have microfossils of cyanobacteria and other eukaryotes going back to as far as 3.5 billion years ago and spanning the entire fossil record since that ancient time. Then, about 600 million years ago, we get the first good evidence of multicellular animals, the Ediacara fauna. They are larger and multicellular but did not have hard shells. The earliest stages of the Cambrian, the Nemakit-Daldynian and Tommotian stages, are dominated not by the little shellies, which were just beginning to develop small mineralized skeletons.

FIGURE 7.5. Examples of the extraordinary soft-bodied middle Cambrian fauna from the Burgess Shale, near Field, British Columbia. Note the exquisite preservation of fine detail, including appendages and other soft tissues. (Photos courtesy of the Smithsonian Institution)

Only after several more steps do we see the full Cambrian fauna. In short, the fossil record shows a gradual buildup from single-celled prokaryotes and then eukaryotes to multicellular soft-bodied animals to animals with tiny shells, and finally by the middle Cambrian, the full range of large shelled invertebrates. This gradual transformation by logical advances in body size and skeletonization bears no resemblance to an instantaneous Cambrian explosion that might be consistent with the Bible but instead clearly shows a series of evolutionary transformations.

All of this information has been known for at least the past few decades, and the first Precambrian microfossils were discovered over 70 years ago. They are published in all the standard geology and paleontology textbooks and have been for decades. But the creationists either don't want to know or cannot understand the implications of these discoveries. Their out-of-context quotations of real scientists puzzling about the Cambrian explosion are all from old sources that do not reflect what we have learned from recent discoveries. Even their most recent books, including the "intelligent design" texts, persist in perpetuating this out-of-date picture. A few years ago, I debated an ID creationist on the KPCC radio station in Los Angeles, and as soon as he mentioned the Cambrian explosion, it was clear he didn't know anything about paleontology and hadn't heard that the Cambrian explosion was a myth.

In 2013, ID creationist Stephen Meyer published an entire book on the topic entitled *Darwin's Doubt: The Explosive Origin of Animal Life and the Case for Intelligent Design.*

Most scientists ignored it, but the few who wasted their time reading it lambasted it (Cook 2013; Marshall 2013). As I wrote in my own review (Prothero 2013), the book is a piece of incompetent scholarship from one end to the other, with mistakes, misstatements, quote mining, cherry-picking of data, ignoring inconvenient facts, and outright lies about the fossil record in nearly every page. This is not surprising, since Meyer has no formal training in paleontology (his Ph.D. is in history of science), and no published research in paleontology, so everything he cites comes across as amateurish and filtered by his creationist biases. There is no space in this short chapter to list all the errors and lies in his book (for details, see Prothero 2013). The most crucial deception that Meyer pulls is that he completely ignores the first two stages of the Cambrian! Nowhere in the book are the "little shellies" or the Nemakit-Dalydinian or Tommotian stages even mentioned! Naturally, if you deliberately leave out the crucial evidence of the intermediate stage of life evolving from large soft-bodied Ediacarans (which he dismisses as irrelevant because we're not sure they are members of modern phyla) to the large shelly trilobites of the third stage of the Cambrian (Atdabanian), it will look more explosive. I even debated him on this topic in Hollywood in 2009, and he dodged the issue entirely, so it cannot be a case of him not knowing any better. No, he's fully aware that this evidence would invalidate his entire book, so he ignores it—and he counts on his readers to not know the difference.

Even if we grant the premise that a lot of phyla appear in the Atdabanian (solely because there are no soft-bodied faunas older than Chengjiang in the earliest Cambrian), Meyer claims the 5–6 million years of the Atdabanian are too fast for evolution to produce all the phyla of animals. Wrong again! Lieberman (2003) showed that rates of evolution during the "Cambrian explosion" are typical of any adaptive radiation in life's history, whether you look at the Paleocene diversification of the mammals after the nonavian dinosaurs vanished, or even the diversification of humans from their common ancestor with apes 6 million years ago. As distinguished Harvard paleontologist Andrew Knoll put it in his book, *Life on a Young Planet* (also cited in the epigraph to this section), it wasn't an "explosion," nor was it "cartoonishly fast."

Finally, one might wonder: what's all the fuss about the "Cambrian explosion"? Why should it matter whether evolution was fast or slow during the third stage of the Cambrian? Some scientists might find this puzzling, but you must understand the minds of creationists. They operate by a god of the gaps argument: anything that is currently not easily explained by science is automatically attributed to supernatural causes. Even though ID creationists say that this supernatural designer could be any deity or even extraterrestrials, it is well documented that they are thinking of the Judeo-Christian god when they point to the complexity and "design" of life. They argue that if scientists haven't completely explained every possible event of the early Cambrian, science has failed and we must consider supernatural causes.

Again and again, creationists persist in presenting a version of the Cambrian that is at least 70 years out of date either because they don't know any better (the "clueless" hypothesis) or because they *do* know better (the "deceiver" hypothesis, of which Meyer is a good example). Either way, it is bad science.

For Further Reading

Ayala, F. J., and A. Rzhetsky. 1998. Origins of the metazoan phyla: molecular clocks confirm paleontological estimates. *Proceedings of the National Academy of Sciences USA* 95:606–611.

Briggs, D. E. G., and R. A. Fortey. 2005. Wonderful strife: systematics, stem groups, and the phylogenetic signal of the Cambrian radiation. *Paleobiology* 31(2):94–112.

Conway Morris, S. 1998. *The Crucible of Creation*. Oxford: Oxford University Press.

Conway Morris, S. 2000. The Cambrian "explosion": slow-fuse or megatonnage? *Proceedings of the National Academy of Sciences USA* 97:4426–4429.

Cook, G. 2013. Doubting "Darwin's doubt." *New Yorker*, July 2, 2013.

Erwin, D., and J. W. Valentine. 2013. *The Cambrian Explosion: The Construction of Biodiversity*. New York: Roberts.

Glaessner, M. F. 1984. *The Dawn of Animal Life*. New York: Cambridge University Press.

Gould, S. J. 1989. *Wonderful Life: The Burgess Shale and the Nature of History*. New York: Norton.

Grotzinger, J. P., S. A. Bowring, B. Z. Saylor, and A. J. Kaufman. 1995. Biostratigraphic and geochronologic constraints on early animal evolution. *Science* 270:598–604.

Knoll, A. H. 2003. *Life on a Young Planet: The First Three Billion Years of Evolution on Earth*. Princeton, N.J.: Princeton University Press.

Knoll, A. H., and S. B. Carroll. 1999. Early animal evolution: emerging views from comparative biology and geology. *Science* 284:2129–2137.

Lieberman, B. S. 2003. Taking the pulse of the Cambrian radiation. *Integrative and Comparative Biology* 43:229–237.

Marshall, C. R. 2013. When prior beliefs trump scholarship. *Science* 341:1344.

McMenamin, M. A. S. 1998. *The Garden of Ediacara*. New York: Columbia University Press.

McMenamin, M. A. S., and D. L. S. McMenamin. 1990. *The Emergence of Animals, the Cambrian Breakthrough*. New York: Columbia University Press.

Narbonne, G. M. 1998. The Ediacara biota: a terminal Neoproterozoic experiment in the evolution of life. *GSA Today* 8(2):1–6.

Peterson, K., M. A. McPeek, and D. A. D. Evans. 2005. Tempo and mode of early animal evolution: inferences from rocks, Hox and molecular clocks. *Paleobiology* 31:36–55.

Prothero, D.R. 2013. Stephen Meyer's fumbling bumbling Cambrian follies: a review of *Darwin's Doubt* by Stephen Meyer. *Skeptic* 18(4):50–53.

Runnegar, B. 1992. Evolution of the earliest animals. In *Major Events in the History of Life*. ed. J. W. Schopf. New York: Jones and Bartlett, 65–94.

Schopf, J. W., ed. 1983. *Earth's Earliest Biosphere: Its Origin and Development*. Princeton, N.J.: Princeton University Press.

Schopf, J. W. 1999. *Cradle of Life: The Discovery of the Earth's Earliest Fossils*. Princeton, N.J.: Princeton University Press.

Schopf, J. W., and C. Klein, eds. 1992. *The Proterozoic Biosphere, a Multidisciplinary Study*. New York: Cambridge University Press.

Seilacher, A. 1989. Vendozoa: organismic construction in the Proterozoic biosphere. *Lethaia* 22:229–239.

Seilacher, A. 1992. Vendobionta and Psammocorallia. *Journal of the Geological Society of London* 149:607–613.

Valentine, J. W. 2004. *On the Origin of Phyla*. Chicago: University of Chicago Press.

Wray, G. A., J. S. Levinton, and L. H. Shapiro. 1996. Molecular evidence for deep Precambrian divergences among metazoan phyla. *Science* 274:568–573.

FIGURE 8.1. A typical assortment of planktonic microfossils from a deep-sea core. Each is about the size of a pinhead or smaller, and many thousands are found in every cubic meter of sediment. They include the foraminifera (larger bubble-shaped shells) and the radiolaria (smaller porous conical shells), along with long, spikelike sponge spicules. (Photo courtesy Scripps Institution of Oceanography)

SPINELESS WONDERS OF EVOLUTION

8

Invertebrate Transitions

There is a story, possibly apocryphal, of the distinguished British biologist J. B. S. Haldane, who found himself in the company of a group of theologians. On being asked what one could conclude as to the nature of the Creator from a study of his creation, Haldane is said to have answered, "An inordinate fondness for beetles."
—J. E. Hutchinson, "Homage to Santa Rosalia
or Why Are There So Many Kinds of Animals?"

Most people are interested only in our own phylum Chordata, which includes vertebrates. They don't know or care about clams, snails, and "bugs" (or they consider them disgusting and don't want to know). Typically, people call most members of the phylum Arthropoda (including not only insects but also spiders, scorpions, "pillbugs," centipedes, millipedes, lice, ticks, and many other unrelated groups) "bugs" and think of most marine invertebrates that are worth eating as "shellfish." Nevertheless, the invertebrates make up more than 99 percent of all living animals on earth. In fact, insects alone outnumber all other groups of organisms in total diversity, and among insects, beetles are more diverse with more species than any other group of animals. Despite the lack of public familiarity, invertebrates are not only the most diverse animals alive today, but they also include the best-fossilized groups by far. We will devote the remaining chapters of this book to looking at more familiar and more popular examples from birds, mammals, and reptiles, but we cannot neglect the excellent fossil record of transitions within the invertebrates. The invertebrates may not be as cute and cuddly as mammals or birds, but they show us far more about evolution than the much less complete fossil record of vertebrates.

The Incredible World of Microfossils

Unknown to Darwin, uninterrupted sedimentation does occur in the open ocean, especially on aseismic ridges and plateaux. These areas experience a continuous rain of particles to the sea bed, and are among the most geologically quiescent places on Earth. A steady build-up of sediment is the result. . . . The sediments in question are composed mainly of the shells of microscopic plankton such as foraminifera, radiolaria, diatoms and coccolithophorids. Large numbers of individuals can easily be extracted. Their evolution can be followed through geological time, simply by comparing one closely-spaced sample with the next. This reveals morphologically isolated and continuous lineages which it is reasonable to infer represent lines of genetic descent. These lineages sometimes split from one another, and often evolve gradually over vast periods of time, or become extinct. . . . Does the fossil record provide a true and accurate record of first and last occurrences of species? Emphatically, the answer is yes! Microfossils are used routinely for biostratigraphic correlation by thousands of specialists the world over. This would not be possible unless the sediment record was good and reliable. We now know (within fairly precise limits) when hundreds of species of mineralizing plankton arose and became extinct, through a history that spans over a hundred million years.

—Paul Pearson, "The Glorious Fossil Record"

Although the evolution of dinosaurs and humans is a far more glamorous subject, by far the best fossil record is found in the microscopic fossils left behind by single-celled organisms in the deep sea. These protistans occur by the trillions in many larger oceanic water masses, and their shells literally carpet the shallow seafloor with hundreds to thousands of individual specimens in a cubic centimeter of sediment (fig. 8.1). Their density can exceed a million specimens per cubic meter of sediment and weigh up to 10 grams per cubic meter. Most of the open ocean floor shallower than 3,000 m is completely covered by "calcareous ooze" composed of the skeletons of microfossils made of calcium carbonate. The sand of many tropical beaches is composed almost entirely of the skeletons of microfossils. In a typical sample of tropical marine sediment, there may be 60 to 70 species. In some groups, such as the foraminifera, there are over 3,600 described genera and perhaps 60,000 species, making them more diverse than any other group of marine animals or plants.

In addition to their great abundance and diversity, microfossils are ideal for evolutionary studies for several other reasons. Cores of the sediments covering the deep-sea bottom have been taken by rotary drilling and by plunging a long tube into the sea bottom ("piston coring"), and both retrieve an almost continuous record of marine sedimentation over that part of the ocean floor. Some cores span many millions of years with no breaks or gaps whatsoever. These cores can be precisely dated by methods such as stable isotope analysis and magnetic stratigraphy, as well as with the biostratigraphy of the microfossil groups themselves. Thus we can trace the history of many microfossil lineages through many millions of years over a single spot in the world, something that is impossible with the much less complete record of shallow marine invertebrates or land vertebrates. Finally, the biogeography of microfossils is relatively simple. Most are confined to a few water masses where the ocean waters are of a given temperature, and these species range over that entire water mass (Prothero and Lazarus 1980). Thus a few cores from an area representing a single water mass will sample all the populations in that water mass, and there will be no small "peripheral isolate" populations that could be missed. As a result, Prothero and Lazarus (1980) showed that if we have cores representing most of the world's major water masses in a given time interval, we can look at a lineage or group and see practically all there is to see about the evolution of their skeletons. Prothero and Lazarus (1980) argued that microfossils are our best "laboratory animal" or "fruit fly" to study evolution in the fossil record.

Naturally, there are a few drawbacks to using microfossils to study evolution. The biggest problem is that we still know relatively little about the biology of the living relatives. Some are difficult to keep alive in the laboratory once they have been caught. Researchers have studied the few species that can be cultured only over short intervals of time. Besides, the most valuable information about their biology relates to how they live in the open ocean, which is hard to simulate in a lab. In addition, microfossils have relatively simple skeletons, without the many levels of anatomical detail that you find in many macroinvertebrates or vertebrates. There is good evidence that some forms have evolved more than once through convergent evolution (Cifelli 1969). In other cases, it is not clear that the skeletal shape is that well constrained during the life of the organism. Finally, what we do know of the biology of these organisms suggests that many of them are asexual through at least part of their life histories and reproduce by cloning, especially when they are trying to multiply quickly to exploit an abundant food resource. Others may hybridize across lineages (Goll 1976), ending the reproductive isolation of species that characterizes multicellular animals. Thus we cannot always be sure that the rules about speciation and evolution developed for sexually reproducing multicellular animals and plants (such as Mayr's allopatric speciation model discussed earlier) can be applied to partially asexual or hybridizing microfossils.

But every research problem has its strengths and limitations, and the strengths of micropaleontology are so enormous that it has proven one of the most fruitful areas of research in all the geologic sciences. Hundreds of micropaleontologists work for oil companies, helping to precisely date and correlate the formations that they drill through to find oil or helping to determine the depth of the water in which ancient oceanic sediment was deposited. Other micropaleontologists work mostly as marine geologists and paleoclimatologists, using microfossils to determine how oceanic currents and climate have changed through time. A small number of micropaleontologists study the biology and shell chemistry of these organisms, and even a smaller number are interested in using microfossils as exemplars of evolution. Despite this trend, however, there are hundreds of well-documented examples of evolution in the microfossil record, of which we will have space to mention only a few.

Although there are many types of microfossils that could be studied (see Prothero 2013a: chap. 12), the most important ones include just a few groups. Two are animallike protistans related to the amoeba, with its oozing, flowing protoplasm—but unlike an amoeba, they have internal mineralized shells. The most diverse and most widely studied are the Foraminifera (fig. 8.2A) or "forams" for short, which secrete skeletons of calcium carbonate (the mineral calcite). Most forams live on or in the sea bottom (*benthic*), but one family, the Globigerinidae, are tiny and buoyant and make up a major part of the marine plankton. The second group of amoeba-like plankton is the Radiolaria (or "rads" in the trade), which secrete skeletons of opaline silica instead of calcium carbonate. These delicate porous glassy skeletons have been compared to miniature Christmas ornaments in their beauty and symmetry (fig. 8.2B).

The other important groups of microfossils are actually planktonic plants, members of the division Chrysophyta, or the golden-brown algae. They include the diatoms, which secrete their skeletons out of silica and are found in marine and fresh waters all over the world (fig. 8.2C), and the coccolithophorids, which secrete hundreds of miniscule (a few microns in diameter) button-shaped plates over their spherical cells (fig. 8.2D). These *phytoplankton* (planktonic algae) are the base of the entire food chain in the world's oceans, and all other organisms (from the foraminifera and radiolaria to megascopic predators like crustaceans to fish all the way up to whales) feed directly or indirectly on them. In addition, phytoplankton are the single largest producer of the oxygen we breathe (much more important oxygen producers than land plants), and in many places in the ocean they are so abundant that they pave the sea bottom with trillions of their shells. In fact, the rock known as chalk is actually made of millions of skeletons of coccolithophorids and a few foraminifera. Phytoplankton are so important to life on this planet that any major crisis in their evolution has caused mass extinctions all the way up the food chain. To a great extent, none of us would be here without the oxygen that phytoplankton release, and the food that they provide for the life of the sea.

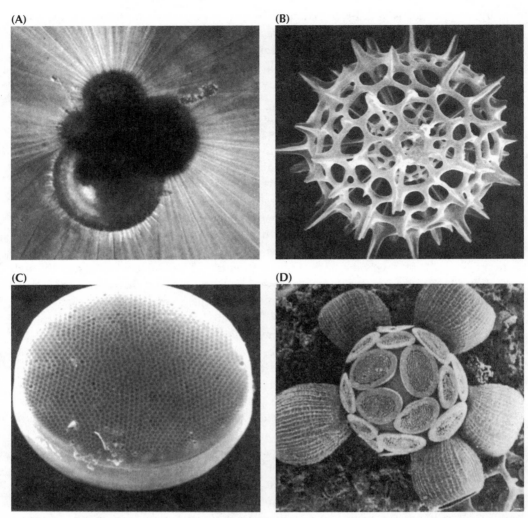

FIGURE 8.2. Examples of common groups of planktonic microfossils. (A) A foraminiferan, showing the bubble-shaped chambers made of calcite surrounded by long fingers of protoplasm known as pseudopodia. (Courtesy J. Kennett) (B) A radiolarian, with its characteristic porous spiny shell made of silica. (From Haq and Boersma 1978) (C) A diatom, with its perforated petri dish–shaped shells made of silica. (Courtesy J. Barron) (D) A coccolithophorid alga, surrounded by multiple plates (coccoliths) made of calcite. (Courtesy W. Siesser)

Let us look at some of many examples of evolution in the well-studied forams first. One of the classic examples of long-term evolution in the foraminifera is provided by a group known as the Fusulinidae (fig. 8.3). Fusulinids were benthic (bottom-dwelling) foraminifera, so they were not constrained to be tiny and float in the plankton. Instead, they secreted shells that ranged from the size of a grain of rice up to 5 centimeters (2 inches) long, which is enormous for a single-celled organism. Like many modern large benthic foraminifera, fusulinids probably harbored symbiotic algae within their tissues to enable them to grow so

FIGURE 8.3. The bottom-dwelling (benthic) foraminifera known as fusulinids were about the size and shape of a grain of rice and extraordinarily abundant in late Paleozoic limestones, where they may number in the trillions and make up entire rock units. (Photography by W. Hamilton, courtesy U.S. Geological Survey)

large. That idea is confirmed by the fact that fusulinids lived in enormous numbers on shallow sea bottom (probably shallow enough for light penetration) during the late Paleozoic (Mississippian to Permian periods, 355–255 million years ago). In many places, huge volumes of limestone are made of nothing but fusulinids, comprising trillions of individuals (fig. 8.3). Then, at the peak of their success, they were wiped out by the greatest mass extinction event in earth history, the Permian catastrophe, which extinguished 95 percent of the marine species on the planet. Whatever the cause of this great mass extinction, the fusulinids were one of the most conspicuous victims.

The fusulinid shell is shaped like a spindle or a grain of rice (figs. 8.3 and 8.4). As the shell grew, more and more shell layers were added, spiraling around the long axis of the spindle, so that when you cut it across the middle, you typically see a spiral pattern. In cross section, you can see that the spiral layers are supported by a dense network of smaller walls and chambers, with an intricate, complex structure. These complex wall structures make each genus and species of fusulinid distinct and easy to recognize for the specialist. Fusulinids evolved rapidly through the late Paleozoic (fig. 8.4) from simple forms with only a few chambers like *Millerella* from the Late Mississippian and earliest Pennsylvanian, to a variety of different lineages that become larger with more and more complex wall structure and interesting variations on the spindle-like symmetry. The enormous variations within this basic body form are apparent even to the nonspecialist and provide a dramatic example of evolution within a single lineage. To the specialist, the different species are so distinctive, and their fossils are so widespread and abundant in late Paleozoic limestones, that they are the principal method of dating rocks of late Paleozoic age. If you want to know the age of any marine limestone from the Pennsylvanian

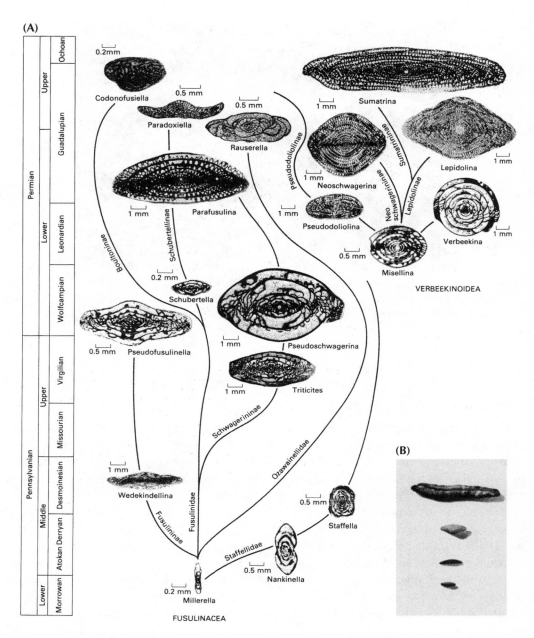

FIGURE 8.4. (A) Fusulinids evolved very rapidly during the late Paleozoic (Pennsylvanian and Permian) by developing increasingly more complex chambers and wall structure, and a variety of shapes based on the fundamental plan of a spiral shaped like a spindle. (B) Some photos of fusulinids at natural size. (Modified from Boardman et al. 1987; courtesy Blackwell Scientific Publications)

or Permian, ask a fusulinid expert to look at it, and you will get the most precise estimate possible.

There are many other examples of dramatic transformations in the forams that could be shown. For example, in the Pliocene, one of the common planktonic forams is *Globigerinoides sacculifer* (Kennett and Srinivasan 1983), which has a shell shaped like a series of porous oblong bubbles clustered together in a spiral arrangement (fig. 8.5A). Through the many cores that sample the Pliocene oceans we can find more and more specimens that develop these long slender fingerlike extensions all over the final few chambers. As you move up the cores, these little "fingers" become longer and more common. These creatures are so distinct from the ancestral lineage that they branched away from that they are given their own species: *Globigerinoides fistulosus* (they do indeed look like little fists). Another common trend is the gradual evolution of foraminifera with flatter chambers and keels along the edges from species with more primitive bubble-shaped chambers. These trends can be seen in the evolution of keeled *Morozovella* from *Praemurica* in the late Paleocene (fig. 8.5B), in keeled forms of *Globoconella* in the early Miocene (fig. 8.5C), and in keeled *Fohsella*, also during the early Miocene (fig. 8.5D).

Let us look at the other amoeba-like group, the radiolarians. As we mentioned already, this group is much like the forams, only their skeletons are made of opaline silica (fig. 8.2B). Unlike forams, which are both benthic and planktonic, all radiolaria are planktonic, so their evolution and ecology closely mirrors the change in water temperature and chemistry where they live and grow. Most of the time, they flourish only in places where nutrients are brought up from the deep ocean to the surface. To the radiolarian, the scarcest and most important nutrient is silica itself, which is depleted in normal surface seawater. When silica is carried up from the deep by upwelling ocean currents, radiolaria and diatoms bloom in enormous numbers and consume almost all of it immediately. After they die, their delicate skeletons rain down on the ocean floor by the millions, so any ocean sediment in areas of upwelling (typically in the boundary currents between water masses) is full of siliceous plankton such as radiolaria and diatoms.

When I was a graduate student at Columbia University and at the American Museum of Natural History in the late 1970s and early 1980s, I decided to learn about micropaleontology to balance out my education in fossil vertebrates and invertebrates. Affiliated with Columbia is Lamont-Doherty Geological Observatory (now Lamont-Doherty Earth Observatory), one of the foremost geologic research institutions in the world. Lamont was the place that led the revolution in geology known as plate tectonics in the 1960s. I rode the shuttle bus up the Hudson River to Lamont so that I could take classes from the giants of plate tectonics, paleomagnetism, and seismology. Lamont has long been one of the pioneers in oceanography and marine geology, with a collection of deep-sea cores from the oceans of the world that is second to none. While at Lamont, I spent most of my time in the core laboratory, where I got to examine microfossils by the thousands and learned about foraminifera from Tsunemasa Saito, radiolaria from Jim Hays, and diatoms from Lloyd Burckle. Soon I was working on a research project with my fellow graduate student Dave Lazarus, counting and measuring hundreds of specimens on slides and trying to decipher the evolutionary patterns in a group of radiolaria known as *Pterocanium* (Lazarus et al. 1985). These cute little "Christmas ornaments" were shaped like a lacy bell (the *thorax*) with a knob (the *cephalis*) and a spike on top, and three long spines sprouting out from the open base (fig. 8.6). From the ancestral form *Pterocanium charybdeum allium* (so named by Dave because the thorax was shaped like a clove of garlic, *allium* in Latin) that lived 7 million years ago, we documented a complex pattern of divergence among different shapes. Some developed larger pores and more robust spines that flared out from the base and lost the distinct "knob" of the cephalis at the top (*Pterocanium audax*). Another lineage developed a more cylindrical, boxy shape with distinct "shoulders" (*Pterocanium prismatium*, an important index fossil for the Pliocene). Another lineage developed huge flaring spines and shrank the size of the thorax to a little ball (*Pterocanium*

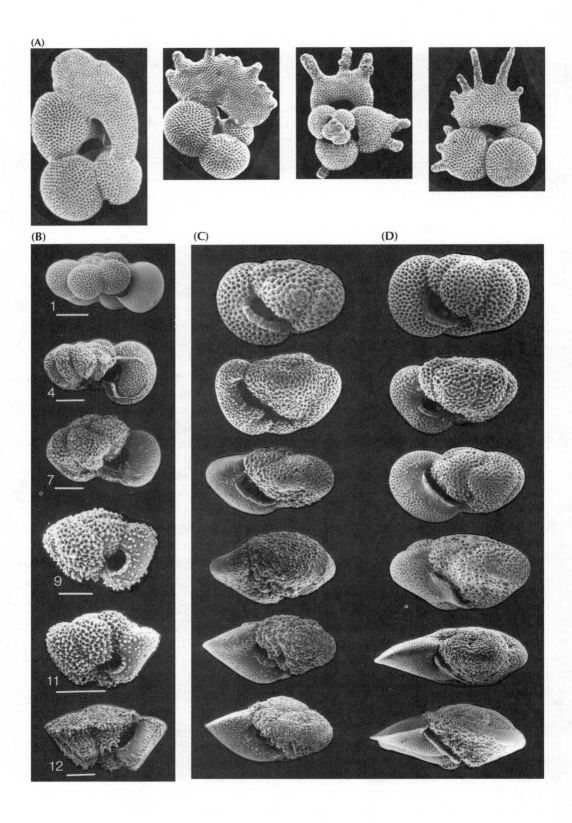

(B)

(C)

(D)

1

4

7

9

11

12

FIGURE 8.5. Evolutionary sequences in some of the planktonic foraminifera. (A) In the Pliocene and Pleistocene, the fingerlike projections of *Globigerinoides fistulosus* evolved from the smooth-shelled *Globigerinoides sacculifer*. (B) The evolution of *Praemurica* (top) to the keeled *Morozovella* (bottom) that occurred between 63 and 59 million years ago. (C) The evolution of *Globoconella* in the early Miocene (20–18 million years ago), from the unkeeled forms at the top to the keeled *Globoconella conoidea at* the bottom, which vanished about 6 million years ago. (D) A similar trend in the genus *Fohsella*, from the ancestral unkeeled forms at the top to the highly keeled species at the bottom. (Photos in (A) courtesy J. Kennett; (B–D) courtesy R. Norris)

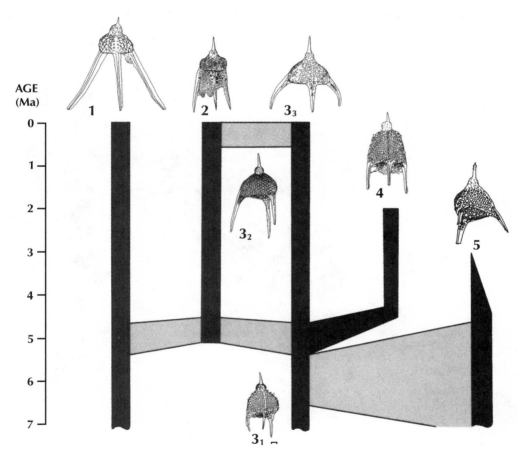

FIGURE 8.6. Evolutionary patterns in the radiolarian *Pterocanium from* the late Cenozoic. Solid bars show the time distribution of the main lineages, while gray zones show intervals of intergradation and possible hybridization of lineages. Time scale along left indicates millions of years before present (Ma). Scale bars = 10 micrometers. 1, *Pterocanium korotnevi*; 2, *Pterocanium praetextum*; 3_1, *Pterocanium charybdeum allium*; 3_2, *Pterocanium charybdeum charybdeum*; 3_3, *Pterocanium charybdeum trilobum*; 4, *Pterocanium prismatium*; 5, *Pterocanium audax*. (From Lazarus et al. 1985: fig. 21; used by permission of the Paleontological Society)

korotnevi). Two lineages that had been distinct for millions of years (*Pterocanium praetextum* and *Pterocanium charybdeum trilobum*) appeared to hybridize during the late Pleistocene. The diagram in figure 8.6 does not capture how the change can be traced gradually centimeter by centimeter through many different cores, so we could see every transitional step between one extreme shell shape and another. However, it was possible to measure many different features, such as the length of the thorax, and see the gradual shift of sizes in each population spanning 7 million years.

This is just one example of evolutionary changes among the radiolaria, and there are hundreds more that could be cited. In fact, radiolaria are still not that well studied (especially compared to forams), so there could be hundreds more as yet unknown. Let us look at just one more classic example, probably the most extreme change in morphology ever documented in the fossil record. If you look at samples of microfossils from the middle Eocene (50 million years ago), you will find distinctive spongy ball-shaped radiolarians known as *Lithocyclia ocellus* (fig. 8.7). As you trace the spongy balls up through the sediments spanning

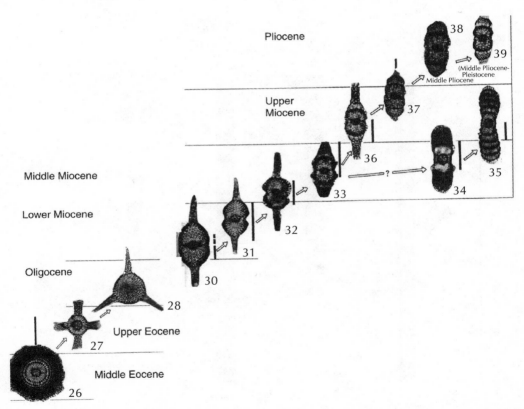

FIGURE 8.7. Evolutionary transformation in the cannartid-ommatartid lineage of radiolaria over the past 50 million years, from spongy balls to four- and then three-armed and finally two-armed bipolar structures, with further variations in the spongy caps later in their evolution. Taxa are as follows: 26, *Lithocyclia ocellus*; 27, *Lithocyclia aristotelis*; 28, *Lithocyclia angusta*; 30, *Cannartus tubarius*; 31, *Cannartus violina*; 32, *Cannartus mammiferus*; 33, *Cannartus laticonus*; 34, *Cannartus petterssoni*; 35, *Ommatartus hughesi*; 36, *Ommatartus antepenultimus*; 37, *Ommatartus penultimus*; 38, *Ommatartus avitus*; 39, *Ommatartus tetrathalamus*. (Modified from Haq and Boersma 1978)

millions of years, you see them gradually lose their spongy outer layers and develop into a small nucleus with four spongy arms (*Lithocyclia aristotelis*), then three arms (*Lithocyclia angusta*), and finally two arms forming a spindle-like shape (*Cannartus tubarius*). The *Cannartus* lineage then gradually develops a "waist" on the central sphere, then the arms get shorter and thicker, and finally, they split into two lineages: *Cannartus peterssoni-Ommatartus hughesi*, which evolves into a form with two arms with multiple spongy layers, and *Ommatartus*, which develops shorter arms and a fatter central sphere. If you look at the two extremes (a spongy sphere turning into a spindle-shaped shell with multiple caps), you could never imagine that they are closely related—yet I have looked at the slides from those cores and seen the gradual transition from one extreme to the other with my own eyes.

There are multiple studies on evolutionary patterns and transitions in other microfossils, including the diatoms and coccolithophorids (see Lazarus [1983] and the articles in *Paleobiology*, volume 9, number 4, fall 1983 for examples). For space reasons, however, we will leave the extraordinary record of microfossils and look at the patterns in the more easily studied macroinvertebrates.

The Story in the Seashells

Why don't paleontologists bother to popularize the detailed lineages and species-to-species transitions? Because it is thought to be unnecessary detail. . . . Paleontologists clearly consider the occurrence of evolution to be a settled question, so obvious as to be beyond rational dispute, so, they think, why waste valuable textbook space on such tedious detail?

—Kathleen Hunt, FAQ, www.talkorigins.org

If you find it difficult to relate to tiny fossils that can only be studied with expensive microscopes, you can go right out to your local fossiliferous outcrop and study patterns of evolution there. Take, for example, the famous cliffs along the shores of Chesapeake Bay (fig. 3.2A and B) that are made of solid shells of mollusks from the Miocene (about 18–5 million years ago). Some of the most common and distinctive fossils are the large scallops known as *Chesapecten*. One of these species, *Chesapecten jeffersoni*, is the state fossil of Virginia. As numerous studies have shown (Ward and Blackwelder 1975; Miyazaki and Mickevich 1982; Kelley 1983), the shells of these scallops change continually through time. The earliest forms are known as *Chesapecten coccymelus* (fig. 8.8); they are abundant in Zone 10 of the middle Miocene Calvert Formation. As you pass up from the Calvert Formation to the Choptank Formation, there are specimens of *Chesapecten nefrens*, which have shells that are longer from front to back than they are high from hinge to opening; this trend continues through the series. As you move up through the series, the number of ribs decreases as well, except in *Chesapecten middlesexensis*, which reverses the trend and develops more ribs. Many more subtle differences in the shells can be detected (Ward and Blackwelder 1975), so many that any competent paleontologist can tell the species apart easily and decide what time in the Miocene or Pliocene is represented just by the scallops.

If you live in the United Kingdom, go out to the White Cliffs of Dover and nearby areas and walk along the base of the cliffs. Weathering out of the soft chalky limestone (made entirely of coccoliths, as we just mentioned) are hundreds of small heart urchin shells from

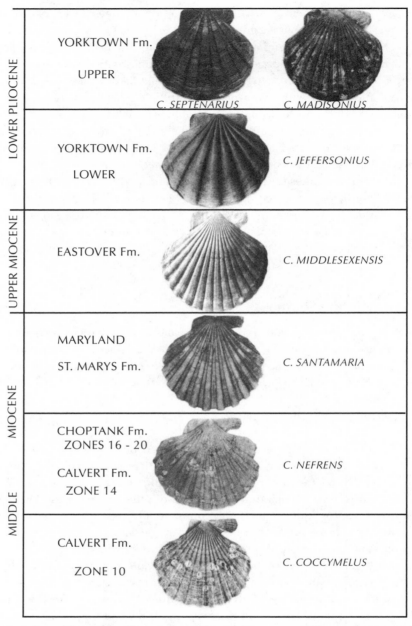

FIGURE 8.8. Evolution of the scallop Chesapecten during the Miocene and Pliocene, as preserved along the Calvert Cliffs in Chesapeake Bay. (After Miyazaki and Mickevich 1982: plate 1; originally based on Ward and Blackwelder 1975; used by permission of Plenum Publishing)

the Late Cretaceous. Originally, they were the subject of a classic study by Rowe (1899), who thought that they provided a clear case of gradual transformation from smaller shells with primitive characteristics (derived from the ancestral form *Epiaster*) to shells that are broader, with the tallest and broadest parts shifted forward. The groove in the front deepens and fills with tiny bumps, and the mouth shifts forward with a more prominent lip. The area for the

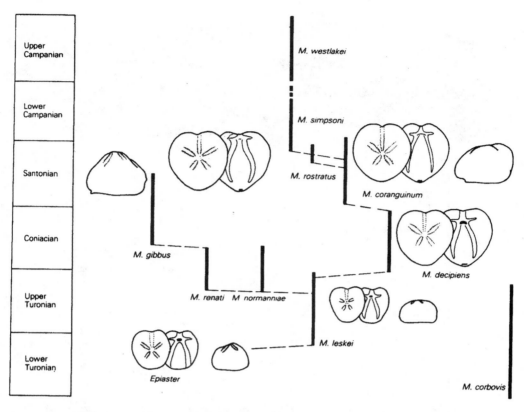

FIGURE 8.9. Evolutionary trends in the heart urchin *Epiaster* and *Micraster* from the Cretaceous chalk beds of the White Cliffs of Dover. (From Nichols 1959; courtesy of the Royal Society of London)

tube feet lengthens and straightens out (fig. 8.9). However, later research showed that Rowe's sampling was inadequate and he did no statistical studies at all. In 1954, Kenneth Kermack (since updated by Nichols [1959], Ernst [1970], and Stokes [1977]) performed a more rigorous analysis and found that, although the sequence was not a single smooth linear trend, there is still a remarkable branching sequence of species that partially overlap in time (fig. 8.9), from *Micraster leskei* to *Micraster cortestudinarum* ("tortoise heart") to *Micraster coranguinum* ("eel heart"), and a side branch leading to *Micraster corbovis* ("cow heart") and *Micraster gibbus*. These fossils are no longer evidence of gradualism within a single unbranched lineage, but there is still a good evolutionary sequence of transitional forms.

In the Jurassic beds of England to west of the White Cliffs, there are many additional excellent fossil sequences. A classic one of these is the weird oyster known as *Gryphaea*. Called "devil's toenails" by the collectors, these creatures had one shell that was shaped like a coiled saucer that lay flat on the bottom (concave side up), and the other shell was much smaller and formed a lid on top (fig. 8.10). In a classic study, Trueman (1922) argued that *Gryphaea* gradually became more and more tightly coiled until the coils actually impeded the opening of the shell. In the thinking of the time, this was evolution run amok, developing to the point of obsolescence and no longer under the control of natural selection. But later, more careful studies by Philip (1962, 1967), Hallam (1968, 1982), Gould (1972), and Hallam and Gould (1974) showed that Trueman had misinterpreted his data. There is a trend toward *less*

(A)

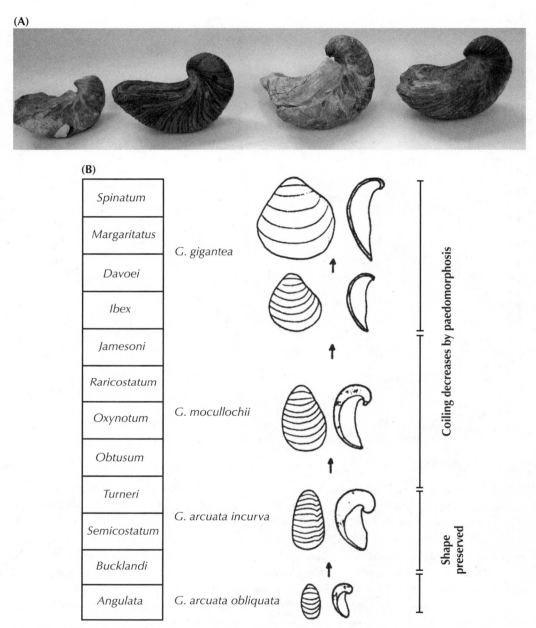

(B)

FIGURE 8.10. Evolution within the Jurassic oyster *Gryphaea*. (A) A series of *Gryphaea* shells, showing their decreasingly tight coiling through time. (Photo by the author) (B) The sequence of less tightly coiled shells found in the Jurassic beds of southern England. (After Hallam 1968: fig. 26; courtesy of the Royal Society of London)

tightly curved and thinner, more dish-like shells that are much wider than long (fig. 8.10), but it happens in a number of different species lineages through the Jurassic, and much of the change in shape is primarily due to growth to a larger size.

Many other examples from the mollusks could be added to this list. For example (Rodda and Fisher 1964), the marine snail *Athleta* is very common in the Eocene (from 55 to 34 million

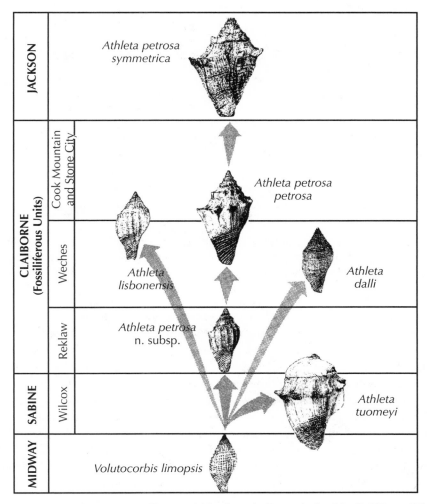

FIGURE 8.11. Evolution in the marine snail lineage *Athleta* from the Eocene beds of the Gulf Coastal Plain. (From Rodda and Fisher 1964; courtesy of the Society for the Study of Evolution)

years old) rocks along the Gulf Coast of Texas, Louisiana, Mississippi, and Alabama (fig. 8.11). They evolved from a simple nonornamented shell of *Volutocorbis limopsis* but quickly branched into several very different lineages. In the Wilcox beds, there is the huge, broad shelled *Athleta tuomeyi*, as well as the more normal lineage of *Athleta petrosa*. In the Weches beds, there are side branches to the nonornamented *Athleta dalli* and the more bulbous *Athleta lisbonensis*. The main *Athleta* lineage concludes in the upper Eocene Lisbon beds with the heavily ornamented *Athleta petrosa symmetrica*.

Or how about everyone's favorite fossils, the trilobites? Many different studies have been published on their patterns of evolution. In most cases, they show trends that are due to very subtle changes, such as the changing number of ribs in the thorax (fig. 8.12), as documented by Peter Sheldon (1987) on a 3-million-year-long sequence of over 15,000 specimens from eight trilobite lineages from the Ordovician of central Wales. Some lineages, such as *Ogyginus*, show almost no net change in rib number (stasis), while the nileids, *Ogygiocarella*, and especially *Nobiliasaphus* show a dramatic increase in rib number over the interval. This example

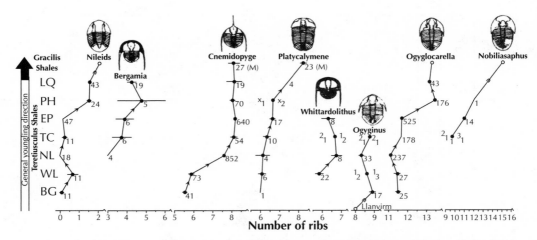

FIGURE 8.12. Evolutionary trends in numerous trilobite lineages from the Ordovician beds of central Wales. Most lineages show a gradual increase in the number of rib segments in the thorax through the different time intervals of the Ordovician. (After Sheldon 1987; copyright the Nature Publishing Group)

shows both stasis and gradual evolution, but the conclusion is clear: these trilobites were not instantaneously created but keep changing through time. Other groups of trilobites show additional subtle changes through time, such as more complex eyes, increased size, more complicated tail segment (*pygidium*), and the development of spines (Eldredge 1977; Fortey and Owens 1990). Most of these examples are not gradual like those documented by Sheldon but show punctuation and stasis—and yet, contrary to creationists' misconceptions, they do change through time, so they are good examples of evolution (just not gradual evolution).

I could go on and on with many more examples such as these, but for space reasons, I will move on to the next topic. I recommend reading about some of the many examples documented in Hallam (1977), Boardman et al. (1987), McNamara (1990), Clarkson (1998), and Prothero (2013a) if you are interested in further documentation.

What About Macroevolution?

There is no need to apologize any longer for the poverty of the fossil record. In some ways it has become almost unmanageably rich, and discovery is out-pacing integration: the growing number of species of Foraminifera that remain undescribed in the cabinets of the oil companies probably is of the order of thousands; and while most other organic groups are not so fully collected the ratio of added finds to palaeontologists studying them is constantly expanding. But what remains to be discovered is likely to be of less and less radical importance in revealing major novelties, more and more of detailed infilling of fossil series whose outlines are known. The main phyla, in so far as they are represented by fossils, now have a long and full history that is made three-dimensional by a repeatedly cladal phylogeny. The gaps are being closed not only by major annectant forms, the "missing links" that Darwin so deplored. . . . Together, the discovery of new fossil forms, the filling out of the details of bioserial change, the interpretation of biofacies, the adoption of new techniques both in fossil morphology

and in fossil manipulation, and the establishment of a progressively refined timescale contribute to a present-day palaeontology offering the strongest support, the demonstrative "proof," of the fact and the process of evolution in terms wholly concordant with the essence of Darwinian theory.
—T. George Neville, "Fossils in Evolutionary Perspective," 1960

When you point out the examples that we just discussed to the creationists, they weasel out of the problem by arguing that these changes are all within the "created kinds" as described in the Bible. As we pointed out in chapter 5, however, the concept of "kind" has absolutely no biological meaning, but creationists use it as a convenient dodge to allow changes on the microevolutionary scale (whatever evolves is within one kind) but to deny that major macroevolutionary changes could occur. Of course, this concedes that a heck of a lot of evolution is taking place, because nearly every group in the fossil record shows some evolution, and they can't all be created kinds—and they couldn't even all fit on Noah's ark, as we already mentioned in chapter 3. But let's play by their absurd rules and look at some radical changes in body form and ecology that are clearly macroevolutionary.

How about sand dollars? These cute little flat shells that are so popular with tourists and beachcombers are actually related to sea urchins and heart urchins and their kin. They are members of the phylum Echinodermata that also includes sea star, brittle stars, and sea cucumbers as well as many extinct groups. Living sand dollars are very flat and spend much of their time just below the sea bottom, buried with a light coating of sand. When they are feeding, however, they maneuver themselves with their short fuzzy spines and tube feet so that they stick out diagonally from the seafloor like a set of shingles (fig. 8.13A). Their mouth are on their undersides and face into the current, allowing them to trap food particles. Looking at a sand dollar, it seems nothing like any other "kind" on the seafloor, and it seems hard to imagine that it could evolve from something else—but it did! As documented by Porter Kier (1975, 1982), sand dollars evolved rapidly in the late Paleocene and early Eocene from the biscuit-shaped urchins known as cassiduloids to the slightly flatter oligopygoids to an even more flattened transitional fossil known as *Togocyamus* from the late Paleocene of Togo, West Africa (fig. 8.13B). In addition to getting flatter and flatter, sand dollars also show a progressive reduction in the size but an increase in the number of spines and bumps on the shell, so they go from rough and spiny to almost "furry" with tiny spines and bumps (which makes burrowing easier). The mouth, which had been on the leading edge of the bottom of the shell, shifts to the middle of the base of the shell, and the anus, which had been on the top side of the shell in primitive forms, shifts to the back edge of the bottom surface. Later in their evolution, they developed little holes and notches on the edge of the shell, which help modify the water flow around the shell as they lie buried. All of these transformations are well documented in specimens from the Paleocene and early Eocene (65–40 million years ago), starting in the central West African region, and eventually spreading worldwide.

Or how about a classic "living fossil," the horseshoe "crab"? They are familiar to anyone who has combed the beaches of the Atlantic Coast of the United States, since they wash up frequently. During particularly high tides once a year, hundreds of them crawl up on the beach, mate, and lay their eggs in the sand in an orgy straight out of prehistory. Yet horseshoe "crabs" are not true crabs but instead are members of the group known as the Chelicerata that includes spiders and scorpions. (True crabs are a family within the Crustacea, a different group entirely.) Surely, they don't look like any other group of arthropods, and

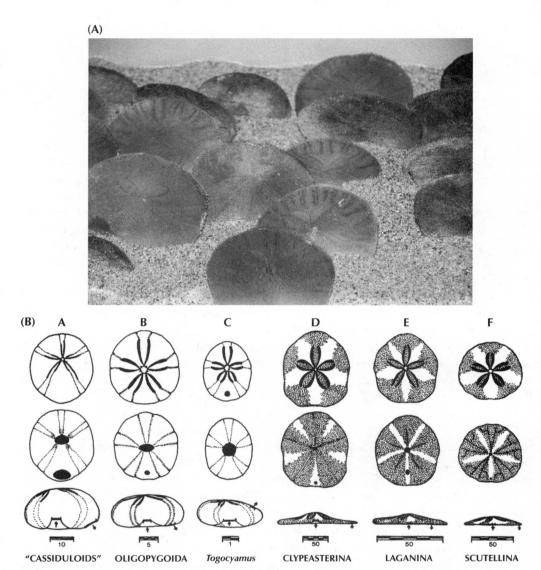

FIGURE 8.13. The evolution of sand dollars. (A) Living sand dollars in feeding position, half-buried obliquely into the sand like shingles, with their mouths on the bottom facing into the food-bearing water currents. (Photo by the author) (B) Evolution of flat sand dollars from the biscuit-shaped cassiduloid sea urchins and slightly flatter oligopygoids through the transitional Paleocene fossil *Togocyamus* from western Africa and concluding with progressively flatter and more specialized sand dollars. The petal-shaped areas on the top of the shell bear the tube feet, while the mouth is on the bottom of the shell and gradually shifts from the center of the bottom to the lower front edge. Meanwhile, the anus shifts from the center of the top of the shell to the back edge of the shell as burrowing ability becomes more specialized and improved. (After Mooi 1990; used with permission of the Paleontological Society.)

if they are supposedly unchanged through millions of years, they would have no transitional forms that could produce the "horseshoe crab kind" from some other kind? Wrong! When I collected in the Upper Cambrian beds of Wisconsin, one of the most spectacular fossils I encountered were the plates of the primitive arthropods known as aglaspids (fig. 8.14A). These large creatures don't look like the modern horseshoe crab, but still they are either close

(A)

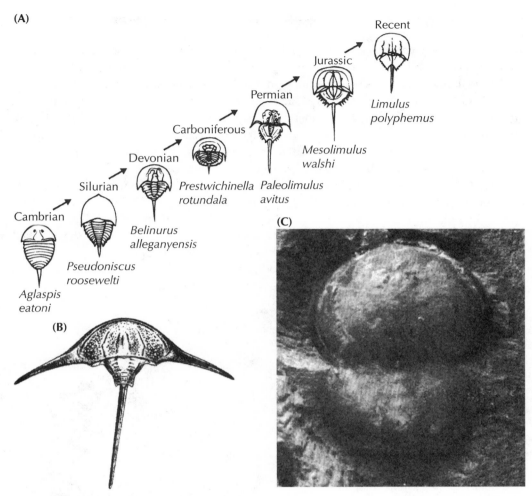

Recent

Jurassic

Permian

Carboniferous

Devonian

Silurian

Cambrian

*Limulus
polyphemus*

*Mesolimulus
walshi*

*Prestwichinella
rotundala*

*Paleolimulus
avitus*

*Belinurus
alleganyensis*

*Pseudoniscus
roosewelti*

*Aglaspis
eatoni*

(C)

(B)

FIGURE 8.14. Evolution of the horseshoe crabs. (A) Evolutionary trends within the lineage from the primitive Cambrian aglaspids through more and more specialized and modern-looking horseshoe crabs. (B) The weird "boomerang-shaped" Australian horseshoe crab known as *Austrolimulus*. (C) The peculiar double-button-shaped horseshoe crab known as *Liomesaspis*. (Part (A) from Newell 1959; by permission of the American Philosophical Society; (B and C) courtesy D. Fisher)

to their ancestry (Newell 1959; Fisher 1982, 1984) or related to both the horseshoe crabs and trilobites (Briggs et al. 1979; Briggs and Fortey 1989), so both trilobites and horseshoe crabs can be traced back to a fossil that looks like neither. Through the rest of the Paleozoic, there are additional species of the horseshoe crab lineage (subclass Xiphosura, "sword tail" in Greek), which develop progressively larger and larger head shields and fewer thoracic segments and, unique to the group, the long tail spine that gave them their name. Finally, by the Jurassic we find specimens of *Mesolimulus walchi*, which are very similar to the living species *Limulus polyphemus*, although their thorax is still not as heavily fused together and is much more spiny than the modern species. Thus xiphosurans haven't changed much in the past 100 million years, but they changed a lot before then. Plus, there are weird experiments such as *Austrolimulus fletcheri* (fig. 8.14B), which has long boomerang-shaped spines on the corners of its head shield (appropriately, it is from the Triassic of Australia), or *Liomesaspis* (fig. 8.14C), which is reduced to two smooth button-shaped shields for the head and thorax and almost no tail spine.

Still not macroevolutionary enough? What about the "missing links" between major body plans and classes and phyla? Here, the problem is more difficult, because the great radiation of animals in the Cambrian was largely among phyla with soft, unfossilizable bodies before large shells finally appeared in the late Early Cambrian, or Atdabanian. We have excellent molecular, anatomical, and embryological evidence demonstrating how the major animal phyla are interrelated (fig. 5.7). This evidence shows that the mollusks are most closely related to the segmented worms among the living animal phyla. How about a transitional form between a worm and a mollusk? Ask and ye shall receive. Among the earliest fossil mollusks known from the Cambrian are simple cap-shaped shells that had been given names (such as *Pilina*), even though we had no evidence of their soft anatomy. Then, in 1952 a dredge brought up specimens from the deep waters off Costa Rica that included another classic "living fossil" named *Neopilina galatheae* (fig. 8.15). *Neopilina* is clearly a mollusk with a cap-shaped shell secreted by a mantle, as well as a mouth, digestive tract, anus, and gills.

But it is unlike any other mollusk alive today because it still retains the segmentation of its wormlike ancestors. Arranged around the body on the margin of the mantle and below

FIGURE 8.15. The "living fossil" *Neopilina*, a relict of the early Cambrian and a transitional form between segmented worms and mollusks. (A) Diagram showing the segmented paired gills on each side of the body (there are also segmented paired retractor muscles as well). This demonstrates that normally unsegmented mollusks evolved from segmented ancestors such as the annelid worms (confirmed by the molecular data). (B) Photograph of the underside of the living animal, showing the relict segmentation. (C) Top view of the cap-shaped shell. (Courtesy J. B. Burch, University of Michigan)

the lip of the shell are segmented gills, kidneys, hearts, gonads, and paired retractor muscles to pull down the shell. You couldn't ask for a more classic transitional form. It is almost completely molluscan, yet it still has features of its segmented worm ancestry. Since 1952, scientists have come to realize that two soft-bodied wormlike groups, the Caudofoveata and the Aplacophora, are actually very primitive mollusks as well, even though they have no shells and look more like worms in their external form. We now have a nice transition from segmented worms to shell-less wormlike mollusks to mollusks with shells but with relicts of segmentation, and finally to the great radiation of unsegmented mollusks, including clams, snails, squids and octopuses, and all their extinct relatives.

How about transitional forms for the largest, most diverse phylum in the world, the Arthropoda, or "jointed legged" animals, which includes insects, spiders, scorpions, crustaceans (crabs, shrimp, lobsters, and barnacles), horseshoe crabs, trilobites, centipedes, millipedes, and many other groups? The tree of life (fig. 5.7) shows that arthropods are more closely related to nematodes (roundworms) and rotifers among living organisms. How could we imagine a transition between a nematode "kind" and an arthropod "kind" such as a millipede? It turns out that the transitional forms are still alive today; they are known as "velvet worms" or phylum Onychophora (fig. 8.16), pronounced "on-ih-KOFF-o-ra." There are about 80 species of these creatures, living mostly in the tropical jungles of the world; they were originally mistaken for "slugs" when they were first described in 1826. But more careful observation (fig. 8.16) shows that although they may look superficially wormlike, they have many features of arthropods as well. Unlike the unsegmented nematode worms, onychophorans are segmented and have legs that resemble those of caterpillars. Their partially segmented legs end in horny hooked "claws." Onychophora have cuticles made of the protein *chitin*, just like arthropods, and periodically have to molt in order to get larger (a feature found elsewhere only in the arthropods). They also have antennae, compound eyes, and mouthparts that are much like those found in arthropods. As outlined by Brusca and Brusca (1990:683), there are many other features that unite onychophorans with arthropods and make them outstanding transitional forms between the phylum Nematoda and the phylum Arthropoda.

The clincher is that we also have them in the Cambrian as well. The Burgess Shale in Canada and the Chengjiang Fauna of China produce amazing fossils of a variety of marine onychophorans known as lobopods, including *Aysheaia* (fig. 8.17A) and *Hallucigenia* (fig. 8.17B) from the Burgess Shale and *Microdictyon* from Chengjiang. *Aysheaia* is almost indistinguishable from some modern onychophorans, but *Hallucigenia* and some of the other Cambrian forms show an amazing array of spines and other features, giving the group much more diversity than we would appreciate from seeing only their living relatives.

FIGURE 8.16. The living transitional form between worms and arthropods, known as the "velvet worms" or phylum Onychophora. Although they have segmented wormlike bodies, they also have jointed appendages and antennae and shed their cuticle like arthropods do. (Photo from IMSI Master Photo Collection)

FIGURE 8.17. Some examples of marine lobopod onychophoran fossils from the middle Cambrian Burgess Shale of Canada. (A) *Aysheaia*. (B) *Hallucigenia*. (Photos courtesy S. Conway Morris)

(A)

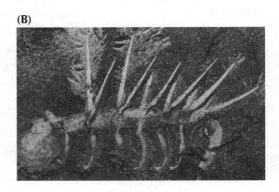

(B)

Once arthropods evolved, they diverged into a wide variety of body forms, from elongate multilegged centipedes and millipedes, to six-legged insects and eight-legged spiders, to the huge diversity of crustaceans. How did the onychophoran "kind" turn into the millipede "kind," the insect "kind," the spider "kind," the crustacean "kind," and so on? As we saw in chapter 4, the answer lies in the Hox genes and their ability to dramatically change body form through small changes in gene regulation. We have already seen how minor changes in the Hox genes produce homeotic mutants such as flies with legs on their heads or with four wings (fig. 4.5). Arthropods are particularly suited to this type of evolution because they have a modular construction with multiple segments, and each segment bears appendages that can be easily changed from a leg to a wing to an antenna to a pincer to mouthparts. Experiments have shown that a few Hox genes cause arthropods to add or subtract segments, and other Hox genes can produce whatever appendage is needed (fig. 8.18). Ronshaugen et al. (2002) put a shrimp *Ubx* Hox gene into an insect larva and showed how this gene was responsible for suppressing the development of limbs in insects (which have 6 legs, compared to the 10 in most crustaceans). Lewis et al. (2000) and Pearson et al. (2005) have shown how by manipulating Hox genes you can get just about any type or number of appendage on each segment of an arthropod, therefore making radical changes in body plan with a simple gene change. And there are many fossils that show primitive arthropods with more than two sets of wings, as in dragonflies, showing that the creature in figure 8.18 is not imaginary (Kulakova-Peck 1978; Raff 1998). Ironically, creationists attacked the first edition of this book by focusing just on one figure, figure 8.18, and claiming that it's "made up." Because of their ignorance, they didn't realize there are fossils of insects with six, eight, and even more wings, so if anything, this multiwinged insect is a bit too conservative.

In addition, arthropods can undergo radical changes in body form each time they shed their exoskeleton during molting. Think about how radically the body is rearranged from

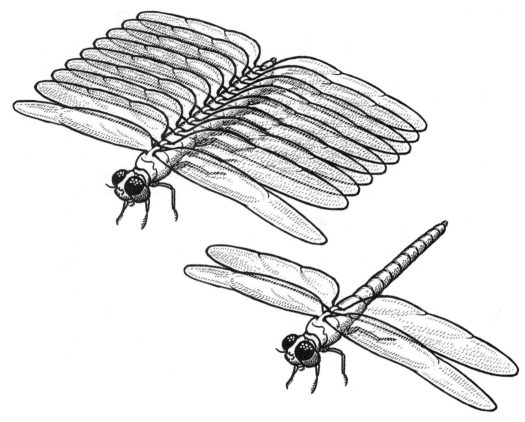

FIGURE 8.18. The evolutionary mechanism by which Hox genes allow arthropods to make drastic changes in their number and arrangements of segments and appendages, producing macroevolutionary changes with a few simple mutations (see fig. 4.6). (Drawing by Carl Buell)

a caterpillar to an adult moth or butterfly, or from a maggot to an adult fly. Thus we can experimentally show that macroevolutionary transition from one body form to another with a completely different number of segments and appendages is a very easy process. No wonder arthropods are the most successful, abundant, and diverse organisms on earth. After we humans are long gone, the cockroaches and other insects will still rule the earth, as they have for over 300 million years so far.

In summary, the molecular-anatomical-embryological phylogeny of the animal kingdom (fig. 5.7) links mollusks and annelids; we have transitional forms, both fossil and living, between these two phyla. It also links nematodes and arthropods, and we have transitional forms, both fossil and living, between these two phyla as well. Using Hox genes, we can demonstrate how radical changes in body plans are controlled by relatively simple genetic mechanisms and allow macroevolutionary changes to take place. As we shall see in the next chapter, we also have abundant transitional forms that link the vertebrates and the echinoderms and show us the earliest ancestry of the vertebrates from soft-bodied ancestors.

For Further Reading

Benton, M. J., and P. N. Pearson. 2001. Speciation in the fossil record. *Trends in Ecology and Evolution* 16:405–411.

Boardman, R. S., A. H. Cheetham, and A. J. Rowell, eds. 1987. *Fossil Invertebrates*. Cambridge, Mass.: Blackwell.

Clarkson, E. N. K. 1998. *Invertebrate Palaeontology and Evolution*. 4th ed. Oxford, U.K.: Blackwell Science.

Eldredge, N., and S. M. Stanley, eds. 1984. *Living Fossils*. New York: Springer-Verlag.

Fisher, D. C. 1982. Phylogenetic and macroevolutionary patterns within the Xiphosurida. *Proceedings of the Third North American Paleontological Convention* 1:175–180.

Gould, S. J. 1972. Allometric fallacies and the evolution of *Gryphaea*. *Evolutionary Biology* 6:91–119.

Hallam, A. 1968. Morphology, palaeoecology, and evolution of the genus *Gryphaea* in the British Lias. *Philosophical Transactions of the Royal Society of London B* 254:91–128.

Hallam, A., ed. 1977. *Patterns of Evolution as Illustrated in the Fossil Record*. New York: Elsevier.

Hallam, A. 1982. Patterns of speciation in Jurassic *Gryphaea*. *Paleobiology* 8:354–366.

Haq, B. U., and A. Boersma, eds. 1978. *Introduction to Marine Micropaleontology*. New York: Elsevier.

Kier, P. M. 1965. Evolutionary trends in Paleozoic echinoids. *Journal of Paleontology* 39:436–465.

Kier, P. M. 1975. Evolutionary trends and their functional significance in the post-Paleozoic echinoids. *Paleontological Society Memoir* 5:1–95.

Kier, P. M. 1982. Rapid evolution in echinoids. *Palaeontology* 25:1–10.

Kukalova-Peck, J. 1978. Origin and evolution of insect wings and their relation to metamorphosis, as documented by the fossil record. *Journal of Morphology* 156:53–125.

Lazarus, D. B. 1983. Speciation in pelagic Protista and its study in the microfossil record: a review. *Paleobiology* 9:327–340.

Lazarus, D. B. 1986. Tempo and mode of morphologic evolution near the origin of the radiolarian lineage *Pterocanium prismatium*. *Paleobiology* 12:175–189.

Lazarus, D., H. Hilbrecht, C. Spencer-Cervato, and H. Thierstein. 1995. Sympatric speciation and phyletic change in *Globorotalia truncatulinoides*. *Paleobiology* 21:975–978.

Lazarus, D. B., R. P. Scherer, and D. R. Prothero, 1985. Evolution of the radiolarian species-complex *Pterocanium*: a preliminary survey. *Journal of Paleontology* 59:183–221.

Malmgren, B. A., and W. A. Berggren. 1987. Evolutionary change in some late Neogene planktonic foraminifera lineages and their relationships to paleoceanographic change. *Paleoceanography* 2:445–456.

Malmgren, B. A., and J. P. Kennett. 1981. Phyletic gradualism in a Late Cenozoic planktonic foraminiferal lineage, DSDP Site 284, southwest Pacific. *Paleobiology* 7:230–240.

Malmgren, B. A., W. A. Berggren, and G. P. Lohmann. 1983. Evidence for punctuated gradualism in the Late Neogene *Globorotalia tumida* lineage of planktonic foraminifera. *Paleobiology* 9:377–389.

McNamara, K. J., ed. 1990. *Evolutionary Trends*. Tucson: University of Arizona Press.

Miyazaki, J. M., and M. F. Mickevich. 1982. Evolution of Chesapecten (Mollusca: Bivalvia, Miocene-Pliocene and the biogenetic law. *Evolutionary Biology* 15:369–409.

Pearson, P. N. 1993. A lineage phylogeny for the Paleogene planktonic foraminifera. *Micropaleontology* 39:193–232.

Pearson, P. N. 1998. The glorious fossil record. *Nature*, November 19. www.nature.com/nature /debates/ fossil/fossil_1.html.

Pearson, P. N., N. J. Shackleton, and M. A. Hall. 1997. Stable isotopic evidence for the sympatric divergence of *Globigerinoides trilobus* and *Orbulina universa* (planktonic foraminifera). *Journal of the Geological Society of London* 154:295–302.

Prothero, D. R. 2013. *Bringing Fossils to Life: An Introduction to Paleobiology*. 3rd ed. New York: Columbia University Press.

Raff, Rudolf A. 1998. *The Shape of Life: Genes, Development, and the Evolution of Animal Form*. Chicago: University of Chicago Press.

Rodda, P. U., and W. L. Fisher. 1964. Evolutionary features of *Athleta* (Eocene, Gastropoda) from the Gulf Coastal Plain. *Evolution* 18:235–244.

Sheldon, P. R. 1987. Parallel gradualistic evolution of Ordovician trilobites. *Nature* 330:561–563.

Smith, A. B. 1984. *Echinoid Palaeobiology*. London: George Allen and Unwin.

Ward, L. W., and B. W. Blackwelder. 1975. *Chesapecten*, a new genus of Pectinidae (Mollusca: Bivalvia) from the Miocene and Pliocene of eastern North America. *U.S. Geological Survey Professional Paper* 861.

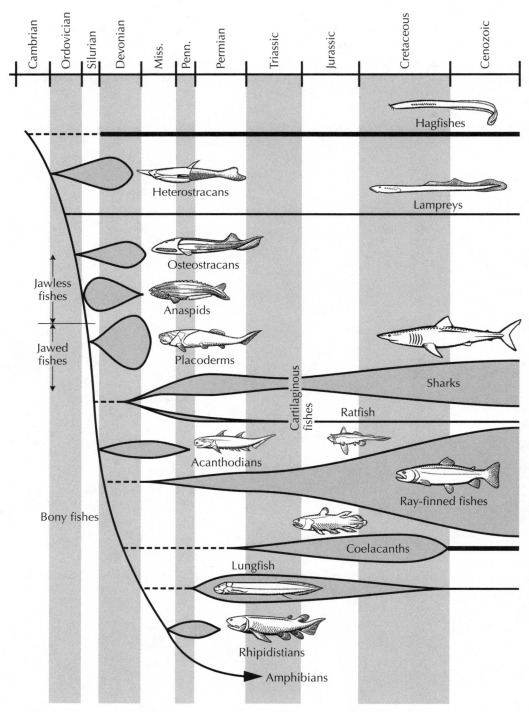

FIGURE 9.1. Evolutionary history of fishes and other early vertebrates. (Drawing by Carl Buell)

FISH TALES

9

Show Some Backbone!

We can trace without a break, always following out the same law, the evolution of man from the mammal, the mammal from the reptile, the reptile from the amphibian, the amphibian from the fish, the fish from the arthropod, the arthropod from the annelid [segmented worms], and we may be hopeful that the same law will enable us to arrange in orderly sequence all the groups in the animal kingdom.

—Walter Gaskell, *The Origin of Vertebrates*

Most people don't get too excited about evolution in sand dollars, snails, scallops, or microfossils. But they are far more interested in where our group, the vertebrates, came from. For this transition, there is abundant evidence not only from the fossil record but also from embryology and from a number of "living fossils" that preserve the steps in the evolution of vertebrates and are still alive today.

Humans are members of the phylum Chordata. This group includes the vertebrates (animals with a true backbone and other kinds of bone as well) such as mammals, birds, reptiles, amphibians, and fish (fig. 9.1). The Chordata also includes a variety of near-vertebrates that have some of the specializations of vertebrates but do not have a backbone. Many of these near-vertebrates have a long flexible rod of cartilage known as the *notochord* instead of the bony backbone; this defines the group known as the phylum Chordata. When you were an embryo, you had a notochord before the cartilage was replaced by the bone of your adult spinal column.

Where do chordates come from? For over a century, all the anatomical and embryological evidence (and more recently, all the molecular evidence as well) clearly shows that our closest relatives among living animals are the echinoderms—the sea star, sea urchins, and sea cucumbers. You may not think of the sea star as your close relative (or even think of it as an animal), but that's what the biological facts clearly show. The most striking demonstration of this comes from our embryology. When you were a simple ball of cells (*blastula*) just a few cleavages after you formed by fertilization, there was a small opening in the ball called a *blastopore*. If you had been an embryo of a worm or an arthropod, your blastopore would have developed into the mouth end of your digestive tract. But in the *deuterostomes* (echinoderms plus chordates), the blastopore becomes the anus, and the mouth develops on the opposite side of the blastula. There are many other embryological similarities as well. The cells in the fertilized egg in most animals cleave in a spiral pattern, but those in deuterostomes do so in a radial pattern. Deuterostome embryos have cells that are indeterminate, meaning that their fates are not determined at the very beginning (as in most animals) but can become part of a new organ or even regenerate an organ if necessary. If you break up the larvae of a sea urchin early in development, each ball of cells can turn into a complete animal. Finally, the internal fluid-filled body cavity (the *coelom*) of the Deuterostomata forms from an outpocketing of the inner layer of cells, or endoderm, rather than from a split in the middle layer of cells, or mesoderm, as occurs in worms and arthropods.

All of these unique specializations show that the echinoderm and chordate larval pattern is the common link between these two very different phyla of animals. And in the past 20 years, every molecular system that has been examined confirms that the Deuterostomata is a natural monophyletic group, so there's no longer any doubt among biologists that sea stars and sea urchins are among our close relatives. From the common larval pattern, one set of developmental commands begins to produce the larvae of echinoderms, and another set of pathways produces the classic chordate embryo. We would not expect to find these fragile embryos preserved in the fossil record, but just a few years ago, a remarkable discovery of late Precambrian embryos was found in Doushantuo, China, which seem to show that our earliest common ancestors lived around 600–700 million years ago (as the molecular clock estimates also suggest).

What defines a chordate besides the notochord? At one end of the basic chordate body plan (fig. 9.2) are the sensory organs (eyes, nostrils) and a mouth opening into a throat cavity known as the *pharynx*. In humans, the pharynx houses the vocal cords, but in fish the pharynx is the region of the gills and gill basket, and in some groups, feeding takes place in the pharynx as well. Chordates are also distinctive in that they have a nerve cord along the back, above the notochord, and the digestive tract along the belly, below the notochord. By contrast, annelids and arthropods have their digestive tract along their backs and their main nerve cord along their bellies. Finally, many chordates have a long row of segmented V-shaped muscles known as *myomeres* that pull and flex the notochord and allow the side-to-side swimming motion found in nearly all fish. Last but not least, chordates differ from worms in that the digestive tract ends with an anus not at the very end of the body (as in worms and arthropods), but only partway back; the tail (composed of notochord and myomeres) usually extends behind the anus.

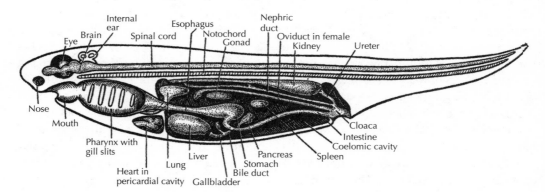

FIGURE 9.2. The basic organization of the chordate body plan. The front part of the body has the sense organs (eyes, nose) and a mouth with a pharyngeal basket for filtering out food and oxygen. The nerve cord and notochord run along the back, while the digestive tract runs along the belly. The anus is not at the rear tip of the body, but midway down the body, with a long tail and V-shaped segmented muscles (myomeres) running the length of the body. (After Romer 1959; reprinted with permission of the University of Chicago Press)

With these basic parts of the chordate body plan outlined, we can now look at the steps that produced a vertebrate from a nonvertebrate deuterostome (fig. 9.3). The most primitive living relatives of the chordates belong to a closely related phylum, the Hemichordata ("half chordates"). Today these include the acorn worms (fig. 9.3) and a group of plankton feeders known as pterobranchs that bear little resemblance to vertebrates. Acorn worms are known from about 80 species that live in U-shaped burrows in the sand and use their muscular proboscis to dig and the collar behind it to trap food particles as they burrow. Pterobranchs, on the other hand, are tiny colonial animals that live on long ringed tubes, and the animal consists largely of a U-shaped digestive tract with a fanlike filter-feeding device at one end. To the casual person walking among the tide pools or the beach sand, neither of these creatures resembles a fish, let alone a human. Yet careful examination of these animals reveals important clues. Although hemichordates do not yet have a notochord, they have the embryonic precursor of the notochord. In addition, both groups have a true pharynx, which occurs in no other group but chordates and their relatives. Finally, they have nerve cords along the back, and the digestive tract along the belly, a configuration that occurs elsewhere only in chordates.

There is also a lot of embryological evidence that hemichordates are our closest relatives. Their distinctive tornaria larva is nearly identical to the larvae of primitive chordates and also very similar to some echinoderm larvae. All the recent molecular analyses consistently show hemichordates as our closest relative other than echinoderms, or slightly closer to echinoderms but also clustered with chordates. Finally, acorn worms don't fossilize, but the extinct relatives of the pterobranchs, known as graptolites, are extremely common in early Paleozoic rocks. Once again, we have convergence of evidence from anatomy, embryology, paleontology, and molecular biology that points to one conclusion. Although we may not like to think of ourselves as having evolved from a creature like the acorn worm or pterobranch, that's where the evidence leads.

How do we get to the next stage? According to one hypothesis (fig. 9.4), the ancestral larvae that we chordates share with echinoderms developed into the filter-feeding pterobranchs (much like the primitive filter-feeding echinoderms). By retaining the embryonic stages of pterobranchs, the filtering arms were lost, and an acorn worm develops from the embryo instead. The next step is known as the "sea squirts" or the tunicates (fig. 9.5), which today are represented by over 2,000 species in the ocean, although they are so tiny and translucent that most people never see one. These delicate little blobs of jelly hardly resemble us, or even a fish for that matter. As adults, they are shaped like a little sac, with an opening at the top through which water is sucked in, then filtered through a basketlike pharynx, and finally out the little "chimney" on the side of their body. The adult sea squirt doesn't suggest much about chordates at all, although the pharynx is a clue. But the best evidence comes from their larvae (fig. 9.5A, left diagram), which look nothing like the adult but instead a lot like a fish or a tadpole. The sea squirt larva has a well-developed notochord, a muscular tail with paired myomere muscles, a nerve cord on the back, and a digestive tract along the belly. This peculiar larva swims around looking for a good rocky surface on which to land. Using the adhesive pad on its snout, it attaches and within 5 minutes the tail begins to degenerate. About 18 hours later the metamorphosis into the adult sea squirt is complete.

The adult sea squirt, of course, is too specialized to have had much to do with our ancestry, but the larva is a different matter. Through a mechanism like neoteny (discussed in chapter 3), the next stage of evolution of chordates would come not from the adults but

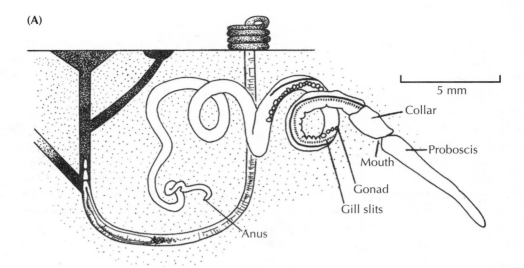

(A)

5 mm

Collar

Proboscis

Mouth

Gonad

Gill slits

Anus

(B)

FIGURE 9.3. The hemichordates include the pterobranchs and the acorn worm (shown here). This creature looks superficially wormlike but has the chordate feature of a pharynx with gill slits, and the precursors of a notochord, as well as a dorsal nerve cord and ventral digestive tract (the reverse of all other true "worms"). Behind the acorn worm is a sketch of its burrow, and the castings that it leaves outside the tail end of the burrow. (Part (A) redrawn after Barnes 1986; (B) courtesy Wikimedia Commons)

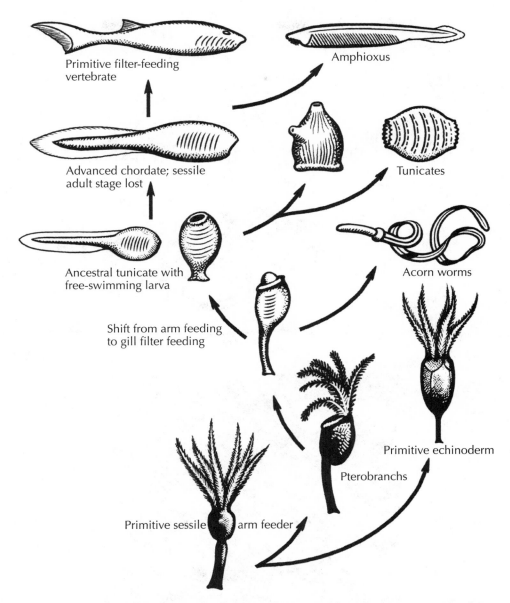

Primitive filter-feeding
vertebrate

Amphioxus

Advanced chordate; sessile
adult stage lost

Tunicates

Ancestral tunicate with
free-swimming larva

Shift from arm feeding
to gill filter feeding

Acorn worms

Primitive echinoderm

Pterobranchs

Primitive sessile   arm feeder

FIGURE 9.4. A diagrammatic family tree showing how chordates evolved from more primitive forms, as postulated by Walter Garstang and Alfred S. Romer. In many cases, such as the transition from hemichordates to tunicates ("sea squirts"), or tunicates to higher chordates, the larval form with its swimming tail enables them to escape the dead end of their highly specialized adult body forms. (After Romer 1959, redrawn by Carl Buell)

from organisms that retain the features found in the juveniles but never metamorphose into adults. And indeed the next step (fig. 9.4) is not much more advanced than the sea squirt larva. Known as the lancelet or amphioxus, it looks more and more like the primitive jawless fish (fig. 9.6). This little sliver of flesh is usually only a few inches in length and swims much like an eel, although it does not have a true head, jaws, teeth, or bones. Adult lancelets

(A)

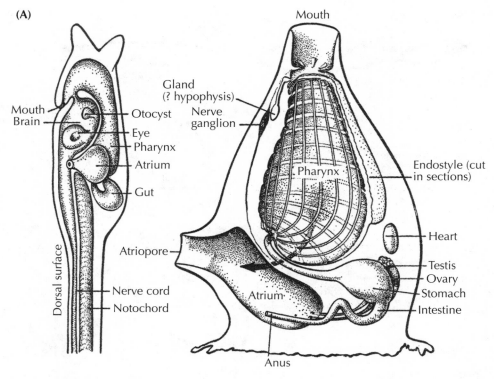

Mouth

Gland
(? hypophysis)

Nerve
ganglion

Mouth
Brain

Otocyst

Eye

Pharynx

Atrium

Gut

Pharynx

Endostyle (cut
in sections)

Dorsal surface

Atriopore

Atrium

Nerve cord

Notochord

Heart

Testis
Ovary
Stomach
Intestine

Anus

(B)

FIGURE 9.5. (A) The tunicates, or "sea squirts," have adult body forms (right diagram) that look nothing like chordates. However, the larvae (left diagram) are free-swimming tadpole-like creatures with tails and a pharynx, that allowed them to escape their adult dead-end body form and evolve into higher chordates. (After Romer 1959, redrawn by Carl Buell) (B) Photograph of the adult tunicate in feeding position. Courtesy Wikimedia Commons)

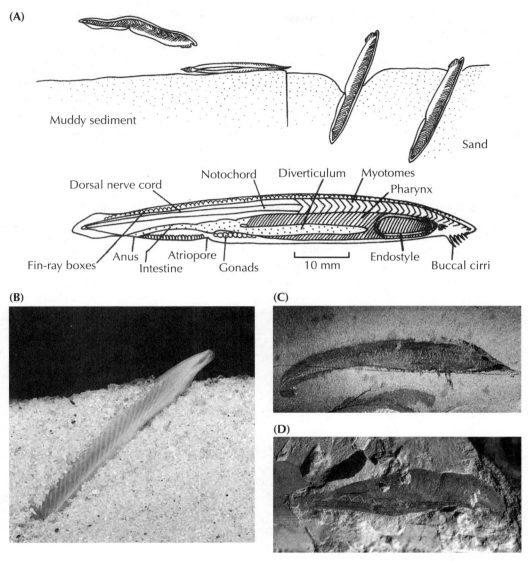

FIGURE 9.6. (A) The lancelets, or amphioxus (*Branchiostoma*) are the most fishlike and specialized of the nonvertebrate chordates. They have a long eellike body with muscles running down the entire length and a notochord supporting the entire body, but the mouth is still a simple filter-feeding pharynx. In life (top diagram) they embed themselves in the sediment with their heads protruding, catching tiny food particles with their mouth filter feeding in the current. (After Barnes 1986, drawn by Carl Buell) (B) Photograph of the living lancelet. (From IMSI Photo Images, Inc.) (C) The middle Cambrian Burgess Shale fossil lancelet *Pikaia*. (D) The Early Cambrian Chengjiang lancelet fossil known as *Yunnanozoon*. ([C and D] photos courtesy D. Briggs)

(fig. 9.6A and B) burrow tail-first into the sandy sea bottom and then use the tentacles around the mouth and pharynx to filter feed. But the anatomy of these animals bears a remarkable resemblance to primitive fish. Lancelets have a well-defined notochord, a nerve cord along the back, a digestive tract along the belly, and many V-shaped myomeres down the length of the body. Their pharyngeal basket is well developed, with over a 100 "gill slits" like those of a primitive fish. In addition, they are the most primitive chordates to have a liver and a kidney, as well as other organ systems not found in hemichordates or sea squirts. They do not have true eyes, but they do have light-sensitive pigment spots on the front of the head to detect light and shadows. The molecular and embryological evidence consistently puts lancelets as our closest nonvertebrate chordate relatives. To top it off, even these delicate animals fossilize occasionally. The Burgess Shale produces a remarkable fossil known as *Pikaia* (fig. 9.6C), which is a very primitive relative of the lancelets. The Cambrian Chengjiang fauna of China yields another form known as *Yunnanozoon* (fig. 9.6D). And in rocks from the Permian of South Africa, there a fossil known as *Palaeobranchiostoma* that looks much like the living lancelet.

It's a Long Way from Amphioxus

Oh a fish-like thing appeared among the annelids one day
It hadn't any parapods nor setae to display
It hadn't any eyes or jaws or ventral nervous chord,
But it had a lot of gill slits and it had a notochord.

It's a long way from Amphioxus
It's a long way to us,
It's a long way from Amphioxus
To the meanest human cuss.
Well, it's good-bye to fins and gill slits, And it's welcome lungs and hair,
It's a long, long way from Amphioxus
But we all came from there.

My notochord shall change into a chain of vertebrae,
And as fins my metaplural folds shall agitate the sea;
My tiny dorsal nervous chords shall be a mighty brain,
And the vertebrae shall dominate the animal domain.

—Philip Pope, to the tune of "It's a Long Way to Tipperary"

From the primitive chordates like the Cambrian lancelets *Pikaia* and *Yunnanozoon*, the next evolutionary step is the jawless fish, which are known from both fossils and living forms. Two groups of living jawless vertebrates are known, and these give us many insights into the fossils. The more primitive of the two is the hagfish (fig. 9.7A), commonly known as the "slime eel" because it can produce copious amounts of mucus to escape predators. Hagfish burrow on the seafloor, slurping up worms, and wriggle into the bodies of dead and dying fish, eating them from the inside out with their rasping teeth. Hagfish are the most primitive chordates that have a definite head region with a brain, sense organs (eyes, nose, and ears), and a full skeleton made of cartilage, not just a notochord. They also have a two-chambered heart and distinctive cells in embryology known as neural crest cells, which are crucial in vertebrate development. But hagfish lack more advanced features, such as bone, red blood cells, a thyroid gland, and many other characters found in the other living jawless vertebrate, the lamprey (fig. 9.7B). Although lampreys look superficially like eels, they are jawless.

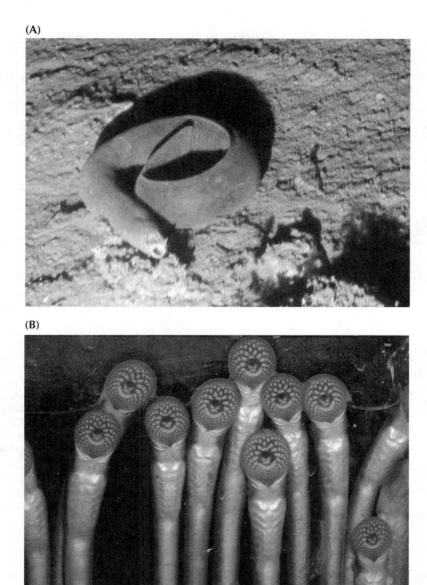

FIGURE 9.7. Two examples of living jawless craniates. (A) The living hagfish, *Myxine*. (Photo courtesy NOAA) (B) The living lamprey, shown here clinging to glass with its sucker-like mouth and displaying its rasping teeth, which it uses to eat a hole in the side of its prey. (Photo courtesy J. Marsden)

They live as parasites by attaching to the side of a fish with their suction cup mouth and using their rasping teeth to suck the fluids out of their host.

These two unsavory characters may not be our favorite cousins, but they are the only jawless vertebrates alive today. However, the fossil record shows that jawless vertebrates had quite a successful run. We can trace them back to the Cambrian, where we find soft-bodied impressions in China (Shu et al. 1999) that have been named *Myllokunmingia*, *Haikouella*, and *Haikouichthys* (figs. 9.8 and 9.9). These recent discoveries push the earliest vertebrates all

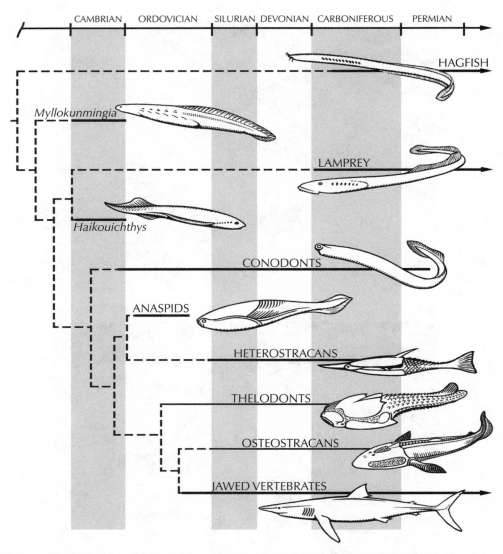

FIGURE 9.8. The evolutionary history of the earliest vertebrates, showing the evolutionary position the new Chinese Cambrian fossils *Myllokunmingia* and *Haikouichthys* with respect to other jawless and jawed fish. (Redrawn by Carl Buell, based on Shu et al. 1999)

the way back to the Early Cambrian, much earlier than previous fossils (which were based on fragments of dermal armor known from the Late Cambrian). Through the rest of the Cambrian and Ordovician, we see nothing more in the fossil record of vertebrates other than isolated plates made of true bone and the microscopic toothlike structures known as conodonts, so apparently the earliest vertebrates remained small and completely soft-bodied for some time. Then in the Early Silurian, about 430 million years ago, we find the first nearly complete armored jawless fish, and this group radiates into an explosion of diversity by the Late Devonian. This Devonian radiation of jawless fish (fig. 9.1) was spectacular, with a wide variety of armored fish that still lacked jaws or a muscular bony skeleton, but nonetheless they were covered in solid bone all over their bodies; some had large curved head shields

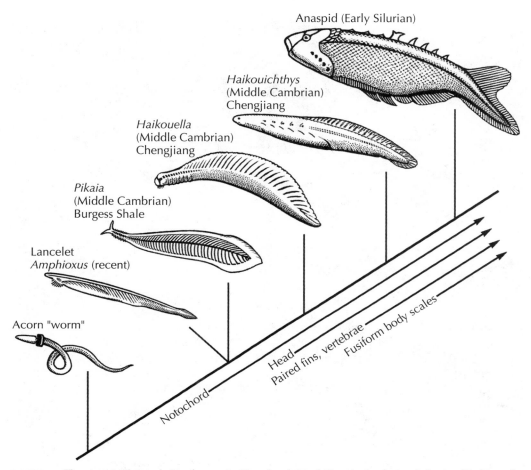

FIGURE 9.9. The steps in the evolution from primitive chordate relatives to the first vertebrates. (Redrawn by Carl Buell, based on Shu et al. 1999)

to protect them and "chain mail" down to their tails. None, however, had a strong pectoral or pelvic fin for steering, as do most modern fish, because they lacked bone to support the fin. Instead, armor covered much of their bodies, although no one is sure what predator they needed all that armor for. Some of these jawless fish, the cephalaspids or ostracoderms, had a large horseshoe-shaped head shield with a flat bottom, and apparently filtered out food from the seafloor. The heterostracans, thelodonts, and anaspids, on the other hand, had simple slit-like mouths and tails with the main lobe pointed downward to keep their heads up while they swam. These fish apparently sucked water through their mouths and filtered it through the gills, as many living fish do today.

Thus, from the primitive acorn worms, we can trace a series of transitional forms that become more and more like vertebrates. Add the notochord and we have lancelets (fig. 9.9). Add the head and the paired fins and we have the early Cambrian Chinese forms *Haikouella* and *Haikouichthys* (figs. 9.9 and 9.10). Finally, by making the body more streamlined and adding bony scales, we have a jawless fish (fig. 9.8). Thus the transition from invertebrates to vertebrates is documented not only by living fossils but also by an increasingly good fossil record.

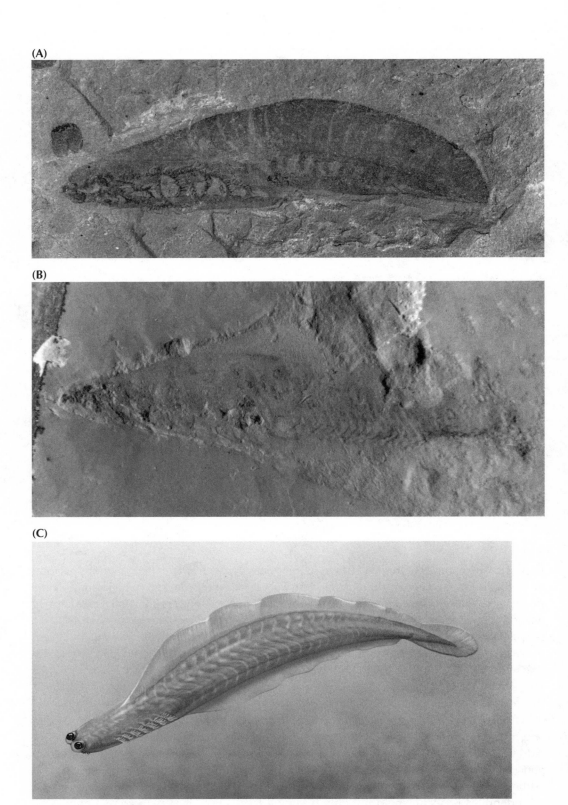

FIGURE 9.10. Early Cambrian soft-bodied vertebrates, the earliest known vertebrate fossils. (A) *Haikouichthys*. (B) *Haikouella*. (Photos courtesy D. Briggs and Jun Yuan Chen) (C) A reconstruction of *Haikouichthys* by Carl Buell.

Jaws: The Evolutionary Story

The term "fish" is of value on restaurant menus, to anglers and aquarists, to stratigraphers and in theological discussions of biblical symbolism. Many systematists use it advisedly and with caution. Fishes are gnathostomes that lack tetrapod characteristics. We can conceptualize fishes with relative ease because of the great evolutionary gaps between them and their closest living relatives, but that does not mean they comprise a natural group. The only way to make fishes monophyletic would be to include tetrapods, and to regard the latter merely as a kind of fish. Even then, the term "fish" would be a redundant colloquial equivalent of "gnathostome" (or "craniate," depending upon how far down the phylogenetic ladder one wished to go).

—John Maisey, "Gnathostomes"

One of the great evolutionary breakthroughs in vertebrate history was the origin of jaws. Before jaws appeared, vertebrates were severely limited in what they could eat (mostly filter feeding or deposit feeding or living as parasites like lampreys and hagfish), and thus in their lifestyles and body size. Jaws allowed vertebrates to grab and break up a food item, which in turn meant they could eat a wide variety of foods, from other fish to plants to mollusks and so on. This then allowed vertebrates to evolve into a great many different ecological niches and body sizes, including superpredators that ate all other kinds of marine life. Eventually, vertebrates used their jaws and teeth for many other things besides eating, including manipulating objects, digging holes, carrying material to build nests, carrying their young around, and making sound or speech.

Once jaws appeared in the Silurian, there was a tremendous evolutionary radiation (fig. 9.1) of different kinds of jawed vertebrates, or more properly, *gnathostomes* (which means "jawed mouth" in Greek). As the quote from John Maisey points out, we have been accustomed to using the word "fish" to describe most of these vertebrates, but that word has no meaning in systematics. Fish are simply vertebrates or gnathostomes that live in the water and are not land animals, or tetrapods. The group is ecological and paraphyletic and not a natural taxon whatsoever. However, for the purposes of this book, we will keep the terminology simple, recognizing that *fish* is a label of convenience and not some real biological entity.

The radiation of gnathostomes began with the dominant group in the Devonian, the extinct *placoderms* (fig. 9.1). These creatures had thick bony plates covering their head and shoulders, but the rest of the skeleton was made of cartilage. Some of them had sharp biting plates on the edge of their mouth shields and reached up to 10 meters (30 feet) long, the largest predator the world had seen up to that time. Others were weighted down with armor over the entire front half of their body (including jointed armor on their pectoral fins that resembled crab legs) and apparently ate small slow prey on the sea bottom. Still others developed flat bodies like rays and skates. All of these different body shapes evolved rapidly in the Devonian and then vanished at the end of that period.

The next group to branch off the family tree of gnathostomes (figs. 9.1 and 9.11) were the sharks, or chondrichthyans ("cartilaginous fish"). We think of the terrors of the movie *Jaws* or the documentaries that feature sharks attacking divers, but sharks are actually much more complex and interesting than that. Most are highly effective predators with rows of razor-sharp teeth, but the largest sharks (whale sharks, megamouth sharks, and basking sharks) have tiny

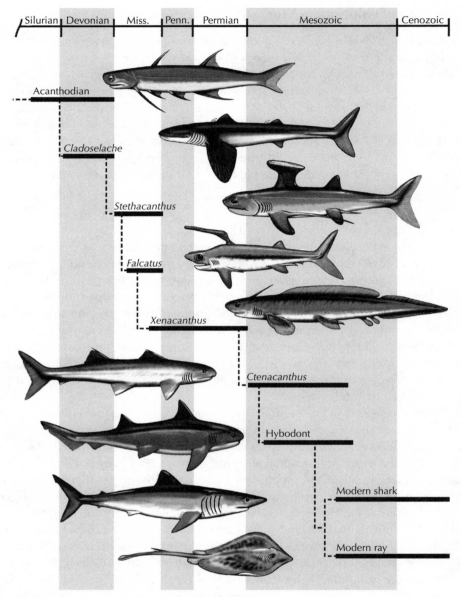

| Silurian | Devonian | Miss. | Penn. | Permian | Mesozoic | Cenozoic |

Acanthodian

Cladoselache

Stethacanthus

Falcatus

Xenacanthus

Ctenacanthus

Hybodont

Modern shark

Modern ray

FIGURE 9.11. The evolution of the sharks. (Drawing by Carl Buell)

teeth and feed on plankton by filtering them through their gills. Other sharks are specialized mollusk eaters with crushing teeth, especially the flat-bodied skates and rays. Because shark skeletons are made of cartilage and do not fossilize well, we know them primarily from their teeth (which are made of bone and enamel) and often from the bony spines that many sharks had in their bodies and fins. Although the basic shark design has been successful for over 400 million years, sharks show considerable evolutionary change through time, contrary to the creationist books and websites (fig. 9.11). The earliest sharks, the cladodonts from the Devonian, had a large, very primitive "skull" (actually a cartilaginous precursor of the skull called a chondrocranium), very broad-based stiff pectoral fins, thick spines in front of their dorsal fins, and a rigid tail that was almost symmetrical. In the Mesozoic, the hybodont sharks are considerably more advanced, with a reduced, more flexible chondrocranium, pectoral fins with narrower bases that allowed more maneuverability, more specialized teeth, and a more flexible tail for powerful swimming. These trends are all continued in the living sharks, or the Neoselachii, which have a chondrocranium that is highly reduced, so they can protrude their upper and lower jaws very far; highly maneuverable pectoral fins; a wide variety of tooth types; and an even more flexible tail. Sharks may have been around for a long time, but they show considerable evolution during this history, and any creationist who claims otherwise has no familiarity with the real fossil record of sharks.

After the sharks and rays split off the family tree (fig. 9.1), the next groups up the clado-gram are all known as *osteichthyans*, the "bony fish." They break into two main groups: the "lobe-finned fish," which include lungfish, coelacanths, and, of course, the tetrapods, to be dis-cussed in the next chapter; and the "ray-finned fish," so called because they support their fins with many long bony rays. Ray-finned fish make up about 98 percent of the species of living fish; only the hagfish, lamprey, lungfish, coelacanth, and chondrichthyans are not members of this group. Like the other main groups of fishes, ray-finned fish appeared in the Devonian, but they have been evolving rapidly ever since then, with hundreds of genera and thousands of species known from both the fossil record and the living world (figs. 9.12 and 9.13). As with sharks, the lineage has been around for 400 million years, but they show remarkable changes over that long history and major changes in the ways in which they feed and swim.

The earliest (mostly Paleozoic) ray-finned fishes (known by the paraphyletic waste-basket name "chondrosteans") have heavy bone surrounding the head region and simple "snap-trap" jaws with limited flexibility and limited room for muscles that close them shut. Although they have a lot of bone in their skeletons, large parts are also made of cartilage,

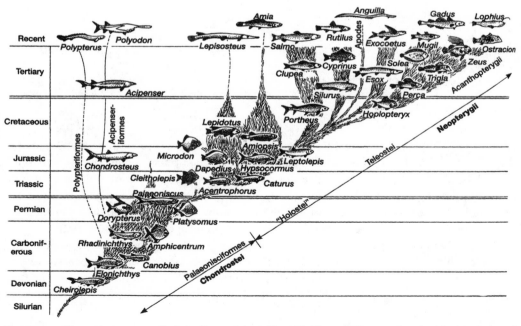

FIGURE 9.12. The evolutionary radiation of bony fishes. (From Kardong 1995; reproduced with permission of the McGraw-Hill Companies)

(A)

(B)

(C)

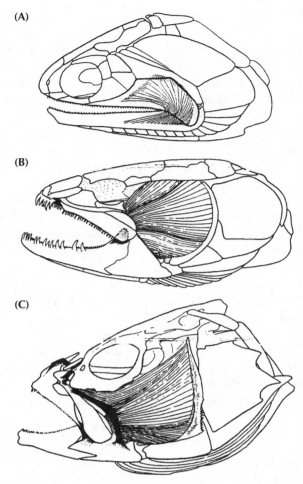

FIGURE 9.13. The transition from (A) primitive bony fish ("palaeoniscoids") with simple robust "snap-trap" jaws and heavy bones throughout their skulls to (B) more advanced "holostean" fish (such as the bowfin *Amia*, shown here), with more protrusible jaws and less ossified skulls, to (C) the modern teleost fish, with a very lightly ossified skull and a protrusible jaw for sucking down prey. (From Schaeffer and Rosen 1961; used by permission)

another primitive sharklike feature. Their bodies are also covered with heavy rhombohe-dral scales, and their tails are very sharklike in having the upper lobe much larger than the lower lobe. By the Mesozoic, these primitive groups had mostly died out with the exception of a few living fossils, such as the sturgeon and the paddlefish. In their place came another great radiation of more advanced ray-finned fish (known by the paraphyletic wastebasket name "holosteans"), which can easily be distinguished from their primitive relatives. Their skulls are still made of fairly solid bone, but the upper jawbones (premaxillary and maxillary bones) are hinged at the front of the skull, allowing them to open their mouths wider and grab larger prey. The back part of the skull is also less solidly bony and more open, so they have enlarged jaw muscles for a stronger bite. Unlike chondrosteans, the holosteans have almost no cartilage in their skeletons but are completely bony. Their scales are thinner and smaller, so they are not as heavily armored as primitive ray-finned fish. The tail is nearly symmetrical, but the vertebral column does curve upward within the upper lobe of the tail (fig. 9.13). Most of these Mesozoic fish groups have died out, but there are a few survivors, such as the garfish and the bowfin.

The final step in fish evolution (fig. 9.12) is the great radiation of the teleost fishes, which make up 98 percent of all the fishes alive today. Nearly every fish you eat or see in the aquarium or in the lakes, rivers, or oceans is a teleost. There are about 20,000 species of teleosts, more than all the amphibians, reptiles, birds, and mammals combined. Teleosts branched off from the more primitive fishes during the Cretaceous and then underwent explosive evolution into hundreds of different families, most of which are still alive today. We mammal chauvinists like to think of the last 65 million years as the "age of mammals," but in terms of diversity, the teleosts were evolving far faster than the mammals, and we could easily think of it as the "age of teleosts."

Teleosts are easily distinguished from the more primitive chondrosteans and holosteans (fig. 9.13). Most have greatly reduced the bone in their skulls, so their heads are supported by a framework of thin bony braces and struts connected by muscles and tendons, not solid walls of bone found in the primitive forms. In particular, the bones of their mouth are much reduced and connected by flexible tendons, so they can protrude and open very easily. Many teleosts have abandoned the old snap-trap jaw mechanism of catching prey by biting with jaws and teeth. Instead they have mouths that open suddenly and create suction, slurping down their prey. (The next time you feed fish flakes to your aquarium fish, notice how they protrude their mouths and suck the food in and do not bite it). Teleosts also continue the trend of reducing bone in the rest of their skeletons as well, so that most of their bones are very light and delicate. Finally, teleosts have a completely symmetrical tail, with only a tiny trace of the upward flexure of the spine near the base.

In summary, vertebrates have come a long way from acorn worms, sea squirts, lancelets, and lampreys to the incredible array of teleosts in the waters of the world. This short chapter does not do justice to their long and incredibly rich evolutionary story. I strongly recommend that you read further on this topic if you are interested.

For Further Reading

Benton, M. J. 2014. *Vertebrate Palaeontology*. 4th ed. New York: Wiley-Blackwell.

Carroll, R. L. 1988. *Vertebrate Paleontology and Evolution*. New York: Freeman.

Forey, P., and P. Janvier. 1984. Evolution of the earliest vertebrates. *American Scientist* 82:554–565.

Gee, H. 1997. *Before the Backbone: Views on the Origin of Vertebrates*. New York: Chapman & Hall.

Long, J. A. 2010. *The Rise of Fishes*, 2nd ed. Baltimore, Md.: Johns Hopkins University Press.

Maisey, J. G. 1996. *Discovering Fossil Fishes*. New York: Holt.

Moy-Thomas, J., and R. S. Miles. 1971. *Palaeozoic Fishes*. Philadelphia: Saunders.

Norman, J. R., and P. H. Greenwood. 1975. *A History of Fishes*. London: Ernest Benn.

Pough, F. H., C. M. Janis, and J. B. Heiser. 2002. *Vertebrate Life*. 6th ed. Upper Saddle, N.J.: Prentice Hall.

Prothero, D. R. 2013. *Bringing Fossils to Life: An Introduction to Paleobiology*. 3rd ed. New York: Columbia University Press.

Schaeffer, B., and D. E. Rosen. 1961. Major adaptive levels in the evolution of actinopterygian feeding mechanisms. *American Zoologist* 1:187–204.

Shu, D.-G., H.-L. Luo, S. Conway Morris, X.-L. Zhang, S.-X. Hu, L. Chen, J. Han, M. Zhu, Y. Li, and L.-Z. Chen. 1999. Lower Cambrian vertebrates from China. *Nature* 402:42–46.

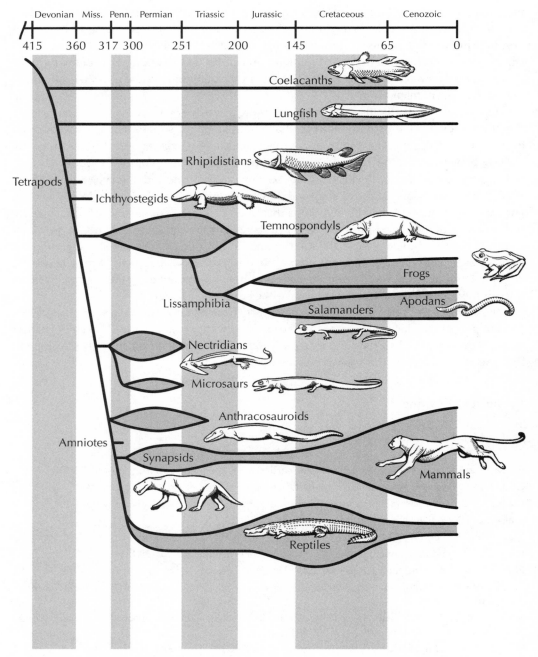

FIGURE 10.1. The relationships of the four-legged animals, or tetrapods. (Drawing by Carl Buell)

FISH OUT OF WATER 10

The Great Leap Upward

What creationists challenge evolutionists to show them, it seems, is a "perfect 10" transitional form, exactly halfway between, say, fish and amphibian. But no such "fishibian," says the Institute for Creation Research (ICR), has ever been found in the fossils.

—Ronald Ecker, *Dictionary of Science and Creationism*

We come now to one of the classic transitions in all of evolution: How did the aquatic vertebrates crawl out on land and become four-legged (*tetrapod*) terrestrial animals? This subject has intrigued paleontologists and biologists for over a century, and naturally plenty of controversy and many mistakes and false leads have occurred (as in any area of science exploring a difficult topic). Creationists, of course, cannot allow themselves to admit that this transition ever occurred, so they attack it with vigor, mostly by citing out-of-date sources (even in more recent books like Gish 1995) and ignoring all the evidence that doesn't fit their point of view. But here the creationists have been left in the dust. Dramatic new discoveries in the past 30 years have completely revolutionized what we once thought about how this transition occurred. You can take the creationist publications and wrap dead fish with them, because they have been completely debunked by what we have learned in the past decade. We don't have every possible transitional form between fishes and tetrapods, but we now have so many steps in the sequence that to deny that this transition occurred is like the neo-Nazis denying the Holocaust—it's a self-evident fact, and there are many fossil witnesses to bear testimony.

Before we discuss this transition in detail, a few semantic issues (some of which are exploited by creationists) need to be clarified. The old Linnaean scheme of animal classification divided the vertebrates into several obvious groups: "fishes," "amphibians," "reptiles," and so on. We are all taught from a young age that amphibians are animals (such as frogs and salamanders) that live both in water and on land (*amphibian* literally means "living both lives" in Greek). But in the context of modern phylogenetic or cladistic classification, *natural groups must include all their descendants*. The lineage that leads to reptiles evolved from one group of amphibians, so amphibians must either include all vertebrates with four legs (tetrapods), or else the amphibians are a paraphyletic "grade" of evolution, halfway between reptiles and fish (fig. 10.1). To get around this problem, most modern cladistic classification schemes do not use the antiquated word "amphibian" anymore but instead use the natural monophyletic group known as tetrapods (all four-legged land animals and their relatives). If you wanted to use "amphibian" as a concept, the three living groups, or "Lissamphibia" (frogs and toads; salamanders; and the apodans, a legless group from the tropics) might be a natural monophyletic clade, and the term could be used there. But then that leaves all the various fossils that have been called amphibians out in the cold. Some extinct groups (fig. 10.1), such as the temnospondyls, may be closely related the living groups and are thus would be true amphibians.

Others, such as the lepospondyls, may or may not be related to the temnospondyl-lissamphibian clade, so their inclusion is more questionable. And then there are the "anthra-cosaurs" (fig. 11.3), a grade of tetrapods that are the sister taxa to the reptiles (more properly, amniotes, as the next chapter will discuss). They don't share the characters that define rep-tiles or amniotes, so paleontologists have traditionally left them in the amphibian waste-basket with the other tetrapods that are not amniotes. In this chapter, we will not use the term "amphibian" further, but will stick with the clumsy but accurate term "non-amniote tetrapod" when we mean frogs and salamanders and what most people call amphibians.

Lobe Fins Lead the Way

Then I turned the page and saw the sketch, at which I stared and stared, at first in puzzle-ment, for I did not know of any fish of our own or indeed of any seas like that; it looked more like a lizard. And then a bomb seemed to burst in my brain, and beyond that sketch and the paper of the letter I was looking at a series of fishy creatures flashed up as on a screen, fishes no longer here, fishes that have lived in dim past ages gone, and of which often only fragmentary remains in rocks are known. I told myself sternly not to be a fool, but there was something about the sketch that seized on my imagination and told me that this was something far beyond the usual run of fishes in our seas. . . . I was afraid of this thing, for I could see something of what it would mean if it were true, and I also realized only too well what it would mean if I said it was what it was not.
—J. L. B. Smith, *Old Fourlegs: The Story of the Coelacanth*

On December 23, 1938, one of the most remarkable scientific discoveries of the past century was made in the mouth of the Chalumna River near East London off the coast of South Africa. Pulled out of a net from the trawler *Nerine* was a huge (almost 1.5 meters [or 5 feet] long, and weighing 58 kilograms [or 127 pounds]) shiny silvery-blue fishlike creature, the likes of which no fisherman had ever seen before (fig. 10.2). The local museum curator, Mar-jorie Courtenay-Latimer, was called and immediately realized it was something of great sci-entific importance, a new species never before caught in the waters off South Africa. As she wrote later, it was "the most beautiful fish I had ever seen, five feet long, and a pale mauve blue with iridescent silver markings." Unfortunately, it was already dead and beginning to rot rapidly in the hot austral summer weather. She did her best to preserve it, but it was so large and rotting so fast that she eventually had to discard most of the innards and saved only the skin. She then sent a sketch of it with her letter to the foremost authority on South African fishes, James Leonard Brierly Smith, whose reaction when he opened her letter and saw her sketch is given in the quote that opens this section. He finally got to see the specimen on January 3, 1939, and as he later wrote,

Coelacanth—yes, God! Although I had come prepared, that first sight hit me like a white-hot blast and made me feel shaky and queer, my body tingled. I stood as if stricken to stone. Yes, there was not a shadow of a doubt, scale by scale, bone by bone, fin by fin, it was a true coelacanth. It could have been one of those creatures of 200 mil-lion years ago come alive again. I forgot everything else and just looked and looked, and then almost fearfully went close up and touched and stroked. (Smith 1956:73)

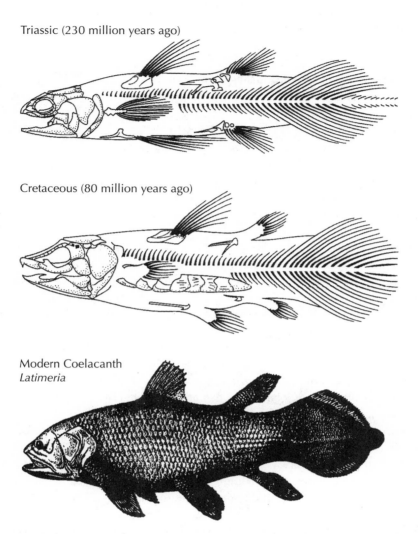

Triassic (230 million years ago)

Cretaceous (80 million years ago)

Modern Coelacanth
Latimeria

FIGURE 10.2. Evolutionary transformation series of the coelacanths, from the primitive Triassic form, which looks much like other early sarcopterygians, to the highly specialized living fossil *Latimeria* (bottom drawing). Even though there is dramatic change in shape, these animals still retain the hallmarks of coelacanths, including the extra lobed fin in the end of the tail, the triangular opercular bone covering the gills, and the distinctive shape of the lobed pectoral, pelvic, and anal fins, but ray-finned dorsal fins. (From Clack 2002; used with permission)

Smith named this astounding find *Latimeria* (after the discoverer) *chalumnae* (after where it was found), and it was the sensation of the scientific world in 1939. After 13 hard years of searching, however, Smith and the fishermen of South Africa had yet to find another and were beginning to despair. So much crucial information had been lost when its guts had been discarded! They sent out a "wanted" poster with a photograph of the fish and a £100 reward and circulated them all over the African coast. Then, in 1952, a lucky break occurred.

Another fisherman, Eric Hunt, had distributed Smith's reward poster up and down the East African coast, and a local fisherman in the tiny Comoros Islands north of Madagascar had found another coelacanth. Soon the fish was flown back to South Africa on the orders of the prime minister himself, and it was well enough preserved that the internal organs had survived.

Since 1952, more than 100 additional specimens have been hauled out of the deep waters around the Comoros and again in South Africa, and a few years ago, another new species was discovered in Indonesia. It turns out that coelacanths live in very deep waters and only come to the surface in the dark of night, which is why they had gone undetected for so long. Unfortunately, they are now so valuable that local fishermen may be hunting them back into extinction, barely more than 75 years since this living fossil was first discovered. Coelacanths have long been known from the fossil record (fig. 10.2), but the last known fossil of a coelacanth dated back to the Cretaceous, while the dinosaurs still roamed the earth. No wonder the world was so astonished to find an animal living that had been thought to be extinct for at least 70 million years.

Coelacanths are part of a group of vertebrates known as "lobe-finned fishes" or Sarcopterygii, which includes not only the coelacanths but also the lungfishes and a number of extinct forms known by the paraphyletic grouping "rhipidistia," as well as their descendants, the tetrapods. Both lobe fins and ray fins share a common ancestor in the Devonian (fig. 9.1). Recent DNA sequencing has shown that coelacanths and lungfish are more closely related to each other than either is to tetrapods. While the ray-finned fish now dominate the world's waters, lobe fins (once common in the Paleozoic and early Mesozoic) have since been reduced to a tiny remnant: the living coelacanths, three genera of lungfish, and, of course, tetrapods. Instead of fins supported by numerous bony rays (as in the fish described in the last chapter), the fins of lobe-finned fish are supported by robust bones and muscles forming a lobe, which is then surrounded by fin rays.

The first living lobe fin to be discovered and described scientifically was the South American lungfish, *Lepidosiren*, which although highly specialized with whiplike fins, has lungs rather than gills. In 1837, a specimen of the African lungfish, *Protopterus* (fig. 10.3), came to the attention of Richard Owen, England's foremost anatomist and paleontologist. As he dissected it, he came across the undisputed evidence that this fish had lungs, although his creationist leanings refused to admit that this gave it tetrapod affinities. Its teeth consisted of large ridged biting plates, which had been known from fossils for years, and here was the source of those mysterious fossils. The clincher came when the Australian lungfish, *Neoceratodus*, was discovered. Not only did it have lungs, but its fins were not as highly modified as those of the African and South American species but still showed the classic lobed form. Since then, many fossil lungfish have been found, and they look much more like the earliest coelacanths (fig 10.2) and the earliest "rhipidistians," so the peculiarities of the living lungfishes and coelacanth disappear as you go back in time.

Further study of the living lobe fins reveals another striking characteristic. Not only are their limbs constructed with robust bones and muscles (like those of a tetrapod) rather than thin bony fin rays (like most fish), but they also use them differently than most living fish. Studies of the fin motion of both coelacanths and lungfish have shown that they move their fins in a "step cycle" similar to the motion of the four limbs of tetrapods. Thus the characteristic leg motion sequence of four-legged animals was already present in lobe fins that never walked on land.

This highly specialized anatomy of the modern forms confuses creationists. They point to the peculiarities of living *Latimeria* or the specialized fins of some lungfish and argue

Protopterus

Neoceratodus

Fleurantia

Dipterus

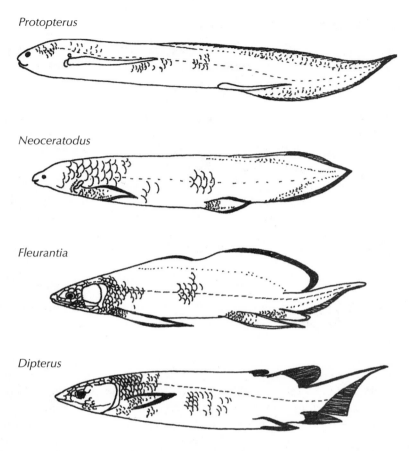

FIGURE 10.3. Evolutionary transformation series of lungfish, from the primitive Devonian fossil *Dipterus*, which closely resembles the earliest coelacanths and rhipidistians, to the highly specialized living forms. (Reproduced by permission of the Royal Society of Edinburgh and P. Ahlberg; from Ahlberg and Trewin 1995)

that they cannot have been ancestors of tetrapods. But once again, they are thinking of ladders when we are talking about branching bushes. None of the living species are ancestral to tetrapods, and no paleontologist ever made that claim. The lungfish, coelacanths, and rhipidistians are distinct side branches, or sister taxa, to the tetrapods, and they share many unique anatomical characters that support that relationship, but they branched off back in the Devonian and have had their own unique history ever since then (fig. 9.1). We can see how close they once were when we compare Devonian fossil forms (fig. 10.3), and it is irrelevant that the peculiar modern descendants have specializations that are not found in tetrapods. Evolution is a bush, not a ladder, and organisms continue to evolve and change, long after they have branched off from their sister groups that lead to humans. It's hard to tell whether creationists can't get this point straight because they are clueless, misinformed, or just don't want to face the truth.

We will leave lungfish and coelacanths aside for now and focus on the fossils that document the transformation from rhipidistians to tetrapods.

Four on the Floor

In my days it was believed that the place for a fish was in the water. A perfectly sound idea, too. If we wanted fish, for one reason or another, we knew where to find it. And not up a tree.

For many of us, fish are still associated quite definitely with water. Speaking for myself, they always will be, though certain fish seem to feel differently about it. Indeed, we hear so much these days about the climbing perch, the walking goby, and the galloping eel that a word in season appears to be needed.

Times change, of course—and I only wish I could say for the better. I know all that, but you will never convince me that a fish that is out on a limb, or strolling around in vacant lots, or hiking across the country, is getting a sane, normal view of life. I would go so far as to venture that such a fish is not a fish in his right mind.

—Will Cuppy, *How to Become Extinct*

As humorist Will Cuppy noted (in the preceding quote), we think of fish as belonging in the water and have trouble with the idea that they could crawl out on land. For years, evolutionary biologists and paleontologists have argued and speculated about the forces that led fishes to finally crawl out on land. The transition seems like a remarkable one. Water-living fishes needed to develop some way of breathing air, supporting themselves on land without the buoyancy of water, preventing their skins from drying out, seeing and hearing on land, and making many other physiological adjustments. Paleontologists used to speculate endlessly about why they made this apparently difficult trek. Some argued that it was to escape drying pools, and others suggested it was to escape predation in the crowded Devonian waters, or to take advantage of new food sources on land (since insects, spiders, scorpions, millipedes, and other arthropods had already colonized the land over 100 million years before vertebrates did). Unfortunately, much of this speculation was predicated on false assumptions and inadequate specimens, and most of it is irrelevant now.

It turns out that crawling up on the land is no big deal. Many different teleost fishes (with flimsy ray fins, not the robust lobe fins) do it all the time. A variety of tide-pool fishes, such as gobies and sculpins, spend a lot of time out of water when the tide goes out, hunting vulnerable prey. In the southeastern United States, the walking catfish (fig. 10.4A) is a legendary pest for its ability to wriggle from one pool of water to another, using only its ray fins for propulsion. Eels, too, are capable of wriggling across the ground for some distance in search of new pools of water. The climbing perch of Africa and Southeast Asia, *Anabas testudineus*, travels in search of water when its ponds dry up. It walks supported by the spiny edges of the gill plates and propelled by the fins and tail and can climb low trees. The most specialized and most amphibious of all the "land fish" is the mudskipper (fig. 10.4B), which is completely adapted to living on the land-water boundary. It even has its eyes up on periscopes so it can look above water while it is swimming. The mudskipper haunts the mudflats in mangrove swamps, catching prey in the mud. It uses its ability to wriggle on land or swim in water to escape predators coming from either direction and can also climb up the exposed roots of the mangrove trees.

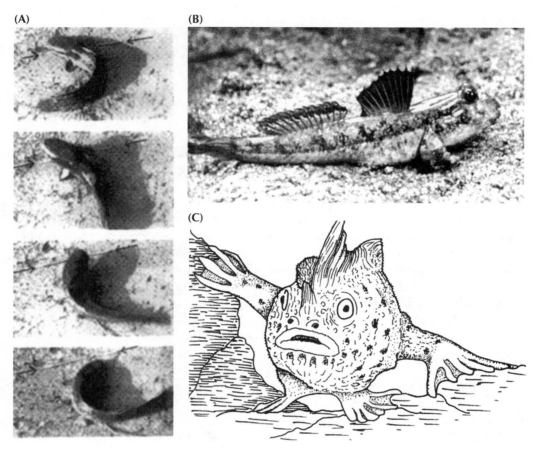

FIGURE 10.4. A number of ray-finned fishes have evolved the ability to live on land and crawl around, or they have modified their rayed fins into walking appendages for use in creeping along the seafloor. (A) The "walking catfish," which wriggles along the ground between ponds when its home pond dries up or becomes too crowded. (B) The mudskipper, which spends most of its life out of water sitting on the mudflats or mangrove roots. ([A] and [B] from Romer 1959) (C) The frogfish, which has modified its ray fins into "fingers" that enable it to creep along the bottom. (From Clack 2002: fig. 4.15; used with permission)

In 2014, scientists published a remarkable study (Standen et al. 2014). They took populations of the African ray-finned fish known as the bichir, or *Polypterus* (distantly related to sturgeons and paddlefish), and made them cross over dry land over and over again. Within 8 months, the bichir that had had been raised on land had modified their ray fins and the muscles to control them so they were more efficient walkers; those that were kept in their normal water habitat did not change. The remarkable footage of these walking fish can be seen online by searching for videos of "walking fish Polypterus."

None of these fishes are perfectly adapted for land, but are "jury-rigged" to live on land just long enough to accomplish certain tasks. They must inhabit humid climates and remain close to water so they don't dry out, and they return to the water frequently to restore their water balance. They don't have lungs but make do with their gills, swim bladders, and the moisture in the air to breathe for extended periods of time without lungs. They don't have robust lobed fins that could become legs and feet, but do the best they can with the relatively flimsy ray fins to push themselves along on their bellies and wriggle across the ground. In fact, there are a variety of completely aquatic teleosts, such as the fingered dragonets, genus *Dactylopus*, and the grunt sculpin, *Rhamphocottus richardsoni*, that have modified their ray fins into separate "fingerlike" features that allow them to creep underwater along the sea bottom in a motion that resembles a spider or a lobster. However, these "fin fingers" are not robust or muscular or as flexible as tetrapod fingers, so they cannot manipulate objects with them. They are jury-rigged features built out of another structure (ray fins) and suboptimally modified to be "semifingers" (yet another blow to "intelligent design"). The frogfish, *Antennarius*, uses its fins with the fingerlike fin rays to walk along the bottom in a motion very similar to that of tetrapods (fig. 10.4C).

When it comes to deciphering how lobe fins crawled out on land, we have some remarkable fossils that demonstrate some of the steps—although preservation of soft features like the nature of their skin and gills or lungs cannot be demonstrated from the bony skeleton. The paraphyletic rhipidistians of the Devonian include a variety of peculiar fish with robust lobed fins such as *Eusthenopteron* (fig. 10.5). Although rhipidistians are fishlike with an aquatic body and multiple fins, the lobe fins have key elements that are homologous, bone for bone, with those in the tetrapod limb. For example, in the pectoral fin, the robust element closest to the body looks very much like the upper arm bone, or humerus, of primitive tetrapods. At the far end of this bone are a pair of bones, which are homologous (and look much like) the radius and ulna of the lower arm in tetrapods. Beyond those bones are a series of smaller rodlike bones that are homologous with the wrist and finger elements. The fin is then surrounded by a series of rays that support the fin membrane itself. If you look at the pelvic fin, the homologies with the thigh bone (femur), shinbones (tibia and fibula), and ankle and foot bones (tarsals and metatarsals) are equally apparent.

The similarities don't stop there. The detailed bone-for-bone structure of the spine of *Eusthenopteron* is a dead ringer for that in primitive tetrapods and totally unlike that in any other group of fish. The detailed patterns of the bones in the skull are also bone-for-bone identical with those in primitive tetrapods; only the relative proportions change in tetrapods, which greatly reduces the bones covering the gills and expands the bones in the front of the snout (fig. 10.5). The lungs or gills themselves don't fossilize, of course, but it's reasonable to assume that *Eusthenopteron* already had lungs because its primitive sister group (the lungfish) and advanced sister taxa (the tetrapods) do. In short, you could not ask for a better fishibian than *Eusthenopteron*. If you study it in detail with the eyes of anatomist (instead of reading superficially about it, as creationists do), you can see all the elements of the tetrapods already in place. Given how easily many modern fish now walk on land, it's not hard to imagine *Eusthenopteron* doing so as well.

The next step is a series of remarkable fossils that demonstrate the step-by-step transition to becoming a tetrapod (figs. 10.5–10.8). These include a variety of more tetrapod-like fish (*Panderichthys*, *Elginerpeton*, *Ventastega*, and *Metaxygnathus*) known from only partial specimens from the Late Devonian. *Panderichthys* (figs. 10.6, 10.7, and 10.10) was a very tetrapod-like lobe-finned fish. Unlike *Eusthenopteron*, these creatures had flattened bodies and upward-facing eyes, and frontal bones, like tetrapods, and a straight tails with a well-developed tail fin. The braincase of *Panderichthys* was originally classified as belonging to a tetrapod, not a fish, until the rest of the body was found. The teeth have the characteristic enfolding of the enamel ("labyrinthodont" teeth) that characterizes the teeth of later

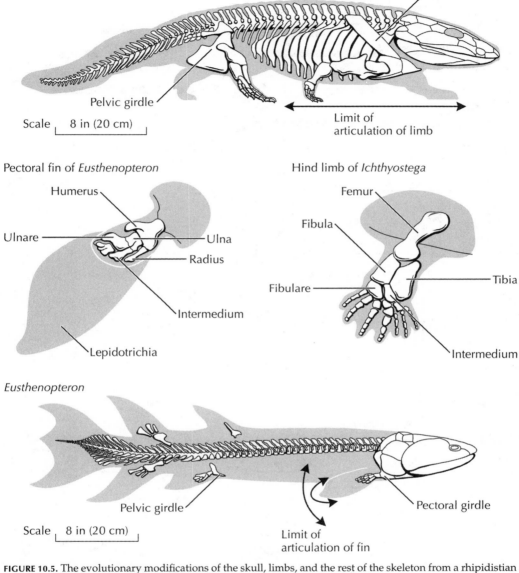

Ichthyostega

Pectoral girdle

Pelvic girdle

Scale   8 in (20 cm)

Limit of
articulation of limb

Pectoral fin of *Eusthenopteron*

Humerus

Ulnare

Ulna

Radius

Intermedium

Lepidotrichia

Hind limb of *Ichthyostega*

Femur

Fibula

Fibulare

Tibia

Intermedium

Eusthenopteron

Pelvic girdle

Pectoral girdle

Scale   8 in (20 cm)

Limit of
articulation of fin

FIGURE 10.5. The evolutionary modifications of the skull, limbs, and the rest of the skeleton from a rhipidistian like *Eusthenopteron* to a tetrapod like *Ichthyostega*. (Drawing by Carl Buell).

tetrapods. *Panderichthys* had both gills and well-developed lungs with nostrils, so it could breathe either way. Most importantly, like tetrapods, *Panderichthys* has lost the dorsal fins and anal fins, leaving only the remarkably footlike lobed pectoral and pelvic fins. This is a classic fishibian: tetrapod-like skull and body and braincase and lungs, all the rest of the fishy fins lost, but still retaining true fins that become the hands and feet.

Another important recent discovery is several nearly complete skeletons of *Acanthostega* (figs. 10.6–10.8) by Jenny Clack, Michael Coates, Per Ahlberg, and others. *Acanthostega* still had a well-developed fin on the tail, large gill openings, and even gills preserved on the

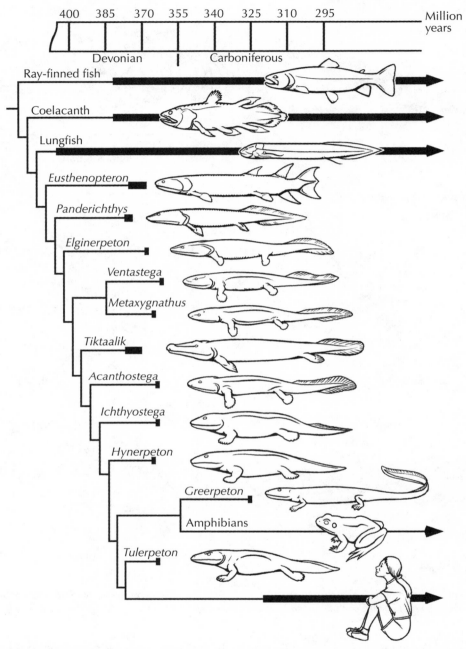

FIGURE 10.6. Phylogeny of the transitional series from rhipidistians through primitive tetrapods. (Drawing by Carl Buell)

inside of the skeleton! It had an ear region adapted for hearing in water, not on land. Yet it also had modified the lobed fin into a tetrapod limb with as many as seven or eight fingers, although the hand and foot were not well adapted for walking, but much better suited for swimming or creeping along the bottom. This is also shown by the proportions of its limbs,

FIGURE 10.7. The transformation of the pectoral fin of lobe fins into the hand and forelimb of primitive tetrapods. Each bone of the lobed fin is homologous with one of the limb bones of the tetrapod, and the main difference is modifications in shape and robustness and the loss of the fin rays, which are replaced by fingers. (From Shubin et al. 2006; fig. 4; by permission of the Nature Publishing Group)

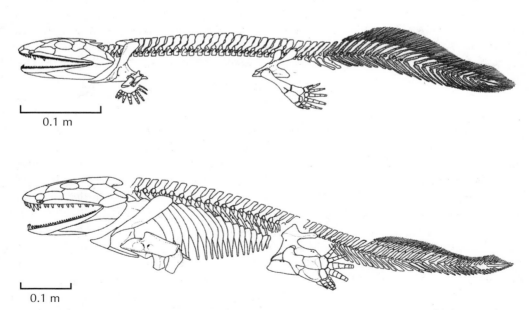

FIGURE 10.8. Sketch of the skeletons of *Acanthostega* (top) and *Ichthyostega* (bottom), showing the mixture of fishlike features (tail fins, lateral line systems, gill slits) and tetrapod features (robust limbs and shoulder and hip bones, reduced back of skull, expansion of snout). (Drawing courtesy M. Coates)

which are more like those of *Eusthenopteron* than of tetrapods, indicating that its limbs were not very good for land locomotion. *Acanthostega* also had the robust vertebrae in the spine that could support its weight out of water. It's a perfect fishibian: gills, fins, ears like a fish, but spine and limbs like a tetrapod. Clack and Coates have convincingly shown that it probably spent most of its time in the water, even though it has the limbs capable of some land locomotion. Like the teleost fish mentioned earlier, it only needed to crawl up on land once in a while, but most of its life was probably aquatic. For that matter, most living salamanders and newts are almost entirely aquatic and use their limbs primarily to swim and push along through the vegetation underwater.

The clinching piece of evidence was announced just over a decade ago (Daeschler et al. 2006; Shubin et al. 2006). Nicknamed the "Fishapod" but formally named *Tiktaalik*, this Late Devonian fossil from Ellesmere Island in the Canadian Arctic (fig. 10.9) was more fishlike than *Ichthyostega* or *Acanthostega*, yet its limbs show the perfect transition between fins and feet (fig. 10.7). It had fishlike scales, lower jaw, fin rays, and palate, but unlike any fish, it had a shortened skull roof and mobile neck (for whipping the head sideways to catch prey), an ear region capable of hearing in both land and water, and a wrist joint that anticipates the condition seen in·terrestrial tetrapods. Thanks to this discovery, we now have a beautiful transitional sequence from fully aquatic lobe-finned fish like *Eusthenopteron* to more amphibian-like forms such as *Panderichthys* and *Tiktaalik* to fully four-legged forms like *Acanthostega* and *Ichthyostega* (which still retain fishlike gills, tail fins, and lateral line systems on the face). This sequence is now so smoothly gradational that it's hard to tell where the fishes end and the amphibians begin—yet is it clear even to a creationist that *Eusthenopteron* is a fish and *Ichthyostega* is an amphibian.

The discovery that early tetrapods had seven or eight fingers came as shock at first. We are so accustomed to the fact that all living tetrapods have only five fingers and toes (or fewer) that we assumed that they always had that number. Early reconstructions of the poorly known hand of *Ichthyostega* often drew in a five-fingered hand, even though there was no evidence one way or another. But in other ways, this large number of toes is not so mysterious. My fellow Columbia student, good friend, and coauthor Neil Shubin at the University of Chicago worked on this problem for his dissertation while he was a graduate student at Harvard. People had always assumed that fingers formed from the branching of the fin rays in lobe fins, symmetrically down the central axis. But Neil studied the embryological development of salamander limbs (Shubin and Alberch 1986) and found that fingers bud off in an arch, from one side of the hand to the other (not symmetrically down the middle axis). This beautifully matches the hand in *Acanthostega* because it apparently had extra fingers (just like extra fin rays) that budded off from this embryonic development pattern. To get the modern pattern of five fingers, the development has to shut down just a bit earlier. Since this discovery, the Hox genes that control this developmental sequence have been identified, so it is clear that changing a limb from seven or eight fingers to just five is not a big problem. It just requires a small adjustment in the developmental timing of limb budding.

Finally, we come to *Ichthyostega* (figs. 10.6, 10.7, and 10.8), the classic transitional form, found in Upper Devonian rocks of Greenland in the 1930s by Danish and Swedish expeditions. It was described briefly by Säve-Söderbergh in 1932, but not fully monographed in detail until 1996 by Erik Jarvik. This creature is very similar to *Acanthostega*, only not as completely preserved. It is also slightly more advanced than *Acanthostega* in the direction of tetrapods, with a smaller tail fin and slightly longer more tetrapod-like limbs. However,

FIGURE 10.9. The newly discovered "fishibian" *Tiktaalik*. (A) Photograph of the nearly complete articulated skeleton. (B) Reconstruction of what the animal looked like in life next to the specimen. (Photos courtesy N. Shubin, University of Chicago, and T. Daeschler, Academy of Natural Sciences, Philadelphia)

it still retains a lot of fishy characteristics, such as the tail fin, the bones covering the gill slits (opercular bones), and especially the series of canals on the face (lateral line canals) for sensing movement and electrical currents under water. This lateral line system is a feature found in most sharks, teleosts, and many other aquatic animals. Once again, we have a classic fishibian: limbs and spine like a tetrapod, but tail fin, gills, and lateral line canals like a fish.

Creationists, of course, cannot admit that this animal is a true transitional form, so they stoop to all sorts of dishonest arguments to deny its fishibian features. Gish (1978; he learned nothing new in his 1995 edition) is typical of the bunch. *Ichthyostega* had limbs and feet, so it must be an amphibian, and then Gish goes on to quote outdated ideas about it. Gish (1978, 1995) keeps claiming that there are no fossils that document the transition from fins to feet, but now with *Panderichthys*, *Tiktaalik*, and *Acanthostega*, we do have fossils that had seven or eight fingers or toes on their limbs, which are clearly still used as fins, not as hands or feet. He never mentions the gill covers, the lateral line system, or the tail fin, aquatic features of *Ichthyostega* that are not found in most tetrapods. This is a clear case of selective citation of evidence and deliberately misleading argumentation. Either Gish can't read the description of the fossil well enough to understand that it has features of both fish and tetrapods, or he is deliberately trying to fool the reader by denying the obvious and well-documented fishlike features of these fossils. Either way, it is extremely poor science and very dishonest.

The creationist arguments usually stop right there: mislead the reader into thinking that *Ichthyostega* is "just" an amphibian and don't mention all its fishy features, and then move on to another topic. Well, that argument is finished for good. The amazing array of new transitional fossils (fig. 10.6) documents the transition in such detail that the creationists can't dodge behind trying to just discredit *Ichthyostega* anymore. The discovery of *Acanthostega* and *Tiktaalik* vastly improves our knowledge of the origin of tetrapods and shows that the earliest tetrapods used their legs not for walking on land but primarily for walking under the water! All of the old arguments about tetrapods needing robust limbs to crawl out to another pool or chase new prey are completely obsolete now that we know that walking underwater was the primary function of the limbs of *Ichthyostega* and *Acanthostega*. And all of their fishy features, such as the ear region, lateral line canals, and tail fins now make sense if we think of them as comparable to modern newts and salamanders that only rarely crawl out on land. As we saw from all the walking teleost fish, crawling out of the water is not such a feat after all, especially if you spend most of your time in the water. For that matter, most amphibians spend most if not all their time in the water, so they haven't made as great an evolutionary leap as the old dogmas once suggested.

How do the ID creationists deal with this extraordinary evidence? They cloud the issue by denying these fossils exist, or by distortion and misstatements. Davis and Kenyon (2004:103, figs. 4–8) show a 55-year-old sketch of *Ichthyostega* and *Eusthenopteron* but make no mention of all the other transitional fossils that were well documented before their book was published in 2004. They show (figs. 4–9, p. 104) the fin and limb bones of each of these creatures but ignore all the beautiful transitional fossils that have been documented in the past 50 years. They make a big deal about how dramatic this transition was, yet falsely claim that "no such transitional species have been recovered." Thanks to *Panderichthys* and *Acanthostega*, and now *Tiktaalik*, that falsehood can be safely laid to rest—but I have no expectation that creationist books will ever acknowledge the existence of these fossils. I'm sure they will simply replay their discredited and outdated arguments.

From the primitive forms like *Ichthyostega* and *Acanthostega*, the tetrapods began a great radiation of more advanced terrestrial forms in the Carboniferous. Some, like *Greererpeton* (figs. 10.6 and 10.10) had long fishy bodies not much different from those of *Acanthostega*. Their limbs and shoulder bones were considerably more advanced and terrestrial than those of *Acanthostega*, yet they still had fishlike features, such as the lateral line canals. By the mid-Carboniferous, we find that tetrapods have branched into many different lineages, including the big flat-bodied flat-skulled temnospondyls (fig. 10.1), the more delicate lepospondyls (some of which became legless and converged on snakes and apodans), and the anthracosaur lineage that leads to amniotes. And, by the middle Carboniferous, we find the first true amniotes as well. We will discuss them in the next chapter.

The "Frogamander"

Creationists frequently taunt scientists by pointing to an image of something as specialized as a frog and saying that there is no way they could imagine a transitional fossil between frogs and other amphibians. But in 2008, a fossil was announced that put this question to rest (Anderson et al. 2008). Formally named *Gerobatrachus hottoni*, it was dubbed "frogamander"

FIGURE 10.10. Drawing of the evolutionary transition of the bones from the skull, shoulder girdle, and forelimb, from fully aquatic rhipidistians like (A) *Eusthenopteron* to the slightly more tetrapod-like (B) *Panderichthys* to the more advanced (C) *Acanthostega* and concluding with a fully terrestrial tetrapod, (D) *Greererpeton*. The fishlike elements of the shoulder girdle, such as the cleithrum and anocleithrum, are gradually reduced, while the clavicles (our "collarbones") become the largest bone in the shoulder girdle. Meanwhile, the major bones covering the gill slits (shaded here in gray) are reduced and then lost, the eyes shift backward as the snout expands, and the cheek regions are reduced. (From Clack 2002: fig. 6.4; used with permission)

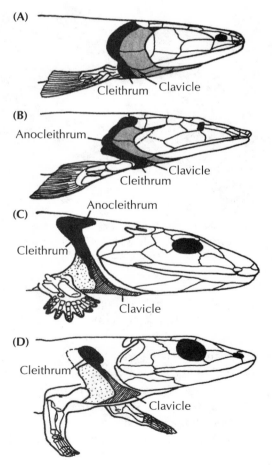

(A)

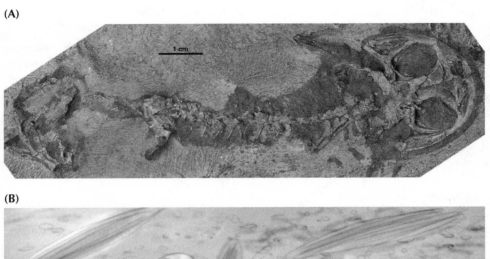

(B)

FIGURE 10.11. (A) The only specimen of *Gerobatrachus hottoni*. (Courtesy Diane Scott and Jason Anderson) (B) A reconstruction of it in life. (Courtesy Nobumichi Tamura)

by the press because it had features of both frogs and salamanders (fig. 10.11). It had a long tail and salamander-like body, but its head is short with a rounded snout like that of a frog. It also had the large eyes and large eardrum found in frogs and not salamanders. Most importantly, its teeth are attached to the jaw on tiny pedestals with a distinct base, a feature that defines the modern frogs and amphibians as a natural group.

There are also fossil frogs from the Triassic such as *Triadobatrachus*. They looked a bit more like living frogs but do not yet have the shortened trunk, reduced number of verte-brae, long hipbones and extremely long jumping hind legs that living frogs do. In short, the transition from a primitive amphibian to a modern frog has now been completely filled by transitional fossils.

For Further Reading

So much new information has now been discovered that I highly recommend reading the books by Zimmer (1998) and Clack (2002) to get the full story. Other useful references are also mentioned (not all of which were published before the new discoveries).

Anderson, J. S., R. R. Reisz, D. Scott, N. B. Fröbisch, and S. S. Sumida. 2008. A stem batrachian from the Early Permian of Texas and the origin of frogs and salamanders. *Nature* 453:515–518.

Benton, M. J. 2014. *Vertebrate Palaeontology*. 4th ed. New York: Wiley-Blackwell.

Carroll, R. L. 1988. *Vertebrate Paleontology and Evolution*. New York: Freeman.

Clack, J. A. 2002. *Gaining Ground: The Origin and Early Evolution of Tetrapods*. Bloomington: Indiana University Press.

Daeschler, E. B., N. H. Shubin, and F. A. Jenkins Jr. 2006. A Devonian tetrapod-like fish and the evolution of the tetrapod body plan. *Nature* 440:757–773.

Long, J. A. 2010. *The Rise of Fishes*, 2nd ed. Baltimore, Md.: Johns Hopkins University Press.

Maisey, J. G. 1996. *Discovering Fossil Fishes*. New York: Holt.

Moy-Thomas, J., and R. S. Miles. 1971. *Palaeozoic Fishes*. Philadelphia: Saunders.

Prothero, D. R. 2013. *Bringing Fossils to Life: An Introduction to Paleobiology,* 3rd ed. New York: Columbia University Press.

Shubin, N. H., E. B. Daeschler, and F. A. Jenkins Jr. 2006. The pectoral fin of *Tiktaalik roseae* and the origins of the tetrapod limb. *Nature* 440:764–771.

Standen, E. M., T. Y. Du, and H. C. E. Larsson. 2014. Developmental plasticity and the origin of tetrapods. *Nature* 513: 54–58.

Thomson, K. S. 1991. *Living Fossil*. New York: Norton.

Weinberg, S. 2000. *A Fish Caught in Time: The Search for the Coelacanth*. New York: HarperCollins.

Zimmer, C. 1998. *At the Water's Edge: Macroevolution and the Transformation of Life*. New York: Free Press.

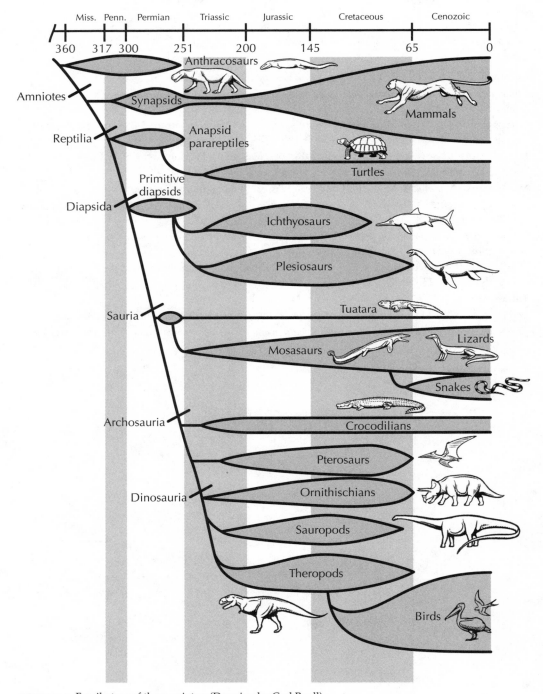

FIGURE 11.1. Family tree of the amniotes. (Drawing by Carl Buell)

ONTO THE LAND AND BACK TO THE SEA: THE AMNIOTES

11

Innovative Eggs

The most fundamental innovation is the evolution of another fluid-filled sac, the amnion, in which the embryo floats. Amniotic fluid has roughly the same composition as seawater, so that in a very real sense, the amnion is the continuation of the original fish or amphibian eggs together with its microenvironment, just as a space suit contains an astronaut and a fluid that mimics the earth's atmosphere. All of the rest of the amniote egg is add-on technology that is also required for life in an alien environment, and in that sense it corresponds to the rest of the space station with its food storage, fuel supply, gas exchangers, and sanitary disposal systems.

—Richard Cowen, *The History of Life*

Like the transition from "fish" to "amphibians," the transition from "amphibians" to "reptiles" has long been confused by misunderstandings and inadequate terminology. If we use the terms "fish," "amphibians," and "reptiles" in the traditional sense, they are not natural monophyletic groups, but "grades" of evolution because they do not include all descendants of a particular group. We have already discussed in chapter 10 how the monophyletic group Tetrapoda is preferred to the archaic term "amphibian." The same goes for the grade we call "reptiles." If you look at the living turtles, snakes, lizards, and crocodiles, they all have scaly skins and sluggish metabolisms, and it is easy to lump them together based on these shared primitive characteristics. But as we will detail in the next chapter, birds are their descendants, so unless you include the birds within the Reptilia, it is an unnatural paraphyletic "wastebasket" group (figs. 5.4 and 11.1). To avoid confusion with this long-standing misuse of the concept, most modern systematists use the term "Reptilia" to mean the clade that includes turtles, snakes, lizards, and crocodiles plus birds. The so-called "mammal-like reptiles" (more properly, the synapsids), on the other hand, branched off from the family tree *before* the branching point of true reptiles, so it is incorrect to label synapsids by that obsolete term (even though people keep using it). Likewise, there are many primitive tetrapods that are neither synapsids nor reptiles but more advanced than the creatures we saw in the previous chapter. To avoid the misleading term "Reptilia" for these animals, most modern systematists prefer to use the term "amniotes" for all vertebrates that are more advanced than traditional amphibians. The Amniota thus includes the classical concepts of Reptilia, Mammalia, and Aves (birds).

The diagnostic character shared by all living amniotes is the land egg, or *amniotic egg* (fig. 11.2). Instead of laying hundreds of tiny soft-shelled eggs in the water (as fish and amphibians do), amniotes lay fewer but larger eggs that can survive out of water. Each egg is covered by a shell (either leathery like a turtle egg or hard like a chicken egg) that protects the delicate tissues and embryo inside against predators and prevents it from drying out.

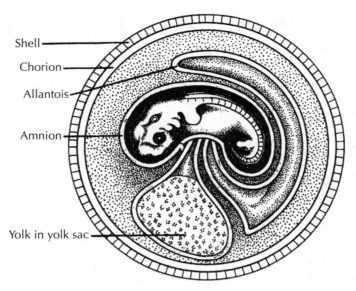

Shell
Chorion
Allantois
Amnion
Yolk in yolk sac

FIGURE 11.2. Diagram showing the parts of the amniotic "land egg." (From Romer 1959; used by permission of the University of Chicago Press)

The embryo is also aided by multiple specialized systems. It is surrounded by a membrane called the *amnion*, which is filled with amniotic fluid to buffer it from shock and temperature change. Attached to the gut of the embryo but outside the amnion is the *yolk sac*, which provides food for the embryo so it can hatch out relatively well developed and ready to face the world. A second sac off the hindgut of the embryo is known as the *allantois*, and it collects wastes and aids in respiration. Finally, the entire assembly—amnion, yolk sac, and allantois—is surrounded by another fluid, *albumin* (the "egg white"), which fills the rest of the volume of the egg. Just beneath the shell is a porous membrane known as a *chorion*, which helps hold in the fluids while allowing oxygen to enter and carbon dioxide waste to escape.

The amniotic egg has other implications. Each egg is more costly to produce, so fewer can be laid, and each embryo will be more completely developed and independent when it hatches. In addition, the egg cannot be fertilized (as is done in most fish and amphibians) by the male swimming near the egg cluster in the water and spraying them with sperm. Instead, the amniotic egg requires internal fertilization, usually involving sexual intercourse. Males and females must copulate so that the sperm can reach the eggs inside the female, and the eggs undergo much of their development inside her, not in the water. Internal fertilization has evolved more than once, of course. Most land-living arthropods (insects, spiders, scorpions, and so on) must copulate for the same reasons that amniotes do—they can't spread their eggs and sperm around in water. In the sea, sharks are among the few non-amniotes that use internal fertilization, and they lay smaller numbers of large eggs, or some even give birth to live young that are fully developed.

How did the transition from primitive tetrapods to the amniotes occur? We cannot tell which extinct organisms known from fossils laid an amniotic egg, because eggs are rarely preserved with their parents. Only a few fossil eggs from the time of the first amniotes are known anyway. Instead, we must work with the features of the skeleton to decide which

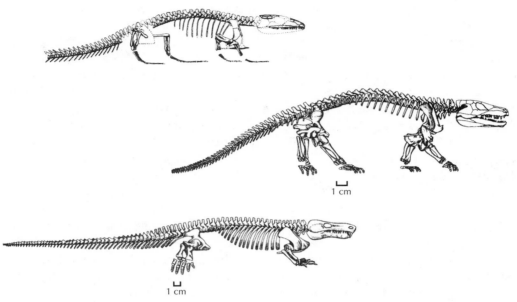

FIGURE 11.3. Skeletons of some of the many "anthracosaurs," which are transitional fossils between "amphibians" and the most primitive amniotes. At the top is the primitive gephyrostegid *Bruktererpeton*. In the middle is the more advanced form *Seymouria*. At the bottom is the highly tetrapod-like *Limnoscelis*. Each one is a mosaic of primitive tetrapod and advanced amniote characters, making it very difficult to decide where the amphibians end and the amniotes begin. (From Carroll 1988: fig. 9-22; courtesy W. H. Freeman and R. L. Carroll)

anatomical characters will diagnose a fossil as an amniote. Once again, we have a long series of transitional forms (the paraphyletic group known as "anthracosaurs") that show progressively more and more amniote-like characteristics, and where to draw the line in this continuous series is problematic (fig. 11.3). We can see several trends in the evolution of anthracosaurs into amniotes. One is the fact that anthracosaurs have skulls that are very deep vertically, with a narrow snout and a shortened region behind the eyes. Their skulls eventually lose the notch in the back for the eardrum. By contrast, most other primitive tetrapods (such as the temnospondyls and lepospondyls) have relatively flat skulls and bodies, with a wide snout, a large region of skull roof behind the eyes, and a well-developed notch for the eardrum. Correlated with this is evidence that limbs of the anthracosaurs were robust and held upright, so they frequently walked with their bellies off the ground. Temnospondyls and lepospondyls had a much more sprawling posture and could rarely move without dragging their bellies. The wrist and ankle bones of anthracosaurs were also modified for much more active motion than the comparable bones in other primitive tetrapods. The neck vertebrae of anthracosaurs were specialized into two bones: the atlas (which holds up the skull, as Atlas once held up the world in Greek mythology) and the second neck vertebra, the axis (the pivot joint on which the skull turns in relation to the neck). These are features found in all later amniotes and allowed anthracosaurs to swivel their heads rapidly to catch prey. Finally, the muscles and bones of the anthracosaur palate were modified so it had a much stronger bite force. By contrast, the palate of temnospondyls and lepospondyls had a much weaker "snapping" motion, with nowhere near as much bite force.

As we move up the sequence of anthracosaurs, we see these features accumulating one by one in a mosaic evolution pattern (fig. 11.3). For example, the more primitive *Solenodonsaurus* has the primitive eardrum notch but advanced jaw muscles. *Gephyrostega* also retains the primitive eardrum notch and primitive vertebrae and limbs but has an advanced ankle joint. *Limnoscelis* and *Seymouria* have lost the primitive eardrum notch and also have more advanced, robust limbs. *Diadectes*, a pig-sized herbivore (the first herbivorous land vertebrate known), is very close to amniotes, with the atlas and axis vertebrae in its neck and strong limbs with advanced ankles, but it still retains the primitive eardrum notch.

Most of these anthracosaurs are apparently end-members of an extinct side branch representing earlier lineages that were not preserved in the fossil record because the oldest fossil that most paleontologists agree is an amniote comes from older beds of the Lower Carboniferous. Nicknamed "Lizzie the lizard" by its discoverer, Stan Wood, this fossil is officially known as *Westlothiana lizziae* (fig. 11.4) and comes from the famous East Kirkton beds of Scotland (Smithson et al. 1994). Slightly later, true amniote fossils known as *Hylonomus* come from the Middle Carboniferous beds of Joggins, Nova Scotia (about 15 million years younger than "Lizzie"). These fossils show a dramatic change from the more primitive tetrapods. They are small (about 20 centimeters long, or the size of a typical lizard) rather than the much bigger dog-sized and even pig-sized anthracosaurs. They were also delicate and slender, with a very long trunk and tail, long slender limbs and toes, and relatively small heads in comparison to most anthracosaurs. Their vertebrae are robust and fused into the arches above the spinal column, a typical feature of amniotes. *Westlothiana* and *Hylonomus* had deep skulls with effective jaw muscles and no eardrum notch in the back of the skull.

(A)

(B)

1 cm

FIGURE 11.4. The oldest known true amniote, *Westlothiana lizziae*, showing both (A) the actual specimen and (B) a reconstruction of its skeleton. (Reproduced by permission of the Royal Society of Edinburgh and R. L. Carroll; from Smithson et al. 1994)

They also had relatively large eyes, suggesting that they may have been night predators, catching insects and other small prey. The best specimens of *Hylonomus* were found preserved inside hollow rotten tree trunks in the famous Joggins locality, suggesting to some that they had been trapped in these deep holes and died there. More recently, paleontologists have argued that the long delicate limbs and fingers suggest that they were good climbers like many modern lizards and probably lived in the hollows of trees, where they were occasionally buried and fossilized.

From early forms like *Westlothiana* and *Hylonomus*, the amniotes radiated into many different groups by the Late Carboniferous (fig. 11.1). One of the earliest lineages to branch off was the synapsids, which eventually gave rise to mammals. They will be discussed in chapter 13. The other lineage is the one we will call the true Reptilia. Its most primitive branch is the group known as Anapsida, which includes the turtles as well as several extinct groups. The next major branch is the Euryapsida, which includes most of the marine reptiles. The third branch is the Lepidosauria, which includes not only the living lizards and snakes but also extinct groups like the marine reptiles known as mosasaurs. The final branch is the Archosauria, or "ruling reptiles," and these include a lot of familiar creatures: crocodilians, pterodactyls, dinosaurs, and birds. For space reasons, we cannot go into detail with each of these groups or look at every transitional form in every lineage. Instead, we will focus on a few examples of the many transitional forms found in certain lineages to demonstrate just how good the fossil record of these creatures has become.

Turtle on the Half Shell

Creationists often mock scientists and proclaim how impossible it is to imagine a creature that is "half a turtle." Once again, the fossil record has answered them with a perfect transitional fossil (Li et al. 2008) that rebukes them in turn. Officially known as *Odontochelys semitestacea*, its name literally means "toothed turtle with half a shell." A number of specimens have been found in Triassic deposits of China, and they are truly remarkable (fig. 11.5). They have a fully developed shell on their belly (the plastron), but their backs are just broad expanded ribs without a back shell (the carapace). They are literally "turtles on the half shell." In addition, they are the last known turtle to have teeth. All more advanced turtles have toothless beaks. Thus, they bridge the gap between more lizard-like reptiles and turtles with fully developed shells on their back and belly.

If a creationist replies by saying, "Where is the transition between this fossil and other reptiles?," we have those fossils too. *Eunotosaurus* is an extinct amniote from the Permian of South Africa that has the broadly expanded back ribs of *Odontochelys* and more advanced turtles, along with some other features of the skull and skeleton, that link it to turtles—yet to the untrained eye it looks like a large, fat lizard. Then, in 2015, my friend Hans-Dieter Sues of the Smithsonian Institution announced *Pappochelys* ("grandfather turtle"), which not only has the broad back ribs of *Eunotosaurus* but also the broad flattened belly bones ("gastralia") that would eventually fuse to become the belly plate (plastron) of *Odontochelys*. This makes it a nice transition from *Eunotosaurus* to all the rest of the turtles—but it has only broad flat ribs on its belly and back, not a fused shell.

Thus, from the very lizard-like *Eunotosaurus*, we now have a complete sequence to *Pappochelys*, *Odontochelys*, and then to fossils that even a creationist could recognize as a turtle. These discoveries were made just in the past decade since the last edition of this book was published. Just imagine how many more transitional fossils we might have in another 10 years!

FIGURE 11.5. *Odontochelys*: (A) the best of the known fossils, showing an incomplete carapace on its back (left), but a complete plastron on its belly (right); (B) reconstruction of its appearance in life. ((A) Courtesy Li Chun; (B) courtesy Nobumichi Tamura)

Evolution of the Great Sea Dragons

There were no real sea serpents in the Mesozoic Era, but the plesiosaurs were the next thing to it. The plesiosaurs were reptiles who had gone back to the water because it seemed like a good idea at the time. As they knew little or nothing about swimming, they rowed themselves around in the water with their four paddles, instead of using their tails for propulsion like the brighter marine animals. (Such as the ichthyosaurs, who used their paddles for balancing and steering. The plesiosaurs did everything wrong). This made them too slow to catch fish, so they kept adding vertebrae to their necks until their necks were longer than all the rest of their body. . . . There was nobody to scare except fish, and that was hardly worthwhile. Their heart was not in their work. As they were made so poorly, plesiosaurs had little fun. They had to go ashore to lay their eggs and that sort of thing. (The ichthyosaurs stayed right in the water and gave birth to living young. It can be done if you know how.)

—Will Cuppy, *How to Become Extinct*

During the "age of the dinosaurs" or the Mesozoic Era, the seas teemed with many different types of life: enormous numbers of plankton, gigantic clams on the seafloor, as well as weird oysters like *Gryphaea* (fig. 8.10), many types of sea urchins and heart urchins (fig. 8.9), squid-like belemnites and ammonites, a great diversity of "holostean" and eventually teleost fish (see fig. 9.12), but the dominant predators of all these creatures were marine reptiles. These include not only huge sea turtles up to 7 meters (22 feet) long and crocodiles known as geosaurs, which had webbed feet and a tail fin, but also three major groups that were unique to the Mesozoic seas: the dolphin-like ichthyosaurs, the long-necked paddling plesiosaurs, and the seagoing Komodo dragons known as mosasaurs.

In marine beds over many parts of the world, we have excellent, often complete skeletons of these creatures. In the western Great Plains (especially South Dakota, Colorado, and Kansas), there are extensive exposures of marine sediments that were deposited in the Cretaceous when a great inland sea covered the Plains region from the Gulf of Mexico to Hudson's Bay. Each of these creatures is so distinctive that it would clearly be a "created kind" to a creationist. And yet we have excellent evidence of the origins and relationships of all three groups both from transitional fossils and from the cladistic analysis of their relationships. If they were not extinct, we might check their molecular phylogeny as well.

The most amazing thing about all three groups is that they were clearly reptiles, so they were descended from terrestrial creatures that developed a land egg, yet these three all independently returned to the oceans (as did crocodilians, sea turtles, sea snakes, seals and sea lions, and whales, of course). Apparently, the food resources were so great in the oceans that land-dwelling reptiles found their way to reap this bountiful harvest in at least five different groups. This seems amazing in itself, yet the fact that it has happened many times shows how powerful the selection forces for this lifestyle must be. Such a radical change in ecology usually caused much convergence in body form as well, so we can see how ichthyosaurs and whales have independently evolved the streamlined torpedo-like shape that is also found in fish. Returning to the ocean makes certain reproductive and physiological demands, in addition to streamlining the body for swimming and modifying the hands and feet into flippers. For example, marine reptiles must still reproduce somehow. We know that sea turtles and saltwater crocodilians crawl out on land and lay eggs in a nest, and presumably mosasaurs

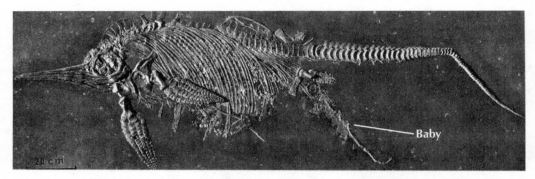

FIGURE 11.6. The famous specimen of a female ichthyosaur giving birth to a live baby, getting caught in the act, and then being fossilized in the Jurassic Holzmaden Shale of Germany. (Image courtesy of the State Museum of Natural History, Stuttgart)

and plesiosaurs could have done so too. But ichthyosaurs are so dolphin-like in body form that they could not have wriggled onto a beach and dug a nest with their flippers. We know that whales and dolphins give live birth, expelling the young from the womb and raising it up to take its first breath, after which it can swim on its own. Apparently, ichthyosaurs could also, since there are several remarkable specimens from the Jurassic Holzmaden shales in Germany that appear to have been in the process of giving birth to live young when they died and were fossilized (fig. 11.6).

Let us focus on the ichthyosaurs first. They represent the greatest challenge because they are the most highly modified and specialized for marine life. They have highly fishlike streamlined bodies with long toothy snouts for catching fish and squid, huge eyes for seeing in dark murky waters, a well-developed dorsal fin, and both the hands and feet are fully modified into flippers. Finally, their tails also have a vertically oriented tail fin. In contrast to the tailfin of most fish, the supporting spinal column of ichthyosaurs flexes downward into the lower lobe of the fin, not upward as in sharks and other primitive fish. A detailed analysis of the skeletal characters of the group shows they belong to the Euryapsida, along with the plesiosaurs, which we'll discuss next (fig. 11.1). The Euryapsida, in turn, are the sister group of the rest of the reptiles.

A number of striking intermediate forms are known from the early Mesozoic (fig. 11.7). First, there is *Nanchangosaurus* from the Triassic of China. Although it has a slightly stream-lined body and a long (but toothless) snout like an ichthyosaur, all of the rest of the features of the skeleton are primitive, including the vertebrae; the limbs, which are not modified into flippers but have normal proportions with all the regular wrist and ankle and toe bones; and a long straight tail with no sign of a tail fin. The original authors were not sure where to put this fossil because it is so primitive, but based on the skull, it seems to be an aquatic lizard on the way to becoming an ichthyosaur.

The oldest known fossil that can definitely be called an ichthyosaur is *Utatsusaurus* from the Early Triassic of Japan (fig. 11.7B). Although it has the general body form of an ichthyosaur, it has a mosaic of primitive features found in its reptilian ancestors. These include a skull with only a short snout and unspecialized teeth, very primitive vertebrae (especially in the neck), hands and feet that are not yet highly modified into flippers but still have discrete fingers and toes, and a long straight tail with no evidence of the downward flexion to

(A)

(B)

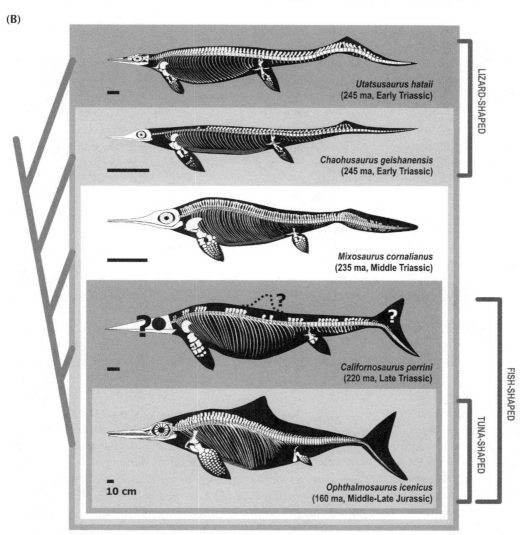

Utatsusaurus hataii
(245 ma, Early Triassic)

Chaohusaurus geishanensis
(245 ma, Early Triassic)

Mixosaurus cornalianus
(235 ma, Middle Triassic)

Californosaurus perrini
(220 ma, Late Triassic)

10 cm

Ophthalmosaurus icenicus
(160 ma, Middle-Late Jurassic)

LIZARD-SHAPED

FISH-SHAPED

TUNA-SHAPED

FIGURE 11.7. The ichthyosaurs evolved from lizard-like forms with asymmetric tails (*Utatsusaurus* and *Chaohusaurus*), primitive amniote skulls, and primitive hands and feet, to the highly specialized ichthyosaurs with symmetrical tails, large dorsal fins, big eyes, long fish-catching snouts with conical teeth, and highly modified flippers. (A) A complete articulated skeleton of the very primitive lizard-like ichthyosaur *Chaohusaurus from* the Early Triassic of China. (B) Diagram showing the evolutionary transformations from more primitive to more specialized ichthyosaurs. (© Ryosuke Motani, http://ichthyosaur.org)

support a vertical tail fin. Another specimen, *Grippia* from the Early Triassic of Spitsbergen, is known primarily from the skull, but it shows a relatively short snout, small eyes, and simple knob-like teeth for crushing mollusks, not the spiky teeth of most fish-eating ichthyosaurs. Yet another Early Triassic form from China, *Chaohusaurus* (fig. 11.7A and B) also has a short snout, simple teeth, primitive vertebrae, and robust limbs that are beginning to form a paddle but still have discrete rows of finger bones with the normal count (not the extra bones of an advanced ichthyosaur paddle—see fig. 11.7A). *Cymbospondylus* from the Middle Triassic of Nevada still retains the primitive hand and foot structure as well and has a relatively short snout with small eyes, and the tail is beginning to show the downward bend that indicates the presence of a small tail fin.

The best known and best preserved of the early ichthyosaurs is *Mixosaurus* from the Middle Triassic of Germany (fig. 11.7B) and many other places. The body has the classic ichthyosaur shape, with the long snout, large eyes, and dorsal fin. The hands and feet are beginning to form flippers, although they have still not multiplied the finger and toe bones as in later ichthyosaurs. And the tail shows just a slight downward bend, with some specimens preserving the body outline and showing that it had a small upper lobe on its tail. Thus it is advanced in many features but still retains the primitive hands and feet and does not yet have the fully bilobed ichthyosaurian tail.

If the ichthyosaurs seem very highly specialized, the plesiosaurs are specialized in a different direction. All of the advanced forms had stout bodies with robust shoulder and hip bones and four large well-developed paddles (fig. 11.8). Some plesiosaurs (especially the elasmosaurs) had long serpentine necks and small heads, while another group (the pliosaurs) had long heads and snouts and much shorter, more robust necks. Instead of employing speedy dolphin-like swimming like ichthyosaurs, plesiosaurs were apparently adapted for slow steady swimming by rowing with their fins (like a sea turtle does) and used their long heads and necks to snap at prey that came within reach.

How could such peculiar "kinds" evolve? We have an even better series of intermediates for plesiosaurs than we do for ichthyosaurs. They start with the Late Permian fossil *Claudiosaurus* from Madagascar (fig. 11.8A). It is so primitive that it may actually be the first known euryapsid, and a sister group to both the plesiosaurs and the ichthyosaurs. However, it shows the condition from which these euryapsids evolved. In most features, it looks just like many other primitive reptiles of the Permian, except that the skull has the characteristic holes, features of the palate, and loss of the temporal bar that earmark it as a euryapsid. In addition, it seems to show the beginning of aquatic adaptations with the loss of the breastbone. This enables it to swim with both limbs moving at the same time in a swimming stroke, not alternating like the gait of a lizard. It also has relatively long limbs and especially long toes, indicative of webbed feet. In fact, its limb proportions closely resemble those of modern aquatic lizards, such as the Galapagos marine iguana. Finally, a large part of the skeleton was reduced to cartilage, another indication of an aquatic lifestyle, since it reduces the weight of bone in a body supported by water and not needing such a robust bony skeleton.

From *Claudiosaurus* we next see a group of Triassic marine reptiles known as nothosaurs (fig. 11.8B). These animals had skulls and bodies that were not noticeably different from primitive euryapsids like *Claudiosaurus*. The biggest difference occurs in the neck, which is much longer and anticipates the long necks of many plesiosaurs. The limbs are not much more specialized for aquatic locomotion than those of its primitive relatives, but they have

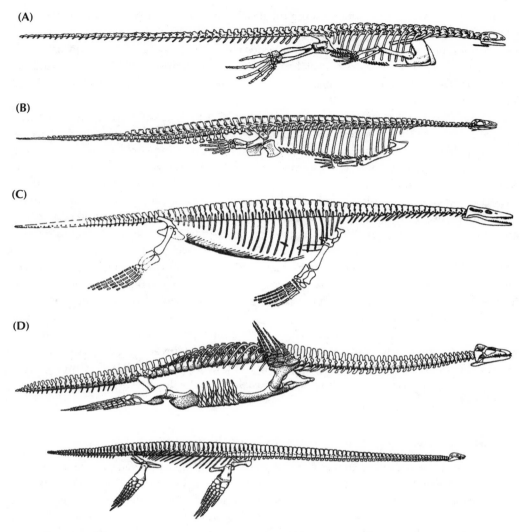

FIGURE 11.8. A transitional sequence of fossils bridging the gap from primitive amniotes to highly derived plesiosaurs. (A) *Claudiosaurus* from the Permian of Madagascar, with the primitive short neck, long tail, and relatively large hands and feet not yet modified into flippers. (B) The Triassic nothosaur *Pachypleurosaurus*, with longer neck, shorter more robust tail, and hands and feet more highly modified for swimming. (C) The Triassic *Pistosaurus*, a relatively primitive plesiosaur, with longer limbs partially modified into paddles, a longer neck, shorter tail, and longer skull. (D) The advanced plesiosaurs *Cryptocleidus* (top) and elasmosaurid *Hydrothecrosaurus* (bottom), with much longer necks, smaller heads, shorter tails, and hands and feet fully modified into flippers. (From Carroll 1988: figs. 12-2, 12-4, 12-10, and 12-12; courtesy W. H. Freeman and R. L. Carroll)

further reduced a lot of the bone to cartilage, another sign of a largely aquatic lifestyle. However, the shoulder girdle and hip bones are becoming much more robust and plate-like in support of the limbs, a hallmark of later plesiosaurs.

Our final step into full-fledged plesiosaurs is *Pistosaurus* from the Middle Triassic of Germany (fig.11.8C). This creature has a relatively primitive head with a slightly longer snout than nothosaurs (but still retaining the nasal bones, which are lost in plesiosaurs), but its palate is more like that of plesiosaurs. The rest of its body is also fairly advanced, with a fairly long neck, deep body, many extra bones along the belly (gastralia), and limbs that are intermediate between the unspecialized nothosaur foot and the highly specialized plesiosaur paddles, which have dozens of extra finger bones (similar to what happened with ichthyosaur paddles—see fig. 11.7).

Finally, we leave these two euryapsid groups and look at the third example of large marine reptiles in the Mesozoic. This group is known as the mosasaurs, and they looked essentially like gigantic Komodo dragons adapted for swimming—which they were. Mosasaurs are members of the family Varanidae, or monitor lizards, which includes not only the Komodo dragon but also all the Australian goannas of *Crocodile Hunter* fame. Although not as highly specialized as plesiosaurs and ichthyosaurs, mosasaurs are completely aquatic, with long bodies, fully developed flippers, and a vertical fin on the tail. Once again, we have beautiful transitional forms, known as the aigialosaurs (fig. 11.9) from the middle Cretaceous of the Adriatic region. As DeBraga and Carroll (1993) and Carroll (1997:324–325) have shown, aigialosaurs are perfectly intermediate between mosasaurs and their varanid ancestors. The aigialosaurs have at least 42 anatomical characters that make them more advanced than varanids, most of which are concentrated in the skull region and the semiaquatic limbs (but otherwise aigialosaurs retain the primitive varanid skeleton). There are another 33 character transformations between aigialosaurs and the most primitive true mosasaurs, most of which involve developing flippers, extending the body, and developing a vertical fin on the tip of the tail. The reader is invited to examine DeBraga and Carroll's (1993) paper in detail and see how a nice transitional sequence of fossils can be documented.

Snakes with Legs and Hopping Crocodiles

Snakes are vertebrates and vertebrates are classified as higher animals, whether you like it or not. I mean you can be a higher animal and still be a snake. This seems to be a rather peculiar arrangement, to be sure. If you can think of a better, let's have it. . . . Snakes in a word, are well worth knowing, unless you'd rather know something else. In closing, I have a little message which I wish you'd relay to some of those people who won't read a snake article because it gives them the jumps: there are no snakes in Iceland, Ireland, or New Zealand. And no snake articles.

—Will Cuppy, *How to Become Extinct*

If ever there were a classic "created kind," it would have to be the snakes. After all, the serpent supposedly tempted Eve in the Garden of Eden and was forever condemned to

FIGURE 11.9. A number of transitional snake fossils with vestigial legs and hip bones are known from the ▶ Cretaceous. (A) *Eupodophis descouensi*, with tiny vestigial hind legs. (B) Detail of the leg bones in the same specimen. (Photos courtesy M. Caldwell) (C) The complete articulated skeleton of the Cretaceous snake with legs known as *Haasiophis*. The large cubes are cork spacers to prevent the fossil from being damaged when it is turned upside down. (D) Detail of the hip region, showing the vestigial hind limbs. (Photos courtesy M. Polcyn, Southern Methodist University) (E and F) The transitional fossil *Adriosaurus*, which had functional hind limbs but vestigial forelimbs, and a long snake-like body. (After A. Palci and M. W. Caldwell, 2007, *Journal of Vertebrate Paleontology* 27:1–7)

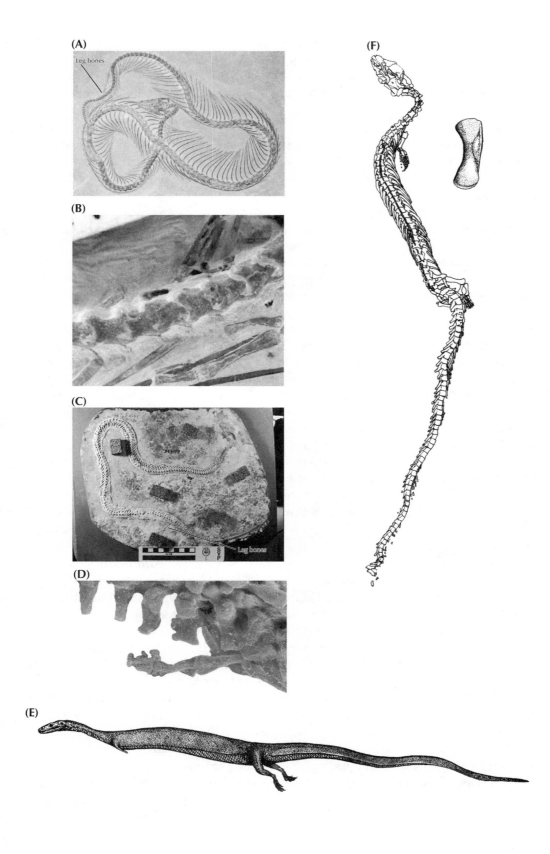

crawl on its belly. Most people have bad feelings about snakes for a variety of reasons on which psychiatrists have long speculated. To a great extent, it is probably because there are many different species of venomous snakes, so all sorts of animals (including ourselves) have evolved a natural fear of them. But snakes can be wonderful, too. They are amazing machines of adaptation, with the ability to completely stretch their skulls and mouths around a prey item much larger than their head, and their incredible adaptations for living in every environment, from the desert to the jungle treetops to the open ocean. Only the cold temperatures of the polar and subpolar regions prevent them from living in those habitats. They are tremendously diverse, too, with over 20 families and dozens of genera and species alive today.

The odds of finding "missing links" for the "snake kind" seem astronomical. Snake skulls and skeletons are made of hundreds of tiny, lightweight bones that break up easily when the snake dies and have a poor potential for fossilization. Only a handful of fossils of complete snake skeletons are known. Most fossil snake species are known only from isolated vertebrae. Thus we have a very patchy fossil record for snakes in general, as would be expected for animals with such poor potential for preservation.

However, occasionally we get lucky. In some extraordinary localities like the Eocene Messel lake beds in Germany, we get complete articulated snake skeletons of essentially modern-looking snakes. And for some reason, we also got lucky when snakes were first branching out from their lizard ancestors in the Cretaceous. A number of fossil snakes with hind limbs (fig. 11.9) are known from the mid-Cretaceous of Israel, Lebanon, and Croatia, including *Eupodophis*, *Pachyrhachis*, and *Haasiophis*. In 2006, a slightly older fossil snake, *Najash rionegrina*, was reported from the mid-Cretaceous of Argentina. It not only had hind limbs but fully functional hip bones as well. Then, in 2007, an even better transitional fossil was reported from rocks in Slovenia that are 95 million years old. Dubbed *Adriosaurus microbrachis*, it was an extremely long-bodied marine lizard with fully functional hind limbs but tiny vestigial front limbs (fig.11.10E and F), showing yet another step between normal lizards and completely legless snakes. Finally, in 2015, the fossil *Tetrapodophis* ("four-legged snake") was found in beds from the Early Cretaceous of Brazil. It has four tiny limbs, but they are clearly too small to have any real function, and must be vestigial (Martill et al. 2015).

This should come as no surprise; we know snakes evolved from lizards that have emphasized locomotion driven by the sinuous motion of their bodies. In fact, a number of lizards, such as the skinks, have highly reduced limbs that are almost nonfunctional. And the legless snakelike body form has evolved several times in the vertebrates, not only in the snakes but also in another living group of reptiles known as amphisbaenids, in a living group of amphibians known as apodans, and in an extinct group of lepospondyl amphibians known as aistopods. Finally, the proof is found in the living snakes as well. A number of groups, such as the boids, still retain the vestiges of their hind limbs and pelvis (fig. 4.9).

Where did snakes come from? What is the nearest relative of snakes? This is a hotly debated topic. Some scientists point to evidence that links them with the marine mosasaurs (Caldwell and Lee 1997), while others cluster them with the amphisbaenid lizards (Rieppel et al. 2003), and the molecular evidence seems to link them with anguimorph and iguanid lizards (Harris 2003). Clearly, the jury is still out, and we have a lot of work to do. The biggest problem is that the snake skeleton is so specialized, with so many elements reduced or lost, that it is hard to make comparisons with any other group of reptiles. But whether we know

(A)

(B)

(C)

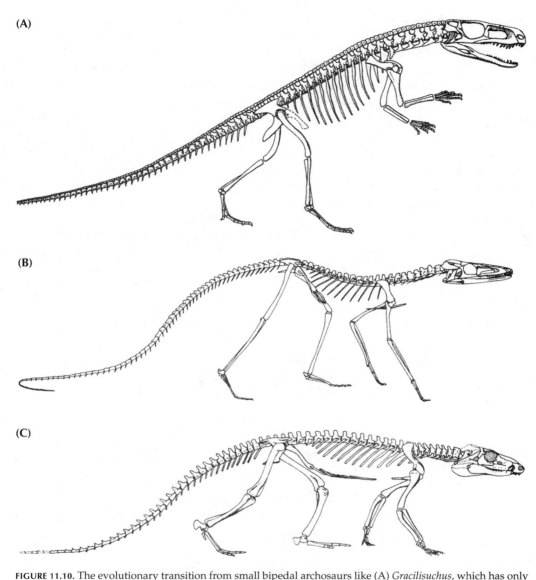

FIGURE 11.10. The evolutionary transition from small bipedal archosaurs like (A) *Gracilisuchus*, which has only a few crocodilian features in its anatomy, and (B) *Terrestrisuchus*, which is more quadrupedal with a longer snout, but still lightly built, to (C) *Protosuchus*, which has more typical crocodilian features, although it is still smaller and much more lightly built than any living crocodilian. (From Carroll 1988: figs. 13-22, 13-24, and 13-25; courtesy R. L. Carroll)

the proper sister group or not, we can see that snakes once had legs and were descended from *some* group of lizards that could walk.

We do not have space to talk about the transitional forms for many other fascinating reptiles, and we will save the dinosaurs for the next chapter. But we should mention one other surprising example: the origin of crocodilians. Today we think of crocodilians as huge, armored, dangerous reptiles that haunt bodies of water, disguised at floating logs until suddenly they lunge at their prey and grab it and drag it underwater in their "death roll." Thanks to movies like *Crocodile Dundee* and the TV show *Crocodile Hunter* with the late Steve Irwin, we have heard a lot more about crocodiles than any other reptile.

Crocodiles have a tremendously diverse fossil record, with many different forms, from 50-foot-long giants that ate dinosaurs for breakfast, to highly specialized marine crocodiles with flippers and tail fins, to the deep-jawed "bulldog" crocodilian *Sebecus*, to the peculiar fish-eating gavials (figs. 11.10 and 11.11). What is most surprising is that crocodiles didn't start out large, heavy, or long-jawed. The earliest crocodiles were nothing like any of their later descendants or like any of the living species. They were small (about a half a meter to a meter, or 2–3 feet long), delicate, and long-legged with a relatively short snout and long, thin tail. In the Triassic, we have delicate creatures like *Saltoposuchus*. Others, like *Gracilisuchus* (fig. 11.10A) were apparently bipedal, walking mainly on their long hind limbs, since their front limbs were so short. Still others, like *Terrestrisuchus* (fig. 11.10B), were quadrupedal but extremely delicate, with long slender legs, and probably were excellent runners. Finally, by the Early Jurassic we have *Protosuchus* (fig. 11.10C), whose name means "first crocodile"

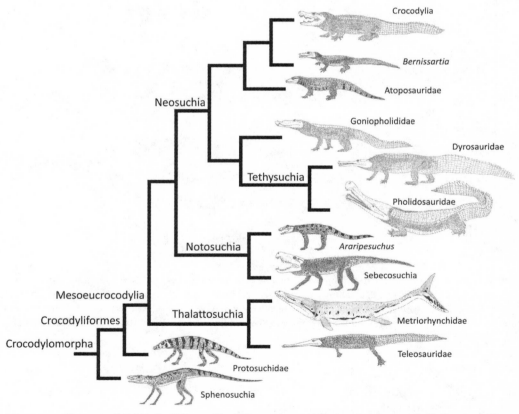

FIGURE 11.11. From small lightly built forms like *Sphenosuchus* and *Protosuchus* (bottom left), the crocodilians have radiated into a wide variety of body and skull shapes and ecological niches, including fully marine crocodiles with flippers and tail fins like the Thalattosuchia, crocodilians with long limbs and deep skulls like dinosaurs (Sebecosuchia), crocodilians with long narrow snouts (Tethysuchia), and the many differently shaped relatives of the living Crocodylia (upper right). (Courtesy D. Naish)

because it was one of the first early crocodilians to be described. This creature has a more classic four-legged stance, but the limbs are still long and delicate, and the snout is still short and slender.

How do we know that these delicate creatures are crocodilians? They sure don't look like it! But the superficial looks are deceiving. Crocodilians have a whole suite of distinctive anatomical features, particularly in the skull and ankle, that are unique and unmistakable. We don't recognize something as a crocodilian fossil by its long snout with lots of teeth but instead by these more subtle and more reliable features.

All of these animals have body proportions not much different from the other primitive archosaurs of the Triassic, to which they are closely related. Thus we can trace the origin of crocodilians from the great branching event of many archosaur lineages in the Early Triassic through a variety of delicately built smaller forms to *Protosuchus*, and by the mid-Jurassic to crocodilians that resemble something we might recognize as such (fig. 11.11).

Finally, you might be wondering why crocodilians remained so small and delicate until the Jurassic. The answer probably lies in competition: during the Triassic, there were huge armored semiaquatic archosaurs, known as phytosaurs, which filled the niche that crocodilians would eventually occupy. In most respects, phytosaurs looked like dead ringers for crocs, from the large body size and armor, sprawling gait, and long tooth-filled snouts. But if you look closely, you can tell in an instant that it's not a crocodilian. All crocodilians have their nostrils on the tips of their snouts. Phytosaurs have shifted the nostril openings to the top of the skull, just in front of the eyes. Apparently, phytosaurs prevented crocodilians from occupying that "croc niche" until they died out in the Late Triassic, and then the delicate *Protosuchus*-like crocodilians quickly evolved into large aquatic predators once the niche was open.

For Further Reading

Benton, M. J., ed. 1988. *The Phylogeny and Classification of the Tetrapods.* Vol. 1, *Amphibians, Reptiles, Birds.* Oxford, U.K.: Clarendon.

Benton, M. J. 2014. *Vertebrate Palaeontology.* 4th ed. New York: Wiley-Blackwell.

Caldwell, M. W., and M. S. Y. Lee. 1997. A snake with legs from the marine Cretaceous of the Middle East. *Nature* 386:705–709.

Callaway, J. M., and E. M. Nicholls. 1996. *Ancient Marine Reptiles.* San Diego, Calif.: Academic.

Carroll, R. L. 1988. *Vertebrate Paleontology and Evolution.* New York: Freeman.

Carroll, R. L. 1992. The primary radiation of terrestrial vertebrates. *Annual Review of Earth and Planetary Sciences* 20:45–84.

Carroll, R. L. 1996. Mesozoic marine reptile as models of long-term large-scale evolutionary phenomena. In *Ancient Marine Reptiles*, ed. J. M. Callaway and E. M. Nicholls. San Diego, Calif.: Academic, 467–487.

Carroll, R. L. 1997. *Patterns and Processes of Vertebrate Evolution.* New York: Cambridge University Press.

DeBraga, M., and R. L. Carroll.. 1993. The origin of mosasaurs as a model of macroevolutionary patterns and processes. *Evolutionary Biology* 27:245–322.

Gauthier, J. A., A. G. Kluge, and T. Rowe. 1988. The early evolution of the Amniota. In *The Phylogeny and Classification of the Tetrapods.* Vol. 1, *Amphibians, Reptiles, Birds.* M. J. Benton, ed. Oxford, U.K.: Clarendon, 103–155.

Laurin, M., and R. R. Reisz. 1996. A reevaluation of early amniote phylogeny. *Zoological Journal of the Linnean Society of London* 113:165–223.

Li, Chun, Xiao-Chun Wu, Olivier Rieppel, Li-Ting Wang, and Li-Jun Zhao. 2008. An ancestral turtle from the Late Triassic of southwestern China. *Nature* 456: 497–501.

Martill, D. M., H. Tischling, and H. R. Longrich. 2015. A four-legged snake from the Early Cretaceous of Gondwana. *Science* 349:416–419.

McGowan, C. 1983. *The Successful Dragons: A Natural History of Extinct Reptiles.* Toronto: Stevens.

Prothero, D. R. 2013. *Bringing Fossils to Life: An Introduction to Paleobiology.* 3rd ed. New York: Columbia University Press.

Rieppel, O. 1988. A review of the origin of snakes. *Evolutionary Biology* 25:37–130.

Rieppel, O., et al., 2003. The anatomy and relationships of *Haasiophis terrasanctus*, a fossil snake with well-developed hind limbs from the mid-Cretaceous of the Middle East. *Journal of Paleontology* 77:536–558.

Schoch, R., and H. D. Sues. 2015. A Middle Triassic stem-turtle and the evolution of the turtle body plan. *Nature* 523:584–587.

Schultze, H.-P. and L. Trueb, eds. 1991. *Origins of the Higher Groups of Tetrapods: Controversy and Consensus.* Ithaca, N.Y.: Cornell University Press.

Smithson, T. R., R. L. Carroll, A. L. Panchen, and S. M. Andrews. 1994. *Westlothiana lizziae* from the Visean of East Kirkton, West Lothian, Scotland, and the amniote stem. *Transactions of the Royal Society of Edinburgh* 84:383–412.

Sumida, S., and K. L. M. Martin, eds. 1997. *Amniote Origins: Completing the Transition to Land.* San Diego, Calif.: Academic.

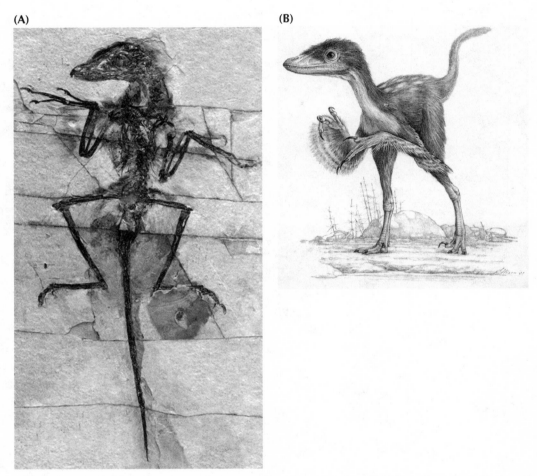

FIGURE 12.1. (A) The famous specimen known as "Dave" or more formally, *Sinornithosaurus*, a feathered nonflying dinosaur from the Lower Cretaceous Liaoning beds of China. (B) Reconstruction of *Sinornithosaurus* in life. (Photo and drawing courtesy M. Ellison and M. Norell, American Museum of Natural History)

DINOSAURS EVOLVE— AND FLY

12

Dinosaur Transitions

Long, long before you and I were born, there were dinosaurs all over the earth, except in New Zealand. Dinosaurs lived and loved in the Mesozoic Era, or Age of Reptiles, which began 200,000,000 years ago and lasted until 60,000,000 years ago. (There are people who know these things. Does that satisfy you?). . . . The brain of a dinosaur was only about the size of a nut, and some think that is why they became extinct. That can't be the reason, though, for I know plenty of animals who get by with less. . . . The Age of Reptiles ended because it had gone on long enough and it was all a mistake in the first place.

—Will Cuppy, *How to Become Extinct*

Dinosaurs are big business these days, with millions of dollars of merchandise featuring their likenesses, four of the highest-grossing movies ever made (the *Jurassic Park-Jurassic World* series), and dozens of documentaries on cable television. Almost every kid between ages 4 and 12 is fascinated with them. Most of the public knows or cares nothing about prehistoric life except for the dinosaurs, and many use the word "dinosaur" for any extinct beast (including prehistoric mammals and many other creatures not even remotely related to dinosaurs). Some people still have the *Flintstones* or the comic strip *B.C.* as their mental image for prehistory and believe that dinosaurs and humans coexisted. With the exception of the birds (fig. 12.1), which are dinosaurs (as we shall soon prove), all the rest of the non-bird ("nonavian") dinosaurs were extinct by 66 million years ago, and our own family did not appear until 5–7 million years ago, so at least 58 million years separate nonbird dinosaurs and humans. Some creationists have tried to perpetuate this myth by claiming that there are human tracks mixed with dinosaur tracks in the Paluxy River bed in Texas, but that has been debunked by creationists themselves, and most of them consider it an embarrassment (Morris 1986; see Glen Kuban's review at paleo.cc/paluxy/sor-ipub.htm).

Nevertheless, creationists know that the public only cares about dinosaur and human evolution, so they feel obligated to trot out examples of cool-looking dinosaurs in their books and debates and at the "Creation Museum" in Kentucky, and then claim that there are no transitional forms for any dinosaur. Not only is this a blatant lie, but it shows that they have not done the least bit of homework, for even the kiddie books about dinosaurs illustrate many transitional forms and primitive members of the major families.

Considering how rare dinosaur fossils are (especially in comparison to the marine invertebrates we discussed in chapter 8), it is remarkable that we have any transitional dinosaur fossils at all. But enough specimens are preserved to show that we now have the transitions between nearly every major group of dinosaurs, plus many other remarkable fossils that show other types of transitions, such as carnivorous dinosaurs becoming herbivorous.

Lifestyles of the Huge and Ancient

Of all creatures that have ever lived, the dinosaurs are of greatest fascination to man, particularly to children. This is perhaps because of their spectacular size in many cases. . . . and because they possessed so many unusual anatomical features. The fossil record of dinosaurs speaks out as clearly for creation as would be possible for creatures now extinct.

—Duane Gish, *Evolution: The Fossils Still Say NO!*

When I debated Duane Gish at Purdue University on October 1, 1983, I had seen his presentation the week before, so I knew he would show some dinosaur slides (mostly from century-old Charles R. Knight paintings, which are grossly out of date) because his audience would find them far more interesting than "mammal-like reptiles" or "fishibians." So I prepared my own segments (the first and third half hours of the first 2 hours of a 4-hour debate) to rebut his misleading information about dinosaurs before he even got to it. Did he change his presentation or acknowledge the fact that I had just shown the transitional forms whose existence he denied? No, like a robot he gave exactly the same slides and same lecture that he had delivered the week before, without any apparent awareness that I had blown his examples to pieces. I don't know what he was thinking, but many of the people who came up afterward and told me that I had won the debate said that his dishonest treatment of dinosaurs convinced them.

Gish (1978, 1995) and the other creationist authors never learn, but (just like their misleading presentation of the second law of thermodynamics) they keep on plugging the same false statements about dinosaurs because dinosaurs are impressive and interesting to their audience, which cares only about dinosaurs and no other prehistoric animals. If you read through the dinosaur chapter in Gish (1995) closely, the entire argument consists of out-of-context quotations from really old, outdated books, especially popular trade books, which are highly oversimplified. Gish never bothered to read more specialized books on dinosaurs, probably because he wasn't trained to do so and couldn't tell one bone from another. And that raises the most important point. Gish had absolutely no qualifications to interpret dinosaur fossils or to make judgments about them. He may have tried to glean what impressions he could from reading children's books and quoting them out of context, but he was no more qualified to make pronouncements based on such simplistic book reports than your average high school kid. (Remember, his Ph.D. was in biochemistry and completely irrelevant). More importantly, reading children's books (rather than studying the actual specimens) is not science nor is it even true research. If Gish really cared to find out whether there were transitional dinosaur fossils or not, he would have gotten the proper training in vertebrate paleontology and gone and studied the fossils for himself. Otherwise, his statements about fossils he has never studied are just baloney.

Let's look at just a few of the many examples of dinosaurs that Gish claimed have no transitional forms. The first one he always shows is the large long-necked sauropod he called "*Brontosaurus*." He apparently had not learned that paleontologists had stopped using that name decades ago; the proper name has been *Apatosaurus* for over a century. (Even most of the children's books have stopped using *Brontosaurus* and correctly use *Apatosaurus* now). In our 1983 debate, he showed a few outdated slides of *Apatosaurus* and *Brachiosaurus* (one of the stars of *Jurassic Park*), then zoomed on to his next examples, claiming that there were no transitional forms between them and other dinosaurs. He even made the same claim in his 1995 book (1995:124).

Apparently Gish never bothers to read even the children's books closely. Nearly every book about dinosaurs illustrates a group of Triassic creatures called prosauropods, whose very name (translated as "before the sauropods") implies that they are primitive relatives of the larger sauropods (fig. 12.2A). The best known of these is *Plateosaurus* from the Triassic of Germany (fig. 12.2B), but there are another dozen genera found in Triassic beds all over the world. Most of these creatures were only about 5–8 meters (15–25 feet) long, one-fourth the size of the giant Jurassic sauropods, but larger than their ancestors. They have the beginnings of a long neck and long tail but not yet the incredible neck or tail of the giant sauropods. The limbs are classic sauropod in the construction of their fingers and toes, yet they are not as robust, and the forelimbs are long and delicate enough that they apparently could alternate between quadrupedal walking and rearing up on their hind legs in a bipedal stance to use their hands. Only when the sauropods reached huge sizes were they forced to walk entirely on all fours, and their limbs also become much more massive in support of their huge body weight.

Even more primitive than *Plateosaurus* was *Anchisaurus* from the Triassic of Connecticut, Arizona, and South Africa. It was only 2.5 meters (8 feet) long, just slightly larger than a human, and had a still shorter neck and tail and even more delicate limbs and feet. In fact, it is the perfect transitional form between the more lizard-like early saurischian dinosaurs, such as *Lagosuchus* (which were much smaller, as we shall soon see). Yet despite its outward appearance, its skull bears all the distinctive hallmarks of sauropods, and it already shows many of the specializations in the vertebrae and especially in the hands and feet that will later come to mark the sauropods. The early saurischians (the "lizard-hipped" dinosaurs) were primarily bipedal, but *Anchisaurus* seems to have been capable of both stances, and *Plateosaurus* was even heavier and more likely to be quadrupedal. We have not only a smooth increase in size from the earliest known dinosaur *Eoraptor* (fig. 12.3B-D) to *Anchisaurus* to *Plateosaurus* to the larger sauropods but also a smooth transition in the anatomical features and in the stance from bipedal to quadrupedal as well.

The other main branch of the saurischians was the theropods, or the predatory dinosaurs. These are all familiar to us from the giants like *Tyrannosaurus rex* and the Jurassic predator *Allosaurus*, but there are dozens of different genera and species of all shapes and sizes. Among the more aberrant were the "ostrich dinosaurs," which had long legs, long necks, and toothless beaked heads much the living ostrich, but they also had a long bony tail (which no living bird has). Gish (1995:124) briefly mentions some of the primitive theropods but clearly has not kept up with the times.

Despite Gish's denials, some of the primitive theropods are actually excellent transitional forms (fig. 12.3). They ranged from only 70 centimeters (2 feet) long (*Compsognathus*) up to 3 meters (10 feet) long (*Coelophysis*), so most were about the size of a chicken up to the size of an adult human (fig. 12.4). Unlike their big theropod descendants, they were lightly built, with small heads, long necks, and slender gracile limbs and tails. Yet their skulls and especially their hands (with their unique combination of only three fingers: the thumb, index, and middle finger) and feet had all the anatomical specializations seen in later theropods.

From primitive theropods like *Coelophysis*, we can trace the theropod lineage even farther back to *Eoraptor* (fig. 12.3B–D), *Staurikosaurus*, and *Herrerasaurus* (figure 12.3E), which are built much like *Coelophysis* but do not yet have all the distinctive specializations of theropods.

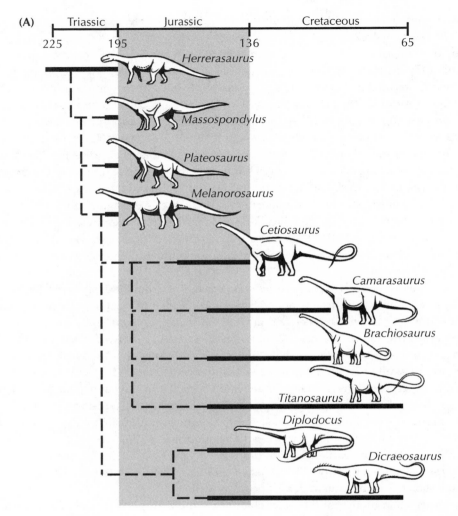

FIGURE 12.2. The prosauropods were transitional forms between smaller primitive bipedal dinosaurs like *Eoraptor* and the sauropods. (A) Family tree of the sauropods. (Drawing by Carl Buell) (B) *Plateosaurus* was capable of both bipedal and quadrupedal locomotion and had a neck and tail intermediate in length between smaller dinosaurs and the huge long-necked sauropods. (Photo courtesy R. Rothman)

(A)

(B)

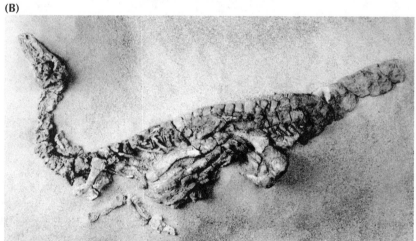

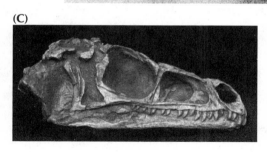

(C)

(D)

FIGURE 12.3. (A) Partial skeleton of a primitive archosaurian relative of dinosaurs known as *Euparkeria* from the Early Triassic of South Africa. (Photo by the author) (B–D) The earliest and most primitive known dinosaur, *Eoraptor* from the Late Triassic of Argentina. (B) A complete articulated skeleton; (C) close-up of the skull; (D) reconstruction of *Eoraptor* in life. (E) Skull of *Herrerasaurus*, another primitive dinosaur from the Late Triassic of Argentina. (Photos (B–E) courtesy P. C. Sereno, University of Chicago)

(E)

FIGURE 12.4. There are many transitional forms between the earliest dinosaurs like *Eoraptor* and the more advanced groups. This is *Coelophysis*, one of the smallest and most primitive carnivorous theropod dinosaurs. (Photo courtesy L. Taylor)

To a casual observer, they look very similar, but to a paleontologist with anatomical training, the differences are clear. *Eoraptor*, *Staurikosaurus*, and *Herrerasaurus* lack the highly specialized three-fingered hand (some still had the full five fingers), the relatively unspecialized vertebrae (both sauropod and theropod vertebrae are very specialized and distinctive), a sliding jaw joint, the fully recurved predatory teeth, and the distinctively theropod modifications of the ankle and foot found in *Coelophysis* and more advanced theropods. Finally, we can trace creatures like *Eoraptor*, *Staurikosaurus*, and *Herrerasaurus* back to primitive nondinosaurian archosaurs like *Euparkeria* (fig. 12.3A), which superficially resembles the earliest dinosaurs but lacks the unique specializations that all dinosaurs have, such as the open hip socket or the distinctive features in the skull.

Thus we can trace both sauropods (through the primitive prosauropods) and theropods (through the primitive creatures like *Coelophysis*) to a common saurischian ancestor along the lines of *Eoraptor*, *Staurikosaurus*, and *Herrerasaurus*, and from there to more primitive archosaurs that were not dinosaurs, like *Euparkeria*. You could not ask for a nicer series of transitional forms. Apparently, Gish never heard of any of these.

And there's a cool final twist to this story of predatory dinosaurs. In 2005, my friend Jim Kirkland and others announced the discovery of a remarkable new fossil known as *Falcarius utahensis* from the Jurassic of Utah (fig. 12.5). This strange creature is a member of an even stranger group known as therizinosaurs, whose exact position *within* the dinosaurs has long been controversial. These beasts have many of the hallmarks of theropods like *Velociraptor*, including long clawed fingers on the hands and a long neck and tail for balancing. But they had toothless beaks and were apparently herbivorous. In recent years,

FIGURE 12.5. Transitional fossils among the dinosaurs. This is the therizinosaur *Falcarius*, which shows the transition between carnivory and herbivory in this peculiar theropod group, shown next to Dr. Jim Kirkland, who discovered and described it. (Photo courtesy J. Kirkland)

the consensus was that therizinosaurs were indeed theropods that had somehow returned to plant eating. The discovery of *Falcarius* provided the "missing link" in this dietary transition because it retains many "raptor" dinosaur features yet is the most primitive therizinosaur with a toothless herbivorous beak. Thus it was a classic transitional form not only in linking therizinosaurs anatomically to raptors but also in showing how they made the remarkable transition from carnivore back to herbivore.

The other main branch of dinosaurs is the Ornithischia, which includes nearly all the herbivorous dinosaurs (except sauropods): the duckbills, the iguanodonts, the turtle-like armored ankylosaurs, the spiky stegosaurs, the bone-headed pachycephalosaurs, and the frilled and horned ceratopsians. The earliest ornithischians include primitive Triassic forms such as *Lesothosaurus*, *Fabrosaurus*, and *Heterodontosaurus*, which are small bipedal dinosaurs that look superficially like *Eoraptor* or *Coelophysis* (fig. 12.6). But on closer inspection, they show all the hallmarks of the ornithischians: part or all of the pubic bone in the hip is rotated back parallel to the ischium; the cheek teeth are deeply inset inside the jaw, suggesting that they had cheeks to confine their food in their mouths while they chewed; and they have a unique extra bone at the tip of the lower jaw known as the predentary bone. All these features are unique to the Ornithischia, yet we can see the transitional forms like *Heterodontosaurus* already had them in the Triassic while they still resembled the other primitive dinosaurs of the time.

The other example that Gish always trotted out is *Triceratops*, one of the last of the horned dinosaurs or ceratopsians. Even in his 1995 book (pp. 119–122) he talked about it and a few other ceratopsians, claiming there are no transitional forms—and completely missing the point of everything he has read! Ceratopsians provide yet another classic case of transitional forms between highly specialized forms, such as the horned and frilled ceratopsians, and much more primitive forms that resemble the common ancestor with other dinosaur lineages.

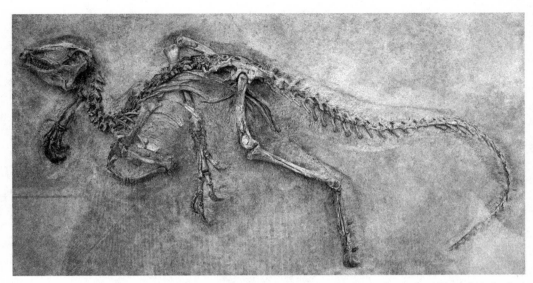

FIGURE 12.6. *Heterodontodosaurus*, the most primitive known ornithischian dinosaur. Although it looked superficially like *Eoraptor* and *Coelophysis* in its small size and bipedal stance, it had the unique predentary bone, backward rotated pubic bones, and other hallmarks of the ornithischians. (Photo courtesy R. Rothman)

In the case of the Ceratopsia, the transition is clear. All of the creatures with horns can be traced back to the very well known *Protoceratops*, which has the frill over the neck and the distinctive bones in the beak and lower jaw but lacks the horns (fig. 12.7). From the very beginning, paleontologists pointed to *Protoceratops* as a nice transition between horned ceratopsians and more primitive dinosaurs, but Gish apparently couldn't figure this out. He cites one out-of-context quotation from the first edition of Weishampel et al. (1990) about the distinctiveness of the Protoceratopsidae (completely missing the point that this does not make them any less a good transitional form), and he also mentions that they occur in the Late Cretaceous, so they can't be ancestral. First, paleontologists are not looking for ancestors but for sister groups (see chapter 5). Second, *Protoceratops* occurred in the *early* Late Cretaceous, millions of years before all of its presumed descendants among the Ceratopsia in the *late* Late Cretaceous.

And we have even better transitional forms. *Bagaceratops* (fig. 12.7) has a slightly smaller frill and beak compared with *Protoceratops*, and its body is not fully quadrupedal. *Archaeoceratops* has an even smaller frill and beak, and a body that is much lighter and most likely bipedal. *Psittacosaurus* (the "parrot lizard"), which was so named because it had a parrot-like beak (composed of the rostral bone unique to the Ceratopsia) and the beginnings of a frill over the neck, is much more lightly built with a bipedal, gracile skeleton rather than the heavier skeleton of *Protoceratops*. This creature not only shows the transition from frill-less skull to one with a small frill to the larger frill of *Protoceratops*, but it also shows the transition from a light bipedal body (typical of nearly all the primitive dinosaurs) to the heavier quadrupedal body of the more specialized groups.

Finally, the most amazing transitional fossil in this sequence was revealed with the 2004 discovery of *Yinlong* (fig. 12.8) from the much earlier beds of the Late Jurassic of China (Xu et al. 2006). Its name means "hidden dragon" in Mandarin, in reference to the

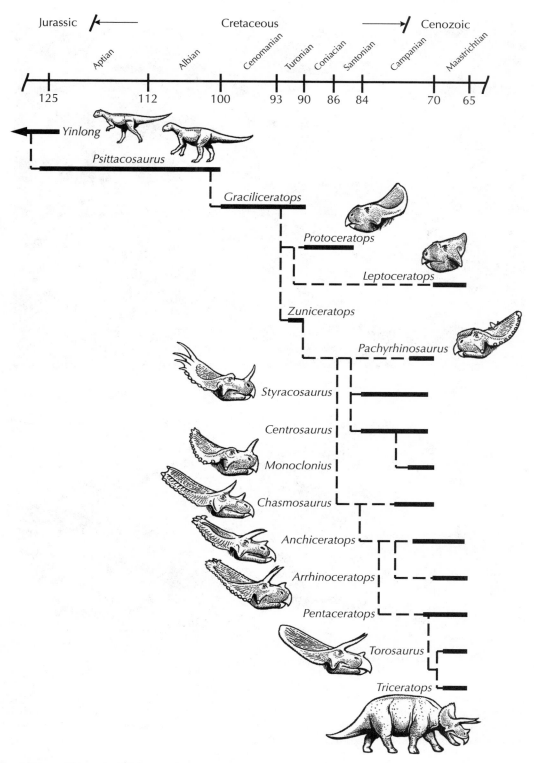

FIGURE 12.7. The evolutionary tree of the horned dinosaurs from *Yinlong* and other primitive marginocephalans. (Drawing by Carl Buell)

(A)

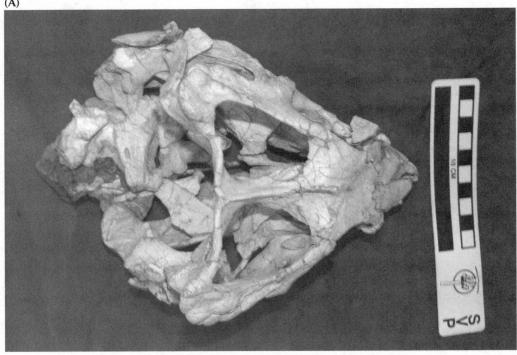

(B)

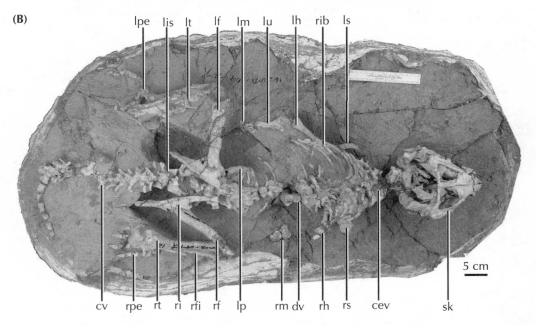

lpe lis lt lf lm lu lh rib ls

5 cm

cv rpe rt ri rfi rf lp rm dv rh rs cev sk

FIGURE 12.8. (A) The primitive marginocephalan *Yinlong*, which shows skull features linking both horned dinosaurs (ceratopsians) and the bone-headed pachycephalosaurs. (B) The complete articulated skeleton of *Yinlong*. For explanation of labels see Xu et al. (2006) (Photos courtesy J. Clark, George Washington University)

popular movie *Crouching Tiger, Hidden Dragon*, which was partially filmed near the locality where the fossil was found. *Yinlong* consists of a beautifully preserved skeleton of a bipedal dinosaur not too different in proportions from *Psittacosaurus*. It has the rostral bone unique to ceratopsians on the tip of its upper beak. However, its skull roof has a unique configuration of bones found in the "bone-headed" dinosaurs, the pachycephalosaurs, which are famous for having a thick dome of bone in their skulls protecting their tiny brains. Paleontologists have long argued that ceratopsians and pachycephalosaurs are closest relatives, based on the fact that they both have a frill of bone around the back margin of the skull (hence their name, "Marginocephala," or "margin heads"). But with *Yinlong*, we have a beautiful transitional fossil that shows features of both ceratopsians and pachycephalosaurs before their lineage split into the two families that every kid recognizes.

As usual, Gish hadn't done his homework or bothered to read more recent sources. The fact that he cited an out-of-context quotation from Weishampel et al. (1990) shows that he could apparently read a more authoritative source, but either he could not read well enough to also discover that the other transitional forms like *Psittacosaurus* are mentioned in the same chapter or his biases were so strong that he can only find short snippets that fit his prejudices. Either way, he completely missed the forest for the trees. And he definitely hasn't looked at the fossils himself or acquired the training necessary to understand what he was looking at. During our debate, I nailed him on this point. After he blathered on and on about no transitional ceratopsians, I showed him not only the examples just discussed but gave my personal testimony. When I was a graduate student, most of the good specimens of *Psittacosaurus* and *Protoceratops* had been cluttering up my office in the American Museum of Natural History for months because my officemate, Dan Chure (recently retired after almost 40 years as the paleontologist at Dinosaur National Monument), had been studying the specimens for his thesis. Not only did I know more about these transitional fossils than Gish, but I have actually studied them as well.

We could go on and on debunking creationist falsehoods about dinosaurs. Anyone with a moderate interest in the subject can peruse the chapters in some of the books listed at the end of this chapter (such as Norman 1985; Weishampel et al. 2004; Fastovsky and Weishampel 2005) and see the beautiful intermediate fossils for nearly every group that Gish denies have transitional forms. There are excellent specimens of primitive relatives of duckbill dinosaurs (they are called ornithopods), of primitive armored ankylosaurs known as nodosaurs (with very limited armor and very primitive delicate skeletons compared to the huge *Ankylosaurus*), and of primitive stegosaurs, such as *Scelidosaurus*, which has limited armor, smaller size, delicate limbs, and a very primitive skeleton compared with *Stegosaurus*. We can trace all of these ornithischian dinosaurs (plus the ceratopsians) back to the most primitive forms, such as *Heterodontosaurus* (fig. 12.6) and *Fabrosaurus* from the Triassic. These creatures, in turn, looked very similar in their external features to *Eoraptor* and *Herrarasaurus*, the earliest dinosaurs, except for a few subtle differences, such as the presence of a predentary bone at the tip of the lower jaw and a primitive ornithischian hip structure. If one reads the literature carefully and with an open mind, these transitions are obvious. If you read with the denial filter of Gish, who only found quotes out of context that seem to support his view, you will get the distorted, misleading version that he presents.

Dinosaurs Are Alive!

And if the whole hindquarters, from the ilium to the toes, of a half-hatched chick could be suddenly enlarged, ossified, and fossilised as they are, they would furnish us with the last step of the transition between Birds and Reptiles; for there would be nothing in their characters to prevent us from referring them to the Dinosauria.
—Thomas Henry Huxley, "Further Evidence of the Affinity Between Dinosaurian Reptiles and Birds"

In 1861, just 2 years after Darwin's book was published, a remarkable specimen was found in the limestone quarries of Solnhofen in Bavaria, southern Germany. These quarries had been excavated for years because they produced nice flat slabs of extremely fine-textured limestone that could be etched with acid to form the lithographic plates that printers used to make book illustrations. Occasionally, these limestones would also yield exquisitely preserved fossils as well, including the tiny dinosaur *Compsognathus* (the "compys" of *Jurassic Park* fame) and some of the first good pterodactyls. But in 1860, an impression of a fossil feather was found, and six months later, workers found a partial skeleton of a peculiar creature that had feathers but bones like a dinosaur.

Naturally, this specimen caused a sensation, and the British Museum in London outbid all the others to acquire it. As soon as it reached London (it is still known as the "London" specimen for where it now resides), it was the responsibility of Richard Owen, curator of the British Museum, and the man who named the "Dinosauria," to describe it. It had already been named *Archaeopteryx* ("ancient wing"), and although Owen basically described it as a bird, he could not help but see all the dinosaurian characteristics of the skeleton. But because he was one of the last reputable biologists to resist evolution, he made no effort to connect this fossil with its relatives.

However, the dinosaurian characteristics did not escape Owen's rival, Thomas Henry Huxley, who by this point had become "Darwin's bulldog" and was making speeches and publishing works that supported Darwin's theory. Having been one of the first to do anatomical studies of modern birds, and having studied a number of dinosaurs like *Compsognathus*, Huxley could not help but notice that *Archaeopteryx* was a classic "missing link" between birds and dinosaurs. At a famous presentation in front of the Royal Society in 1863, he proposed that birds were descended from dinosaurs and listed 35 features shared only by nonavian dinosaurs and birds (17 of these are still used by modern paleontologists). By 1877, an even better fossil was found, the classic "Berlin specimen" (see page 151), which is the best preserved of the 12 known specimens. By then, the Germans had come to realize the importance of *Archaeopteryx*. German industrialists bought it and made sure it stayed in Berlin, where it is now on display in the Museum für Naturkunde. It even survived the bombing during World War II. I have seen both the original London and the Berlin specimens close up, and it is like a pilgrimage to the Holy Grail to see such amazing and historic fossils rather than photographs or casts.

After Huxley's efforts, the dinosaur-bird hypothesis declined in popularity as another paleontologist, Harry Govier Seeley, challenged it. In 1926, the artist Gerhard Heilmann proposed that birds originated from more primitive archosaurs (then known as "thecodonts"), and his influence dominated for half a century. Heilmann did not have much evidence to contradict Huxley's hypothesis, except that he argued that none of the known theropod dinosaurs that could be related to birds had clavicles or collarbones, yet all birds have their clavicles fused into the "wishbone," which serves as an important "spring" in the flight stroke. (Clavicles have since been found in a number of theropods, removing this objection; they are just delicate and rarely preserved). Of course, this is pure ancestor worship. Heilmann's preferred candidates (such as *Euparkeria*) among the archosaurs are simply primitive with respect to all dinosaurs. In the process, he ignored all the derived similarity between dinosaurs and birds that Huxley had demonstrated, largely because most paleontologists thought of dinosaurs as big and specialized and couldn't imagine birds originating from these huge land animals. Heilmann's thinking was also dictated by scenarios about how birds evolved flight by gliding from branch to branch (the "trees down" hypothesis), and large terrestrial dinosaurs don't fit that scenario.

The "birds as dinosaurs" hypothesis remained unfashionable until the 1970s, when Yale paleontologist John Ostrom (a good friend of mine, he died in 2005) looked at the specimens in Europe again. He found one specimen in the Teyler Museum in the Netherlands that had been misidentified as a pterosaur in 1855, but when he looked closer, he saw faint feather impressions and knew he had another specimen of *Archaeopteryx*. Meanwhile, a specimen in the Eichstatt Museum found in 1951 had been misidentified as the Solnhofen dinosaur *Compsognathus*, until F. X. Mayr found feather impressions on it 20 years later. The fact that a small dinosaur and *Archaeopteryx* could be so easily mistaken for each other was a revelation for Ostrom. He revived Huxley's hypothesis and added a long list of evidence to support it further. He was influenced by the fact that in the 1960s he had discovered and described the highly specialized dinosaur *Deinonychus* (misidentified as "*Velociraptor*" in the *Jurassic Park* movies), which shows amazing anatomical similarities to the earliest birds.

After Ostrom's initial papers, the controversy over the "birds are dinosaurs" hypothesis raged for several years but quickly resolved because the evidence soon became overwhelming. Hundreds of specialized shared-derived characters support the hypothesis (Gauthier 1986), and there are no competing hypotheses with even a fraction of that support. All but a tiny minority (less than 1 percent) of paleontologists are convinced by the data, especially with the discovery of feathered nonavian dinosaurs from China that we shall discuss shortly. Those few who insist that "birds are not dinosaurs" have no competing hypothesis with any strong support, and their case consists largely of trying to nitpick individual characters that occur both in birds or dinosaurs. They do not address the overwhelming evidence of most of the characters. Indeed, they don't use or seem to understand cladistics, which is part of the problem. Another part of the problem is that they are wedded to the "trees down" hypothesis of the origin of flight and cannot imagine terrestrial dinosaurs evolving flight from the "ground up." Ken Dial (2003) showed that their emphasis on the "trees down" origin of flight is misguided. Many birds, such as chukar partridges, use the lift of their wings to help them run up steep inclines but seldom use them for flight. It is easy to see how dinosaurs (which already had feathers for insulation, as we will see later) could have adapted these structures to make climbing steep inclines easier, and from there began to develop short glides and eventually true flight.

But evolutionary scenarios must not drive the analysis, and scientists still have to abide by the rules of science and provide positive evidence and well-supported alternative hypotheses, not "just-so" stories. As detailed by Norell (2005:215–229), their attacks on the widely accepted hypothesis mostly amount to ridiculous and often self-contradictory sniping, without proposing an alternative hypothesis of their own. In that respect, they resemble the creationists, who attack one tiny detail of the subject without addressing all the rest of the evidence.

New discoveries have further reinforced the point that theropod dinosaurs were bird-like not only in their anatomy but also in their behavior. In the 1990s, expeditions from the American Museum of Natural History to the Gobi Desert in Mongolia made some remarkable discoveries, including nests of eggs of the dinosaur *Oviraptor*. These eggs were so common in Mongolia that the original American Museum expeditions in the 1920s had attributed them to the most common dinosaur of these beds, the primitive horned *Protoceratops* (fig. 12.7). When bones of a small theropod were found near some nests, they were given the name *Oviraptor* ("egg thief"). But the recent expeditions show that this name is slanderous: *Oviraptor* wasn't stealing the eggs—it was their mother! In some cases, the female *Oviraptor* skeleton was entombed in brooding posture right on top of the eggs as they were both buried in a sandstorm and fossilized. The details of this brooding posture and the way in which it was preserved show that many theropod dinosaurs acted more like birds than like reptiles.

Fastovsky and Weishampel (2005:261) point to another problem: the media. The debate has been unnaturally prolonged by media attention. The origin of birds has been a topic of great public interest for the past 20 years, so much so that the leading proponents are frequently interviewed for newspaper articles and TV specials. The rules of journalism require that "equal time" be given to representatives of each viewpoint. So the supports of the basal diapsid origin of birds often have as much airtime as the supporters of birds as dinosaurs, even though the latter represent probably more than 99 percent of working vertebrate paleontologists.

Before we deal with the creationist distortions about *Archaeopteryx*, let us review the evidence that convinced 99 percent of legitimate scientists that birds are dinosaurs. Much of this evidence is visible in *Archaeopteryx* itself (fig. 12.9A) and pointed out by Huxley from the beginning. Darwin could not have asked for a better transitional form than *Archaeopteryx*. As we saw earlier, most of its skeleton is so dinosaurian that one specimen was mistaken for the little theropod dinosaur *Compsognathus*. Like most theropod dinosaurs (but no living birds), it had a long bony tail, a highly perforated skull with teeth, theropod (not birdlike) vertebrae, a strap-like shoulder blade, a pelvis midway between that of typical saurischian dinosaurs and later birds, gastralia (rib bones found in the belly region of dinosaurs), and unique specializations in the limbs. The most striking of these are in the wrist. Birds and some theropod dinosaurs, such as the dromaeosaurs (*Deinonychus* and *Velociraptor* and its kin), all have a half-moon-shaped wristbone formed of fusion of multiple wristbones known as the *semilunate carpal* (fig. 12.9B). This bone serves as the main hinge for the movement of the wrist, allowing dromaeosaurs to extend their wrists and grab prey with a rapid protraction and retraction. It so happens that exactly the same motion is part of the downward flight stroke of birds. *Archaeopteryx* had the same three fingers (thumb, index finger, and middle finger) as most other theropod dinosaurs, and the middle digit (the index finger) is by far the longest. In addition, the claws of *Archaeopteryx* are very similar to those of theropod dinosaurs.

The hind limbs of *Archaeopteryx* also have many dinosaurian hallmarks. The most striking of these is in the ankle (fig. 12.9C). All pterosaurs, dinosaurs, and birds have a unique ankle arrangement known as the *mesotarsal joint*. Instead of the typical vertebrate ankle, which hinges between the shin bone (tibia) and the first row of ankle bones (as your ankle does), pterosaurs, dinosaurs, and birds developed a hinge between the first and second row of ankle

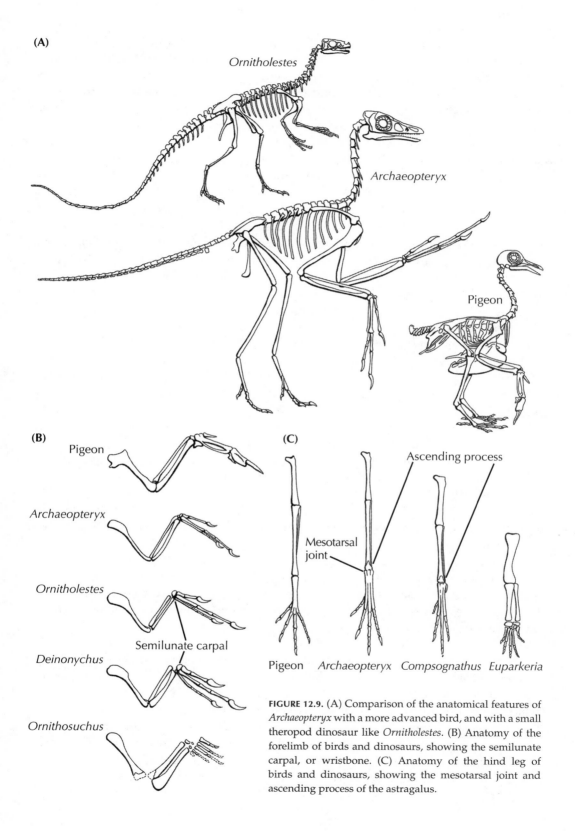

(A)

Ornitholestes

Archaeopteryx

Pigeon

(B)

Pigeon

Archaeopteryx

Ornitholestes

Semilunate carpal

Deinonychus

Ornithosuchus

(C)

Ascending process

Mesotarsal joint

Pigeon   *Archaeopteryx*   *Compsognathus*   *Euparkeria*

FIGURE 12.9. (A) Comparison of the anatomical features of *Archaeopteryx* with a more advanced bird, and with a small theropod dinosaur like *Ornitholestes*. (B) Anatomy of the forelimb of birds and dinosaurs, showing the semilunate carpal, or wristbone. (C) Anatomy of the hind leg of birds and dinosaurs, showing the mesotarsal joint and ascending process of the astragalus.

bones, or within the ankle (mesotarsus). The first row of ankle bones thus has little function except as a passive hinge, and in many taxa, it actually fuses on to the end of the shin bone as a little "cap" of bone. The next time you eat a chicken or turkey drumstick (which is its tibia bone), notice that inedible cap of cartilage at the less meaty "handle" end of the drumstick is actually a relict of the dinosaurian ancestry of birds! In addition, part of this first row of ankle bones has a bony spur that runs up the front of the tibia (the ascending process of the astragalus), another feature unique to saurischian dinosaurs and birds. Finally, the details and the structure of the toe bones and the short hallux, or big toes, are unique to theropod dinosaurs and birds as well, although *Archaeopteryx* did not have the opposable big toe that would have enabled it to grasp branches well.

With all this evidence that *Archaeopteryx* is basically a feathered dinosaur, why call *Archaeopteryx* a bird at all? In fact, it has only a few uniquely birdlike features not found in other theropods: the big toe is fully reversed, the teeth are unserrated, and the tail is relatively short but the arms are long compared to most other theropods. All the other features of *Archaeopteryx*, including the feathers and the fused collarbones or "wishbone," have now been found in other theropod dinosaurs, although some say that the feathers of *Archaeopteryx* are more advanced than those in theropods and have the asymmetry that suggests that *Archaeopteryx* could fly.

In light of all this overwhelming evidence, it is bizarre to read how the creationists distort and misrepresent *Archaeopteryx*. In their minds, the created "kinds" have to be distinct, and transitional forms cannot exist, therefore they will do whatever it takes, no matter how dishonest and unscientific, to try to discredit *Archaeopteryx*. Because *Archaeopteryx* had feathers, in their minds it must be part of the bird "kind," so the creationists will simply say that it's a bird and either distort or not even address the long list of dinosaurian features in the specimen. For example, the creationist books by Davis and Kenyon (2004:104–106), Sarfati (1999:57–68; 2002:130–132), Wells (2000:111–135), and Gish (1995:129–139) mostly quote out-of-date sources to try to discredit *Archaeopteryx* and the "birds are dinosaurs" hypothesis. Or they quote old papers by the tiny handful of crank scientists like Alan Feduccia and the late Larry Martin, who disagreed with 99 percent of the profession—but they don't mention the devastating counterarguments against the ideas proposed by Feduccia and Martin. Gish (1995) and Sarfati (1999) argue that *Archaeopteryx* teeth are not like those of theropod dinosaurs but like those of other toothed birds (which is not true, by the way—they have both primitive similarities with theropod teeth and their own derived features), but this whole argument misses the point: no living birds have teeth, yet if fossil birds like *Archaeopteryx* had them, it links birds and dinosaurs. (Recall from chapter 4 that birds still have the embryonic genes for teeth, but these are normally suppressed during development). Gish (1995) and Sarfati (1999) briefly mention the long tail of *Archaeopteryx* and then blithely say that some reptiles and some birds have long tails and some have short ones. The point is that no living bird has a long *bony tail* (all the tailbones of living birds are fused into the "parson's nose," or *pygostyle*, and the tail is supported by feather shafts instead), yet *Archaeopteryx*, which even Gish admits is a bird, has the long bony tail of dinosaurs. Gish attempts to discredit the three bony clawed fingers of *Archaeopteryx* by pointing to the hoatzin birds of Central America, which also have these three fingers while they are chicks (although he fails to mention that their configuration is entirely different). But one isolated atavism does nothing to discredit the fact that the hand of *Archaeopteryx* is fundamentally dinosaurian. No other living birds besides the hoatzin chicks have this type of hand, which is highly specialized and in no way resembles the hand of *Archaeopteryx*. In short, every time Gish, Sarfati, and the other creationists mention a feature that makes *Archaeopteryx* a dinosaur, they distort the evidence, show their ignorance of the anatomical details, or fail to mention counterarguments or details that would discredit their case. Nowhere do the creationists discuss the other 100 or so anatomical characters, including such unique dinosaurian features as the mesotarsal joint and the semilunate carpal. On those grounds alone, their arguments are worthless and only show how poorly creationists understand the anatomy of these creatures.

The creationists are so wedded to the idea of distinct "kinds" that they cannot even conceive of intermediate forms. Gish was embarrassed during several debates in just this way. When his opponent put up an image of the forelimb of a modern bird and of a theropod dinosaur (fig. 12.19B) and challenged Gish to sketch the likely intermediate anatomy, Gish declined (because he knew it was a trap). Sure enough, his opponent then revealed the forelimb of *Archaeopteryx* as a perfect intermediate between birds and dinosaurs. Gish then mumbled something irrelevant and tried to change the subject. Even more dishonestly, he went on with the same deception at the next debate venue, never correcting mistakes when he was shown to be lying.

The intelligent design creationist authors are even more subtle and misleading. They use a few out-of-context quotations that do not apply to this case and fall back on the old misconception that evolution must be a smooth gradual "chain of being" within a single lineage. Davis and Kenyon (2004:106) write that *Archaeopteryx* "is transitional only if it is part of lineage—one of series of generations in which in-between stages led gradually from one group to another" (illustrated clearly in their figs. 4–11, p. 106). In one sentence, they have shown their complete misunderstanding of the fundamental concepts of evolution. *Archaeopteryx* does *not* have to be part of single, gradually evolving lineage to be a transitional form—those are all misunderstandings about evolution discredited decades ago. It only needs to be one of many species that show transitional features on the bushy, branching tree of life. And in this respect, *Archaeopteryx* could not be a better intermediate transitional form. Wells (2000) claims that paleontologists have "quietly shelved" *Archaeopteryx* and that it is not an "ancestor" because modern birds are not descended from it. This completely misses the point. *Archaeopteryx* does not have to be an actual ancestor to show us how birds evolved from dinosaurs. It has all the transitional features that one might expect from the sister group or "collateral ancestor" of birds (and has no unique specializations that would preclude it from being the actual ancestor). And nobody has "quietly shelved" it—it is still being published on and studied and mentioned in the ever-burgeoning field of Mesozoic bird paleontology.

Finally, Davis and Kenyon, Sarfati, Wells, and Gish all argue that because *Archaeopteryx* and its dinosaurian kin are the same age (or some of the dinosaurian sister taxa appear later than *Archaeopteryx*), then the dinosaurs cannot be ancestral to the birds. As we have said over and over, evolution is a bush, not a ladder. *Archaeopteryx* and other theropods are sister taxa, and their relationships are supported by shared derived characters. Age relationships are irrelevant (especially in such rarely fossilized animals as small dinosaurs and birds). They shared common ancestors back in the Middle Jurassic, when we have a very poor record of terrestrial vertebrates worldwide, so that by the Late Jurassic, the lineages have just split apart, and both theropods and *Archaeopteryx* were living side by side.

But all these arguments of the creationists, as well as the "birds are not dinosaurs" minority like Martin and Feduccia, are now rendered entirely obsolete by an amazing array of new discoveries that have occurred in the past 20 years. If *Archaeopteryx* were still the

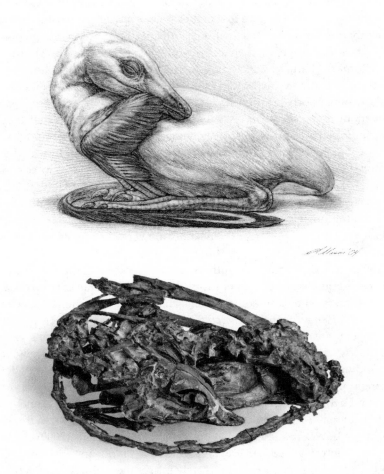

FIGURE 12.10. *Meilong*, a feathered dinosaur from the Liaoning beds that is preserved as an extraordinary three-dimensional specimen coiled up in a sleeping position. (Courtesy M. Ellison and M. Norell, American Museum of Natural History)

only transitional dinosaur-bird fossil, it would be sufficient, but it is not alone any more. An amazing array of new transitional bird fossils and feathered nonbird dinosaurs have been discovered and described (figs. 12.1, 12.6, 12.10, and 12.11) that fill in most of the gaps between theropods and advanced birds, so now we have a wealth of transitional forms, of which *Archaeopteryx* is just one link.

The most earth-shaking discoveries come from the famous Lower Cretaceous Liaoning fossil beds of China, which have now become one of the world's most important fossil deposits. These delicate lake shales preserve extraordinary features in fossils, including body outlines, feathers, and fur, as well as complete articulated skeletons with not a single bone missing. In the past 20 years, a major new discovery has been announced from these deposits every few months, and almost all previous ideas about birds and dinosaurs were quickly rendered obsolete by these discoveries (for a summary, see Norell 2005). The most amazing fossils of all were a number of clearly nonflying, nonavian dinosaurs with well-developed

(A)

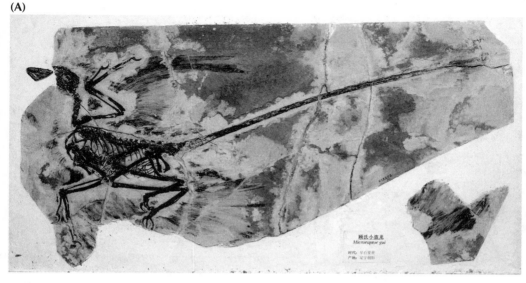

(B)

FIGURE 12.11. Feathered nonflying dinosaurs from the Lower Cretaceous Liaoning beds of China, which show the early stages of evolution from dinosaurs to birds. (A) *Microraptor*, which had feathers on both its hands and legs, although it is still controversial whether it flew. (B) *Sinosauropteryx*, the first feathered nonavian dinosaur to be discovered, with its beautiful preservation of hairlike feathers (especially visible along the spine). (Photos courtesy M. Ellison and M. Norell, American Museum of Natural History)

feathers (figs. 12.1, 12.10, 12.11, and 12.12). These include incredible complete specimens such as *Sinosauropteryx*, *Protarchaeopteryx*, *Sinornithosaurus*, *Caudipteryx*, the large theropod *Beipiaosaurus*, and the tiny *Microraptor*.

Most of these non-bird dinosaurs clearly do not have flight feathers or other indications that their feathers were used for flight. Instead, they show that feathers were apparently a widespread feature among theropod dinosaurs (and perhaps in other dinosaurs and

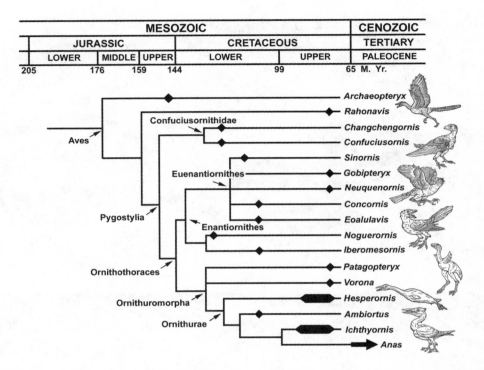

MESOZOIC						CENOZOIC
JURASSIC			CRETACEOUS			TERTIARY
LOWER	MIDDLE	UPPER	LOWER		UPPER	PALEOCENE
205	176	159 144		99		65 M. Yr.

FIGURE 12.12. The family tree of Mesozoic birds, emphasizing some of the recent fossil discoveries. (Courtesy L. Chiappe)

archosaurs as well, especially pterosaurs). Feathers, then, did not evolve for flight but were already present in theropod dinosaurs, presumably for insulation, and were later modified to become flying structures.

Prum and Brush (2003) have completely rethought the origin of feathers and showed that they are not modified scales (as once believed) but from a similar embryonic primordium with different Hox genes controlling development. Type 1 feathers (fig. 12.13) are simple hollow pointed shafts, which appear in the primitive theropod *Sinosauropteryx*. Type 2 feathers are simple down with no vanes, and type 3 feathers have a vane and shaft, but no barbules linking them together like Velcro. Both of these types are found in the large therizinosaur *Beipiaosaurus*, suggesting that they were present in almost all theropods (fig. 12.13). Type 4 feathers have barbules that link the vanes of the feather into a continuous surface, but the shaft is symmetrically aligned down the middle of the feather. This kind of feather appears in *Caudipteryx*, which suggests that these occurred in higher theropods (including *Tyrannosaurus rex*) as well. The classic asymmetric flight feather with the shaft near the leading edge of the vane first appears in *Archaeopteryx*, and for this reason many scientists think that *Archaeopteryx* was one of the first to modify the long heritage of feathers for true flight.

Moving up from *Archaeopteryx* on the cladogram of birds (fig. 12.12), we come to *Rahonavis* from the Cretaceous of Madagascar (Forster et al. 1998). About the size of a crow

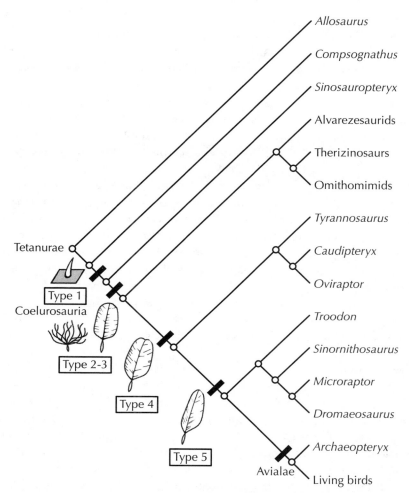

FIGURE 12.13. The evolution of feather types from simple pinshafts to down plumes to complex flight feathers with asymmetric vanes and shaft. On the basis of their appearance in various feathered nonflying dinosaurs from Liaoning, we can demonstrate that most predatory dinosaurs (including *T. rex*) probably had feathers of some sort. (Modified from Prum and Brush 2003)

(fig. 12.14A), it had the primitive sickle-like claws on the hind feet, the long bony tail, teeth, and many other theropod features, but it also had birdlike features such as the fusion of its lower back vertebrae with the pelvis (the *synsacrum*), holes in its vertebrae for all the blood vessels and air sacs found in living birds, fingers with quill knobs, suggesting that it was feathered and could fly (no surprise here), and the fibula (the smaller shin bone) that does not reach the ankle. Birds have reduced the fibula to the tiny splint of bone that you bite into when you are eating a chicken or turkey drumstick, but *Archaeopteryx* has a fully developed fibula like that of dinosaurs.

The next step is marked by *Confuciusornis* and its relatives (fig. 12.14B), which has a unique feature found in all higher birds: the pygostyle, formed by the fusion of all the old dinosaurian tail vertebrae into a single "parson's nose." These higher birds have also increased the number of lower back vertebrae fused to the synsacrum and elongated the bones that reinforce the shoulder, which improved flight. They also are the first birds with a toothless beak. Following this transitional form is another branch point that leads to the extinct Enantiornithes, or "backwards birds" (so named because their leg bones ossify in the reverse direction from that found in modern birds). These include *Iberomesornis* from the Cretaceous Las Hoyas locality in Spain, *Sinornis* from China (fig. 12.14C), *Gobipteryx* from Mongolia, *Enantiornis* from Argentina, and several others. All of these birds are more special- ized than *Archaeopteryx*, Rahonavis, or Confuciusornis in that they have reduced the number of trunk vertebrae, have a flexible wishbone, made the shoulder joint better for flying, fused the hand bones into a bone called the carpometacarpus, and the finger bones into a single element (the meatless bony part of the chicken wing that you never eat).

Continuing up the cladogram, we come to several Cretaceous birds such as *Vorona* from Madagascar, *Patagopteryx* from Argentina, and the well-known aquatic birds *Hesperornis* and *Ichthyornis* from the chalk beds of Kansas. These birds are united by at least 15 well-defined characters, including the loss of the belly ribs or gastralia, reorientation of the pubic bone to the modern birdlike position parallel to the ischium, reduction in the number of trunk verte- brae, and many other features of the hand and shoulder that improved flight performance. *Ichthyornis* is even closer to modern birds in having a keel on its breastbone for the flight mus- cles and a knob-like head on the upper arm bone that made the wing more flexible. Finally, the clade that includes all modern members of class Aves is defined by the complete loss of teeth and a number of other anatomical specializations, such as the fusion of the leg bones to form a tarsometatarsus.

How do creationists respond to this flood of new discoveries? Most of the time, they don't. Even recent books like Sarfati (1999, 2002) and the constantly updated creationist web- sites completely ignore them. Wells (2000) ignores nearly all of them except for one specimen, named "*Archaeoraptor*," which was a composite forged out of two real fossils by an unknown Chinese fossil dealer. Smuggled out of China, the specimen was bought and made into a big deal by amateur dinosaur illustrators (and by *National Geographic*, which wanted to get a scoop without waiting for the specimen to be tested by peer review). As soon as well-trained paleontologists looked at the specimen, they quickly detected that it was a composite of two different specimens put together to enhance its sale price, and the specimen was never even formally published in a peer-reviewed journal. Wells (2000) slanders the entire profession by suggesting that one artful hoax (which was quickly exposed as soon as real paleontologists looked at it) implies that *all* the fossils from China are faked or that qualified paleontologists are easily suckered by fakes. As the facts of the story show, Wells is wrong on all counts.

FIGURE 12.14. In addition to *Archaeopteryx*, there are now dozens of new transitional birds from the Mesozoic, ▶ each of which shows a mosaic of evolutionary changes from more dinosaur-like creatures like *Archaeopteryx* to forms that are similar to modern birds in many ways. (A) *Rahonavis* from the Cretaceous of Madagascar, which still has the teeth, long clawed fingers, and long bony tail of *Archaeopteryx*, but the hip vertebrae are fused to the hip bones (*synsacrum*) as in modern birds. (After Forster et al. 1998; copyright © 1998 Association for the Advancement of Science) (B) *Confuciusornis* from the Cretaceous of China, which has fused the tail vertebrae into a pygostyle, and lost its teeth but still has the long dinosaurian fingers. (After Hou et al. 1995; used by permission of the Nature Publishing Group) (C) *Sinornis*, a primitive enantiornithine bird from the Cretaceous of China, which still has teeth, an unfused tarsometatarsus, and an unfused pelvis but had shorter fingers, a fully opposable big toe for perching, a broad breastbone for flight muscle attachment, and an even shorter pygostyle in the tail. (From Sereno and Rao 1992: fig. 2. Used by permission of the Nature Publishing Group)

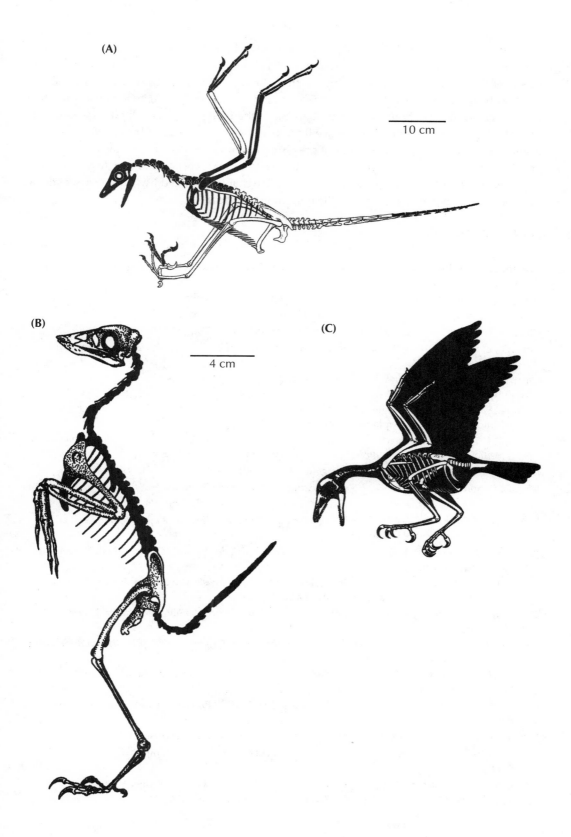

(A)

10 cm

(B)

4 cm

(C)

If this welter of new bird fossils and anatomical characters seems a bit overwhelming, your impression is correct: the past two decades have produced such an explosion of new fossils and new ideas that everything we thought we knew about Mesozoic birds before 1990 is obsolete. Each year brings astonishing new specimens that further transform what we thought we knew about avian evolution. Since the final picture is still taking shape, we cannot tell how many more changes we'll make in our cladograms of birds before the discoveries start repeating themselves. But one thing is abundantly clear: we now have dozens of beautiful transitions from dinosaurs to birds. The creationist books that focus only on *Archaeopteryx* and distort the fossil record are so laughably outdated by the new discoveries that their writings are only fit to line the bottom of a birdcage.

For Further Reading

Benton, M. J., ed. 1988. *The Phylogeny and Classification of the Tetrapods.*Vol. 1, *Amphibians, Reptiles, Birds*. Oxford, U.K.: Clarendon.

Benton, M. J. 2014. *Vertebrate Palaeontology*. 4th ed. New York: Wiley-Blackwell.

Carroll, R. L. 1988. *Vertebrate Paleontology and Evolution*. New York: Freeman.

Chiappe, L. M. 1995. The first 85 million years of avian evolution. *Nature* 378:349–355.

Chiappe, L. M., and G. J. Dyke 2002. The Mesozoic radiation of birds. *Annual Review of Ecology and Systematics* 33:91–124.

Chiappe, L. M. and L. M. Witmer, eds. 2002. *Mesozoic Birds: Above the Heads of Dinosaurs*. Berkeley: University of California Press.

Chiappe, L. M., and Meng Qingjin. 2016. *Birds of Stone: Chinese Avian Fossils from the Age of Dinosaurs*. Baltimore, Md.: Johns Hopkins University Press.

Currie, P. J., E. B. Koppelhus, M. A. Shugar, and J. L. Wright, eds. 2004. *Feathered Dragons: Studies on the Transition from Dinosaurs to Birds*. Bloomington: Indiana University Press.

Dial, K. 2003. Wing-assisted incline running and the evolution of flight. *Science* 299:402–405.

Dingus, L., and T. Rowe. 1997. *The Mistaken Extinction*. New York: Freeman.

Dodson, P. 1996. *The Horned Dinosaurs*. Princeton, N.J.: Princeton University Press.

Fastovsky, D. E., and D. B. Weishampel. 2005. *The Evolution and Extinction of the Dinosaurs*. 2nd ed. New York: Cambridge University Press.

Fastovsky, D. E. and D. B. Weishampel. 2016. *Dinosaurs: A Concise Natural History*. 3rd ed. New York: Cambridge University Press.

Forster, C. A., S. D. Sampson, L. M. Chiappe, and D. W. Krause. 1998. The theropod ancestry of birds: new evidence from the Late Cretaceous of Madagascar. *Science* 279:1915–1919.

Gauthier, J. A. 1986. Saurischian monophyly and the origin of birds. *California Academy of Sciences Memoir* 8:1–56.

Gauthier, J. A., and L. F. Gall, eds. 2001. *New Perspectives on the Origin and Early Evolution of Birds*. New Haven, Conn.: Yale University Press.

Hou, L.-H. Z., Zhou, L. D. Martin, and A. Feduccia. 1995. A beaked bird from the Jurassic of China. *Nature* 377:616–618.

Long, J., and H. Schouten. 2008. *Feathered Dinosaurs: The Origin of Birds*. New York: Oxford University Press.

McGowan, C. 1983. *The Successful Dragons: A Natural History of Extinct Reptiles*. Toronto: Stevens.

Naish, D., and P. Barrett. 2016. *Dinosaurs: How They Lived and Evolved*. Washington, D.C.: Smithsonian Books.

Norell, M. 2005. *Unearthing Dragons: The Great Feathered Dinosaur Discoveries*. New York: Pi.

Norman, D. 1985. *The Illustrated Encyclopedia of Dinosaurs*. New York: Crescent.

Ostrom, J. H. 1974. *Archaeopteryx* and the origin of flight. *Quarterly Review of Biology* 49:27–47.

Ostrom, J. H. 1976. *Archaeopteryx* and the origin of birds. *Biological Journal of the Linnean Society* 8:91–182.

Padian, K., and L. M. Chiappe. 1998. The origin of birds and their flight. *Scientific American* 278:28–37.

Pickrill, J. 2014. *Flying Dinosaurs: How Reptiles Became Birds*. New York: Columbia University Press.

Prothero, D. R. 2013. *Bringing Fossils to Life: An Introduction to Paleobiology*. 3rd ed. New York: Columbia University Press.

Prum, R. O., and A. H. Brush. 2003. Which came first, the feather or the bird? *Scientific American* 288:84–93.

Schultze, H.-P., and L. Trueb, eds. 1991. *Origins of the Higher Groups of Tetrapods: Controversy and Consensus*. Ithaca, N.Y.: Cornell University Press.

Shipman, P. 1988. *Taking Wing: Archaeopteryx and the Evolution of Bird Flight*. New York: Simon & Schuster.

Weishampel, D. B., P. Dodson, and H. Osmolska, eds. 2004. *The Dinosauria*. 2nd ed. Berkeley: University of California Press.

Xu, Xing, C. A. Forster, J. M. Clark, and J. Mo. 2006. A basal ceratopsian with transitional features from the Late Jurassic of northwestern China. *Proceedings of the Royal Society of London B* 273:2135–2140.

FIGURE 13.1. The transformation from primitive synapsids like *Ophiacodon* and the fin-backed *Dimetrodon* to the predatory gorgonopsians to the weasel-like *Thrinaxodon* and finally to true mammals is one of the best transitional series in the entire fossil record. (Drawing by Carl Buell)

MAMMALIAN EXPLOSION 13

From Amniote to Synapsid to Mammals

Of all the great transitions between major structural grades within vertebrates, the transition from basal amniotes to basal mammals is represented by the most complete and continuous fossil record, extending from the Middle Pennsylvanian to the Late Triassic and spanning some 75 to 100 million years.

—James Hopson, "Synapsid Evolution and the
Radiation of Non-eutherian Mammals"

Of all the transitional series that we have examined between major groups of vertebrates, one of the best documented is the transition from primitive amniotes to mammals via the synapsids, formerly known as the "mammal-like reptiles." As we explained previously (fig. 11.1), however, the synapsids that evolve into mammals are not reptiles and never had anything to do with the lineage that leads to reptiles. Both the earliest true reptiles (*Westlothiana* from the Early Carboniferous—fig. 11.4) and the earliest synapsids (*Protoclepsydrops* from the Early Carboniferous and *Archaeothyris* form the Middle Carboniferous) are equally ancient, demonstrating that their lineages diverged at the beginning of the Carboniferous. Older pre-cladistic interpretations had synapsids evolving from a paraphyletic wastebasket of primitive amniotes known as the "anaspid reptiles." This idea is now completely discredited, and anyone who still uses the obsolete and misleading term "mammal-like reptile" clearly doesn't know much about the current understanding of vertebrate evolution.

Focusing on the lineage of synapsids, we can put together an almost continuous series of well-preserved fossils (figs. 13.1, 13.2, 13.3, and 13.4) that span the Carboniferous and Permian and straggle into the Triassic before most lineages died out (possibly in competition from the newly emergent dinosaurs), and the remaining lineages gave rise to true mammals. Each taxon along the way shows a mosaic of mammalian characters, with some advanced features appearing early in the series, while others appeared quite late. The most primitive group is the paraphyletic wastebasket known as "pelycosaurs" (fig. 13.2A and B), which includes not only the oldest and most primitive taxa (such as *Protoclepsydrops* and *Archaeothyris*) but also the spectacular "finbacks" *Dimetrodon* and *Edaphosaurus*, which were among the largest land animals of the Early Permian (fig. 13.2B). Although the earliest forms are almost indistinguishable from the earliest true reptiles in most features, creatures such as *Protoclepsydrops* and *Archaeothyris* still show a number of unique synapsid specializations, including a hole in the side of the skull (temporal opening) beneath the postorbital and squamosal bones, the beginnings of true canine-like teeth, and a number of other subtle features in the skull and palate. Even more advanced "pelycosaurs" like *Dimetrodon* are still primitive in most features but clearly have the lower temporal opening in the skull and the large canines in the front of the mouth.

(A)

(B)

(C)

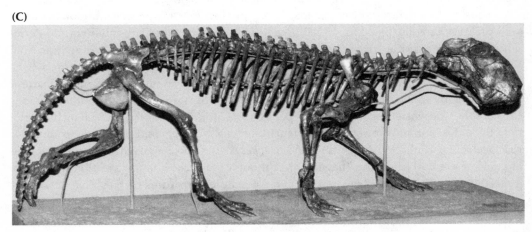

FIGURE 13.2. Skeletons of the various transitional fossils among the synapsids. (A) The very primitive early Permian "pelycosaur" *Ophiacodon*. (B) The finbacked pelycosaur *Dimetrodon*. (C) The predatory gorgonopsian *Lycaenops*, with its large wolf-like skull, big canines, and more upright posture. (D) The highly mammal-like cynodont *Thrinaxodon*, which was the size of weasel. (Photos courtesy R. Rothman)

(D)

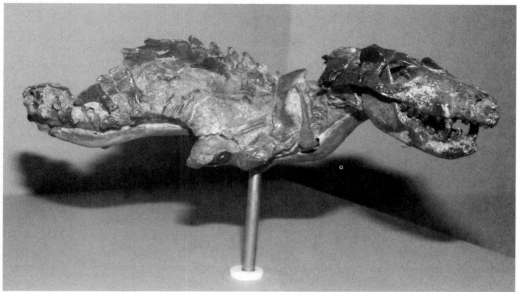

FIGURE 13.2. (*Continued*)

As we move up the cladogram through the diverse groups of synapsids (fig. 13.3), more and more mammalian characters appear in a piecemeal fashion. The next grade up from the pelycosaurs is the "therapsids," which dominated the Late Permian landscape and evolved into a number of large wolf-sized creatures with big saber-toothed canines (fig. 13.2C), as well as lineages of huge herbivorous synapsids, some with beaks and almost no teeth, and others with thickened skulls and ugly bony knobs on their faces for display and head-to-head butting. In short, these therapsids dominated many different niches in the Late Permian terrestrial ecosystem. They were also considerably more advanced than typical pelycosaurs like *Dimetrodon* (fig. 13.2B). The temporal opening on the side of the skull was now much larger, presumably for expansion of jaw muscles, therefore enabling them to have a stronger bite force and even some chewing motion. If you look at the palate of the skull, there is the beginning of a secondary palate, a roof of bone that grows out from the edge of the upper jaws and encloses the original reptilian palate in a tube. This enables advanced synapsids and mammals to breathe and eat at the same time, something that reptiles (except crocodilians, which independently evolved a secondary palate) cannot do. Reptiles, with their slow metabolism, can hold their breath for a long time while they swallow a large prey item, but the presence of a secondary palate shows that synapsids must have been developing an active, "warm-blooded" metabolism and needed to process their food quickly to survive. In addition, the old single ball joint that hinged the skull to the first neck vertebra (known as the occipital condyle) is now split into a double ball joint, presumably for more flexibility in moving the head.

In therapsids, the canines are much larger, and some of the rest of the teeth are now more specialized as well, with serrated edges like steak knives. There are also striking differences in the limbs (fig. 13.4), with a much stronger more flexible shoulder girdle and an increased number of lower back vertebrae fused to the hip bones. Finally, they no longer have the

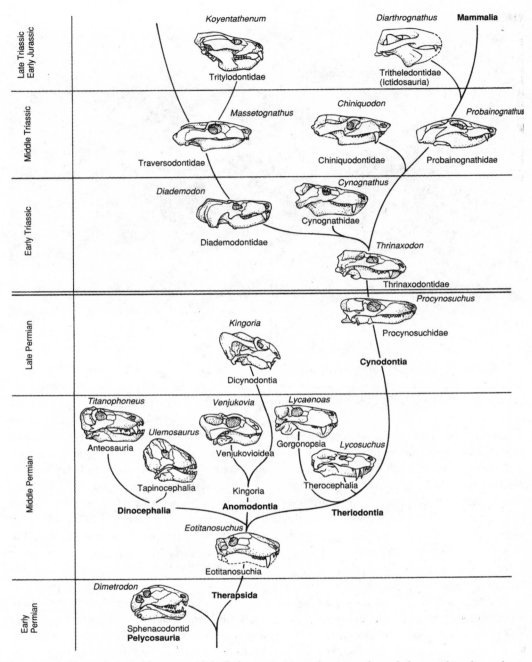

FIGURE 13.3. The evolution of the synapsid skulls from primitive pelycosaurs through therapsids and cynodonts to true mammals. (From Kardong 1995; reproduced by permission of the McGraw-Hill Companies)

Early mammal *(Megazostrodon)*

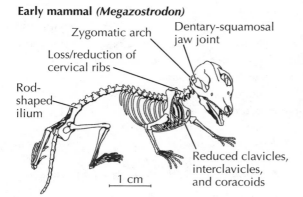

Zygomatic arch

Dentary-squamosal jaw joint

Loss/reduction of cervical ribs

Rod-shaped ilium

Reduced clavicles, interclavicles, and coracoids

1 cm

FIGURE 13.4. The transformation of the synapsid skeleton from primitive pelycosaurs like *Dimetrodon* through cynodonts to true mammals. (Drawing by Carl Buell).

Cynodont therapsid *(Thrinaxodon)*

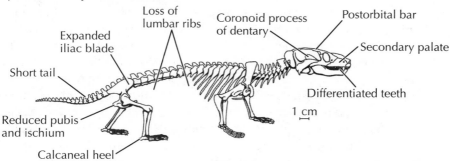

Loss of lumbar ribs

Coronoid process of dentary

Postorbital bar

Expanded iliac blade

Secondary palate

Short tail

Differentiated teeth

Reduced pubis and ischium

Calcaneal heel

1 cm

Noncynodont therapsid *(Lycaenops)*

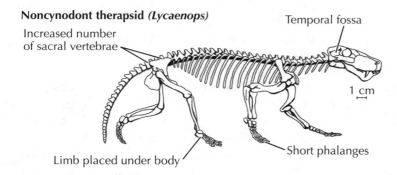

Temporal fossa

Increased number of sacral vertebrae

Short phalanges

Limb placed under body

1 cm

Pelycosaur *(Haptodus)*

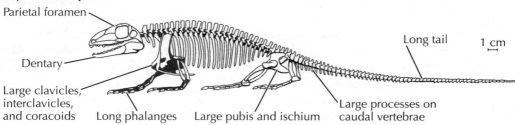

Parietal foramen

Dentary

Large clavicles, interclavicles, and coracoids

Long phalanges

Large pubis and ischium

Large processes on caudal vertebrae

Long tail

1 cm

sprawling limb posture seen in primitive synapsids but instead held their limbs vertically and walked fully upright. Consequently, there are many subtle changes in the shape and musculature of all the limb bones, and the finger bones are much shorter because these animals no longer sprawled and walked flat-footed with toes splayed out like a lizard's but instead began to walk on the tips of their toes like most mammals do.

The next grade up is a group called the "cynodonts," which originated in the late Permian, survived the greatest mass extinction event in earth history at the end of the Permian, and then dominated the world of the Early Triassic. Cynodonts (the name means "dog tooth" in Greek) are very much like mammals in lots of ways (figs. 13.2D and 13.3). The temporal opening at the back of the skull is even more enlarged, allowing for several jaw muscles to develop, which in turn made it possible for cynodont jaws to chew as well as bite. The canines were large, and the postcanine teeth (premolars and molars) are now specialized and multicusped for chewing, not the simple conical piercing teeth of reptiles and primitive synapsids. The secondary palate was now almost fully developed, with the internal air passage opening not in the mouth (like in primitive synapsids) but in the back of the throat (as in mammals). Instead of sprawling, the limbs (fig. 13.4) are even more upright and specialized for rapid running, with a lightweight shoulder girdle and a reduced set of hip bones (except for the expansion of the iliac blade of the hip along the spine). In the heel, the ankle bone known as the calcaneum develops a long extension for the attachment of the Achilles tendon, a sign of much more efficient and faster running. The tail is shorter, too. The rib cage disappears from the lower back, and in some specimens, the ribs are locked together with flanges of bone. This suggests that they did not breathe by expanding the ribs in and out (like reptiles do) but instead had a solid rib cage and used a muscular wall between the lung cavity and abdominal cavity, the diaphragm, to pump air in and out of the lungs. Some advanced cynodonts, such as *Thrinaxodon* (fig. 13.2D) actually have small pits on their snouts that suggest the presence of whiskers. (Hair does not normally fossilize, so it is hard to know whether any given fossil had hair or not.) If so, then cynodonts probably had hair elsewhere on their bodies as well, and hair probably appeared early in synapsid evolution with the evolution of the smaller therapsids.

The most advanced cynodonts, such as the tritylodonts and the trithelodonts of the Early Triassic (fig. 13.3), were small weasel-like forms with highly specialized skulls, large temporal openings and several sets of jaw muscles, highly specialized molars and premolars, and very doglike skeletons with almost all the typical mammalian features. In these and many other features, they are so mammal-like that they have often been called mammals. Indeed, the transition from the most primitive synapsids all the way to mammals is so smooth that it is rather arbitrary where to break the continuous sequence and begin calling advanced synapsids mammals. However, most paleontologists agree that creatures like *Adelobasileus*, *Sinoconodon*, *Megazostrodon* (fig. 13.4), and *Morganucodon*, from the Late Triassic of New Mexico, China, and South Africa, are bona fide mammals, because they are dramatically smaller in body size (rat-sized or smaller), had lost the ribs from their neck vertebrae, had reduced the reptilian elements in their shoulder girdle (clavicles, interclavicles, and coracoids), and had reduced the ilium portion of the hip bones to a simple rod running along the spine. Most importantly, they had a jaw joint between the dentary bone of the jaw and the squamosal bone of the skull, the defining character of Mammalia.

Through this entire gradual transition is an even more amazing story in the jaws and ears of these animals (fig. 13.5). Early synapsids like *Dimetrodon* had a classic primitive amniote jaw.

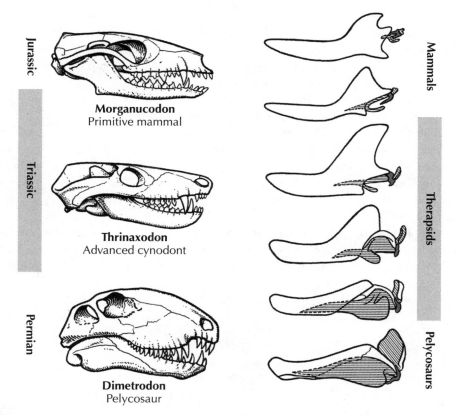

Jurassic

Triassic

Permian

Morganucodon
Primitive mammal

Thrinaxodon
Advanced cynodont

Dimetrodon
Pelycosaur

Mammals

Therapsids

Pelycosaurs

FIGURE 13.5. The gradual transformation of the jawbones within the synapsids, as all the nondentary jawbones (shaded bones: angular, surangular, articular, coronoid, splenials, and so on) are gradually reduced to tiny splints in the inside back part of the jaw, and the dentary bone (unshaded bone) takes over as the principal jawbone. Eventually, all the nondentary jaw elements are lost in mammals, except for the articular bone, which becomes the "hammer" (malleus) bone of the middle ear. (Drawing by Carl Buell)

The front part is composed of the dentary bone (which contains the teeth) but behind it were many other bones: the coronoid bone where the jaw muscles pulled up on the jaw, the articular bone in the jaw hinge, the angular bone in the lower back part (angle) of the jaw, and several other accessory jawbones. The articular bone of the lower jaw hinged against the quadrate bone of the amniote skull, and both were abutting against the middle ear and the "stirrup" bone (the stapes), helping transmit sound from the lower jaw into the ear. But as we go through the series of more and more advanced synapsids, we see some remark-able changes in the jaw. All of the nondentary components (shaded bones in right column of figure 13.5) get smaller and smaller, until by the time we see advanced cynodonts, most of these nondentary bones are just tiny splints in the inside back of the jaw, and the dentary bones have expanded to become almost the entire jaw. The probable reason for this trans-formation is that a single bone (the dentary) is much stronger than a series of bones sutured together, and as synapsids became more active in chewing, they needed a jaw that could handle all the stresses. A portion of the dentary, known as the coronoid process, expanded upward and took over the attachment points of the temporalis muscles and replaced the primitive amniote coronoid bone. Likewise, the primitive articular jaw joint with the quadrate bone of the skull also became highly reduced. Meanwhile, a portion of the dentary bone expanded upward, meeting the skull in the squamosal bone region and initiating a new jaw joint. Eventually, the dentary/squamosal jaw joint will take over completely, and the primitive articular/quadrate jaw joint will be superseded.

There is even one remarkable fossil, a trithelodont known as *Diarthrognathus* (fig. 13.6), which shows how this transition took place. Its name means "two jaw joint," and indeed that's what it has: both the old amniote articular/quadrate jaw joint still attached on both sides of the skull alongside the new dentary/squamosal jaw joint. We could not have asked for a more perfect transitional fossil, which has been caught in the act of making the transition from one set of jaw joints to the other. Eventually, the articular/quadrate jaw joint gets smaller and smaller, and no longer serves as a jaw joint, as the dentary and squamosal take over completely.

So what happened to the quadrate and articular? They could have vanished completely, as did most of the nondentary bones of the lower jaw. But remember the point we discussed earlier? Primitive amniotes hear with their lower jaws, and the sound is transmitted from

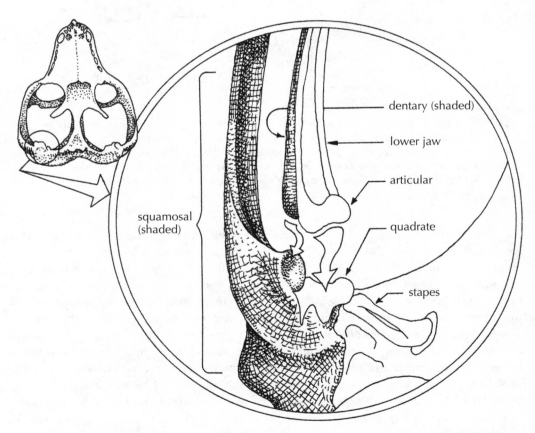

FIGURE 13.6. *Diarthrognathus* actually has the old articular/quadrate jaw joint of primitive synapsids in operation side by side with the mammalian dentary/squamosal jaw joint on both sides of the skull. (Drawing by McLoughlin 1980; Viking, New York; used with permission)

the jaw joint into the middle ear. For example, snakes cannot hear when they rear up to face the snake charmer because their jaw is not in contact with the ground and so cannot pick up vibrations. They are responding to the body motions of the snake charmer, but the sound of the flute is just for the tourists—the snake cannot hear a thing. Sure enough, once the quadrate and articular became tiny and disconnected from the jaw joint function, they did not vanish—they are in your middle ear right now (fig. 13.7)! The quadrate bone turned into the *incus* or "anvil" bone, which transmits sound to the stirrup or *stapes*. The articular bone turned into the *malleus* or "hammer" bone, which transmits sound from the eardrum to the incus. So as you listen to sounds, they are being transmitted through bones that started out as part of the jaw and skull articulation. If that story seems too incredible, just think of this: when you were an embryo, your ear bones were represented by cartilage in the lower jaw and skull, and through embryonic development they shift until they reached the middle ear, retracing the path they took during evolution!

The most amazing clinching fossil, however, is *Yanoconodon*, described by my friend Luo Zhexi and his colleagues (Luo et al. 2007) from the Lower Cretaceous Yixian Formation of Hebei Province. It is equivalent in age to the beds in the Liaoning Province that produced all the birds described in the last chapter. It is a beautiful complete specimen (fig. 13.8) with all the bones articulated in a death pose as it died and was preserved in the delicate lake

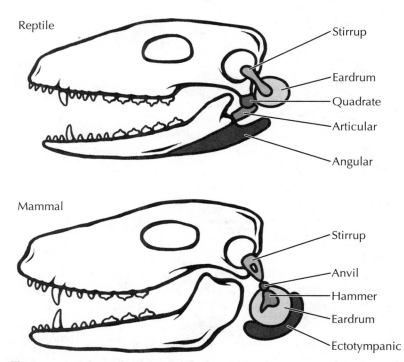

FIGURE 13.7. The ear region also undergoes a dramatic transformation, as the articular bone of the lower jaw hinge and the quadrate bone of the jaw hinge in the skull shift to the middle ear and become the incus and the malleus ("anvil" and "hammer"). This same transformation can be seen not only in fossils but also during the embryology of a mammal. When you were an embryo, your middle ear bones started out in your jaw.

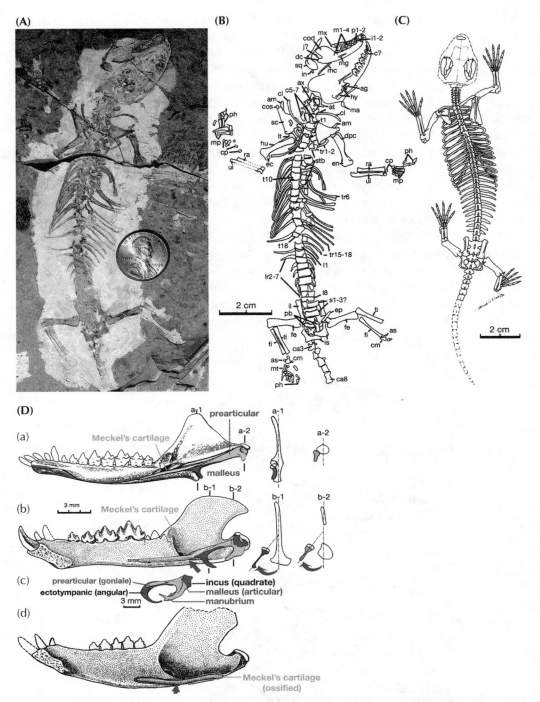

FIGURE 13.8. The primitive triconodont mammal *Yanoconodon allini* from the Lower Cretaceous of China, which still retains its ear bones attached to its lower jaw. (A) Photograph of the original articulated specimen. (B and C) Sketch of the skeleton, labeling the bones, and a reconstruction of the skeleton. (D) Detailed diagrams of the lower jaw, showing the ring of ear bones still attached to the lower jaw while functioning for hearing. Specimen (a) is the jaw of the primitive Triassic mammal *Morganucodon*; (b) is *Yanoconodon*, with a detail of the ear bones shown in (c) Diagram (d) is the lower jaw of the triconodont *Repenomamus*. For further details, see Luo et al. (2007:288–293). (Courtesy Zhexi Luo/Carnegie Museum of Natural History)

sediments. The most amazing thing about the specimen is that the middle ear bones *are still connected to the lower jaw!* This animal could hear with its quadrate/articular (incus/malleus) like any other mammal, yet the bones had still not migrated to the middle ear!

All of this amazing evidence, of course, is hard for creationists to stomach. Some, such as Duane Gish, tried to ridicule the whole idea by joking about these creatures "chewing and hearing while rearticulating their jaws." He never explained to his audience how most reptiles indeed hear with their lower jaw, or that we have fossils like *Diarthrognathus* with both jaw articulations operating simultaneously, or that human ear bones were originally in their jaws during their early embryonic stages. Gish (1995:147–173) attempted to discredit this beautiful evolutionary sequence of synapsids by his usual method: quoting people out of context or quoting outdated sources that do not reflect what we know now. He mines Tom Kemp's (1982) 35-year-old book for quotes that seem to say there was no transitional sequence in synapsids—but if you read the quotes closely, what Kemp is saying is that we don't have many gradual transformations between each of the synapsid genera (most of which are known only from a few fossils). But that does not mean that we can't line up the genera (as we did here) and produce a beautiful evolutionary sequence *among* the genera (fig. 13.3). In other cases, Gish claimed that there is a black hole of missing fossils in the Mesozoic—but that gap has long ago been filled by some amazing fossils. Most of the rest of Gish's criticisms reveal that he had absolutely no firsthand knowledge of these fossils or their anatomy but simply mines other people's work for quotes that seem to support his biases and then pulls them out of context so they seem to say something that the author never intended. Once again, this is dishonest and unscientific. If Gish had been really interested in the truth, he would have done his homework, learned some anatomy and paleontology, and studied the fossils for himself—and he would have found one of the best macroevolutionary transitions between two major groups of animals ever documented.

How do the intelligent design creationists handle this extraordinary transitional series? Wells (2000) doesn't address it, nor does Sarfati (1999, 2002). Davis and Kenyon (2004:100–101) quote a few evolutionists out of context and even concede, "Without a doubt, the Therapsids are highly suggestive of a Darwinian lineage." But then they betray their complete lack of understanding of evolution and try to discount the entire example by arguing that it is not a single ancestral lineage but many different lineages. That is *exactly* how most evolutionary transitions work in a bushy, branching system—not as "missing links" on a nonexistent "chain of being" (the common creationist misunderstanding) but as multiple, closely related lineages that each show progressively more mammalian characteristics.

Furballs in the Age of Dinosaurs

With malleus Aforethought
Mammals
Got an earful
Of their ancestors' Jaws
—John Burns, *Biograffiti*

When some people think of life during the Mesozoic, or "age of dinosaurs," they assume that the dinosaurs dominated the planet, and mammals had not yet evolved. In fact, the earliest mammals evolved from cynodonts in the Late Triassic at exactly the same time as the early dinosaurs (which may have outcompeted the last of the larger synapsids that once ruled the Triassic). But while dinosaurs soon came to dominate the globe for the next 130 million years, mammals remained small and inconspicuous, seldom getting much larger than the size of a house cat. Most mammals apparently lived in the nooks and crannies of the world of the "terrible lizards," hiding in the vegetation, and probably coming out mainly at night. In fact, the first two-thirds of mammalian history was the story of these tiny Mesozoic mammals. Only after the nonavian dinosaurs vanished at the end of the Cretaceous did the world open up for mammals so that they could dominate the planet.

For over a century, very little was known about Mesozoic mammals. Because they were so tiny and delicate, the best we could find were tiny pinhead-sized teeth and partial jaws from animals the size of shrews. None were known from almost any other part of the skeleton. When I first began to work on Jurassic mammals for my master's thesis in 1977, this state of affairs was still true after a century. I had access to a new collection of jaws and teeth from the famous Upper Jurassic Morrison Formation dinosaur quarries at Como Bluff, Wyoming. This allowed me to survey all the early Mesozoic mammals known at the time and conduct the first cladistic analysis of all these poorly understood species. I did several different papers on that research and published them. One result was that I was able to show that the long-abused paraphyletic group "Pantotheria" was a hopeless wastebasket and needed to be abandoned. Indeed, most paleontologists since then have stopped using this obsolete concept. Around the campfire in the long field season of 1977, my graduate advisor Malcolm McKenna, my fellow graduate students, and I daydreamed about what it would be like if we had skulls or even partial skeletons of these critters, instead of tiny, frustratingly incomplete teeth and jaws. For a century before us, everyone else who worked on Mesozoic mammals must have felt the same way. But they did the best they could with what they had.

I moved on from Mesozoic mammals shortly thereafter because there were no more new specimens to study at the time. I also preferred working on larger and more easily studied mammals like camels, horses, and rhinos, which do not require a microscope to see or photograph. Since then, there has been a veritable explosion of new Mesozoic mammal fossils. Not only do we have many more new species based on jaws and teeth, but there have been some extraordinary finds that have decent skulls (fig. 13.9), and in a few cases, even articulated skeletons for many of the different groups. What these specimens show is that most Mesozoic mammals were small, insectivorous creatures, living much like the modern shrew in most of their habits. A few were slightly larger in body size, and one specimen from the Cretaceous of China, known as *Repenomamus*, was over a meter long and actually has a baby psittacosaur dinosaur in its stomach. Generally speaking, however, it appears that mammals avoided the dinosaurs and were not in any position to compete with them, let alone eat them.

The most amazing specimens of Mesozoic mammals come from the same Lower Cretaceous Liaoning lake beds that yielded the many feathered dinosaurs and early birds discussed in the previous chapter. These include a complete specimen of the oldest known marsupial, *Sinodelphys szalayi* (fig. 13.10), which preserves not only the bones in articulation but even the impressions of the fur and soft tissues. This fossil shows that marsupials (the pouched mammals, which today include opossums, kangaroos, and koalas) had already split off from the main mammalian stem at 120 million years ago, and opossum-like marsupial teeth are

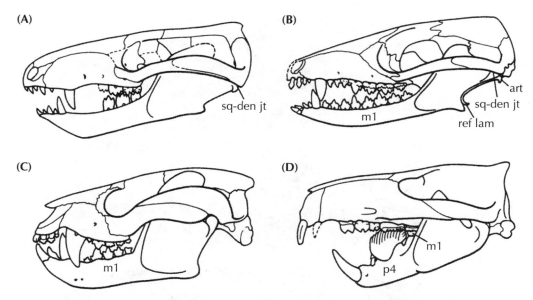

FIGURE 13.9. The skulls of some of the better known Mesozoic mammals. (A) Early Jurassic *Sinoconodon*. (B) Early Jurassic *Morganucodon*. (C) Early Cretaceous *Vincelestes*. (D) The Paleocene multituberculate *Ptilodus*, representative of the long-lived Mesozoic radiation of multituberculates. Abbreviations: sq-den jt, squamosal-dentary joint; ref lam, reflected lamina; art, articular bone; m1, first lower molar; p4, fourth lower premolar. (From Hopson 1994: fig. 9; courtesy J. Hopson)

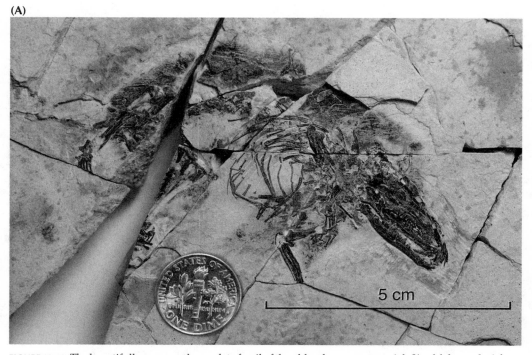

FIGURE 13.10. The beautifully preserved complete fossil of the oldest known marsupial, *Sinodelphys szalayi*, from the Lower Cretaceous of China. (A) The complete skeleton of the type specimen. (B) Artistic reconstruction by Carl Buell of the appearance of *Sinodelphys*. See Luo et al. (2003: 1934–1940). (Courtesy Zhexi Luo/Carnegie Museum of Natural History)

(B)

FIGURE 13.10 (*Continued*)

(A)

FIGURE 13.11. The beautifully preserved fossil (complete with hair) of one of the oldest known placental mammals, *Eomaia scansoria*, from the Lower Cretaceous of China. (A) Photograph of the type specimen. (B) Sketch of the skeleton and reconstruction of the skeleton as it appeared in life. See Ji et al. (2002:816–822). (Courtesy Zhexi Luo/Carnegie Museum of Natural History)

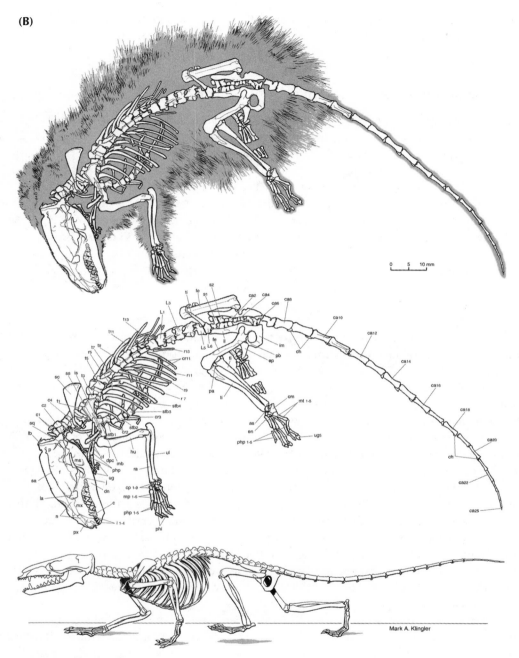

(B)

0   5   10 mm

Mark A. Klingler

FIGURE 13.11. (*Continued*)

known from most of the Cretaceous. The Liaoning beds also yield one of the oldest known placental mammals, *Eomaia scansoria*, and the fossil also preserves the hair and soft tissues (fig. 13.11). An even earlier placental fossil is *Juramaia*, from the Jurassic of China. Thus the split between marsupials and placentals (the mammals that give live birth to their young, including most living mammals) occurred much earlier in the Cretaceous than we thought just a few years ago. By the Late Cretaceous, both marsupials and placentals were evolving very fast, and most of the archaic groups of Mesozoic mammals were gone.

Radiation in the Aftermath

The placental or eutherian mammals comprise about twenty living orders and several extinct ones. The morphological and adaptive range of this group is extraordinary; diversification has produced lineages as varied as humans and their primate relatives, flying bats, swimming whales, ant-eating anteaters, pangolins, and aardvarks, a baroque extravagance of horned, antlered and trunk-nosed herbivores (ungulates), as well as the supremely diverse rats, mice, beaver and porcupines of the order Rodentia. Such adaptive diversity, and the emergence of thousands of living and fossil species, apparently resulted from a radiation beginning in the late Mesozoic between 65 and 80 million years ago. This explosive radiation is one of the more intriguing chapters in vertebrate history.
—Michael J. Novacek, "The Radiation of Placental Mammals"

At the end of the Cretaceous 66 million years ago, the nonavian dinosaurs vanished from the planet. One group of scientists argue that the impact of a big rock from space did them in, while another group points out that there were too many survivors who could not have outlived such an extreme catastrophe. Instead, they suggest that the extinction was due to more gradual changes, and this is supported by evidence of the fossil record (for a review of the topic, see Prothero 2016). Whatever the cause, by the early Paleocene, the terrestrial realm was devoid of large animals, and there were plenty of vacant ecological niches for any opportunistic creature to occupy. Within a million years after the end of the Cretaceous, mammals began an explosive evolutionary radiation, with many new groups appearing for the first time in the fossil record, and the mostly shrew-sized Mesozoic mammals evolving into much larger dog-sized and even cow-sized animals. By the middle Eocene, only 15 million years after the nonavian dinosaurs had vanished, almost all the living orders of mammals (rodents, rabbits, bats, whales, carnivores, primates, and so on) had appeared, although they were very primitive members of those families that look nothing like their living descendants. Paleontologists frequently point to the evolutionary radiation of Cenozoic mammals as a classic example of what life can do when competition is suddenly removed and there are many new ecological resources and adaptive zones left vacant.

Deciphering this evolutionary explosion has been one of the major challenges for paleontologists for over a century. The major problem is that for a very long time the fossil record of mammals was very poor in Cretaceous and Paleocene rocks, so paleontologists had only fragmentary teeth and jaws to work with, and only from a few places such as North America and Europe. Complete skeletons were extremely rare, skulls were scarce and often badly distorted or damaged because deposits this old are usually crushed under the weight of millions of tons of rocks that were deposited on top of them and often deformed by later mountain-building events as well. Paleontologists did the best they could by matching the patterns of teeth from the Paleocene with those from the Cretaceous and attempting to construct ancestor-descendant sequences. Because the hard enamel on teeth is often the most durable part of the skeleton, the teeth and jaws are usually the only parts to survive the beating caused by scavengers and river currents and trampling. Basing our understanding of mammals largely on their teeth may seem inadequate, but fortunately teeth are the most diagnostic part of most mammals, even when we do have the luxury of fossilization of the rest of the skeleton. Teeth not only preserve patterns of their ancestry in the intricate details of their cusps and crests, but they also reflect (to varying degrees) the diet of the animal as well. Thus, if we had to choose only one part of the animal skeleton to be preserved, we are lucky that the most useful part happens to be what fossilizes. Some vertebrate paleontologists joked (after seeing one talk after another on the protocones, paracones, metacones, and other cusps on mammalian teeth) that we "protoconologists" seem to think that one tooth gave rise to another tooth ad infinitum. But this is not by choice, but by necessity. By contrast, a fish or reptile paleontologist usually cannot do much with teeth or other fragments and typically only works with nearly complete specimens. This is one of the reasons that the mammalian fossil record is the most complete and detailed and densely fossiliferous of all vertebrate groups. We can do many things with fossil mammals that cannot be done with any other group of vertebrates.

When I was an undergraduate, my final project in my vertebrate paleontology class was to identify a collection of early Eocene mammal teeth from the Bighorn Basin of Wyoming and then try to research their origins in the scientific literature. I did my best on the project using what was published at the time. But when I came to the American Museum of Natural History in 1976 to begin graduate school at Columbia University, I was in for a shock. Here were the very best minds in the business working with the best fossils of every group of mammals. They were all feverishly studying specimens and drawing cladograms, using this new approach to decipher the century-long puzzle of the relationships of the major groups of fossil and living mammals. I gave a copy of my humble undergraduate thesis to my friend, Earl Manning, the collections manager and a former graduate student, and he tore it to shreds because the new cladistic approach (plus his better knowledge of the actual specimens) gave him a perspective that I could never have obtained as an undergrad reading older published literature. (This is a lesson for creationists: you can't do research by reading other people's work in the literature. Until you do the research with the real specimens yourself, you have no right to talk about these things). Soon, I was discovering for myself the groundbreaking new thinking about mammalian relationships, and the century-old problem of the relationships of the orders of mammals was soon to be solved.

Just a year before I arrived in New York, my graduate advisor Malcolm McKenna published the first cladistic analysis of fossil mammals. At the time he did it, it caused howls of shock and outrage. Cladistics was already second nature to entomologists and ichthyologists by then, but vertebrate paleontologists and mammalogists tended to be more conservative. Although many had heard of cladistics, and some had tried it out on their own group of organisms, none had ever used it to decipher the higher-order relationships of mammals before. But it was just the tool that was needed to address this complex problem that had eluded solution for over a century. Instead of trying to find more primitive ancestral teeth in earlier beds, paleontologists could now use cladistics to take advantage of the anatomy of the entire organism, not just the teeth. This worked especially well for living groups that had little or no fossil record but had all this anatomical data in the soft tissues and skeletons of the living members of the group. In many cases, these living groups with limited fossil records had not entered into the analysis, because paleontologists were only concerned with "connecting the dots" between the teeth that were actually preserved in their local basin.

By the late 1970s and early 1980s, mammalian paleontologists had been studying specimens and collecting reams of anatomical data on every system in living mammals, as well as the important information about the skeletons of fossil mammals. Malcolm McKenna, Mike Novacek, and Andy Wyss published some of the first cladograms that covered all the placental mammals, while I collaborated with Earl Manning on the first cladogram (originally prepared by him in 1977) that covered all the hoofed mammals, or ungulates. All of these cladograms were published in the mid-1980s (Novacek and Wyss 1986; Novacek et al. 1988; Novacek 1992, 1994; Prothero et al. 1986, 1988), especially in the landmark volume on amniote relationships from a 1986 London symposium (Benton 1988a, 1988b), and in a later volume based on a 1990 American Museum symposium on mammalian interrelationships (Szalay et al. 1993).

By the mid-1990s, almost all these cladograms on mammalian relationships seemed to converge on a common topology, with some minor unresolved differences of opinion (fig. 13.12). Some things, however, were very clear and were based on multiple well-supported, well-corroborated analyses. As McKenna had predicted in 1975, the most primitive group of placentals is not the insectivorous mammals (as paleontologists had long thought), but the xenarthrans (sloths, armadillos, anteaters, and their relatives). The rodents and rabbits are closely related after all (most paleontologists thought they were convergent) and formed a natural group called the Glires (pronounced "GLY-reez"). The closest relatives of primates were the tree shrews (no surprise to anthropologists here) and also the colugos (a bizarre Asian species incorrectly referred to as "flying lemurs," even though they are not lemurs and they glide but do not fly). These groups together were called the Archonta. Carnivorous mammals formed their own clade, as did the true insectivores (shrews, moles, and hedgehogs). Finally, the hoofed mammals (subject of the next chapter) also formed a distinct well-supported clade, the Ungulata. Together, these groups seemed to cut the Gordian knot of mammalian interrelationships that had puzzled the great minds of paleontology for decades.

But science is never this simple. Once we seemed to have achieved consensus, another source of information had to be considered: molecular data. In the 1980s and early 1990s, all we had were a few protein sequences for a few examples of each order, and most of the data seemed to be consistent with the topology that the anatomical cladograms were producing. But by the late 1990s, molecular biologists were sequencing the mitochondrial and nuclear DNA of mammals directly, and some of their results are still not consistent with what the anatomy and fossil record seems to indicate (Springer and Kirsch 1993; Stanhope et al. 1993, 1996; Madsen et al. 2001; Springer et al. 2004, 2005; Murphy et al. 2001a, 2001b). For example, tenrecs don't seem to be related to other insectivores, elephant shrews don't cluster with Glires nor do elephants, aardvarks, and hyraxes cluster with Ungulata. Instead, they are united (along with some other insectivorous African groups) in a molecular clade known as Afrotheria, and their branch point is even more ancient than that of the xenarthrans. Ungulates (excluding elephants and their relatives) still clustered together, but with carnivores as their sister group, followed by the bats, and then the insectivores, forming a clade the molecular biologists call Laurasiatheria. Rodents and rabbits still cluster as the Glires, with

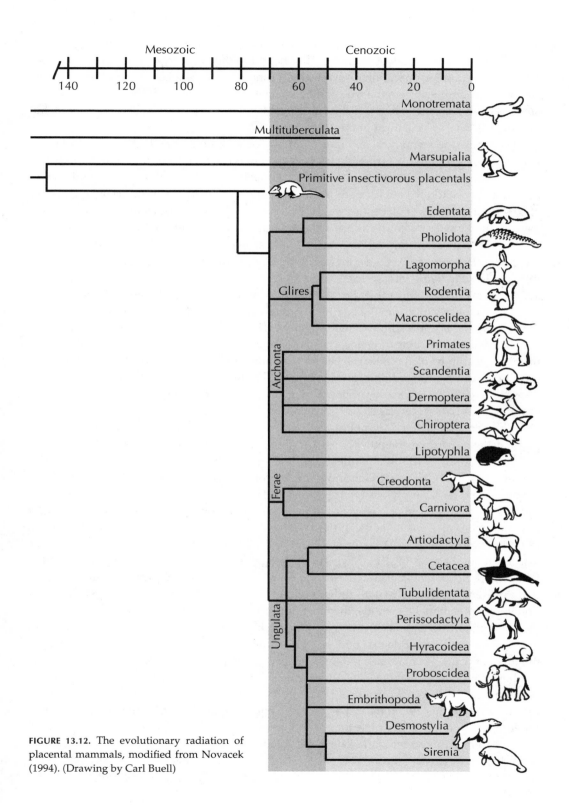

FIGURE 13.12. The evolutionary radiation of placental mammals, modified from Novacek (1994). (Drawing by Carl Buell)

the Archonta (primates, colugos, and tree shrews, but *not* bats) as their sister group, forming a group known as the Euarchontoglires. Thus nearly all the topology of the anatomical tree is retained, a strong corroboration from these two completely independent sources of data that we must be getting close to the true pattern of relationships of placental mammals. Eventually, these inconsistencies between anatomical and molecular data sets will be ironed out, but for now we have at least most of the "big picture."

What both the anatomical and molecular phylogenies of mammals show consistently is that the great radiation was already underway before the nonavian dinosaurs vanished at the end of the Cretaceous. We knew that the most primitive fossils of placentals and marsupials dated back to the Late Jurassic, about 140 million years ago, but the branching history of all the major orders of placentals was difficult to decipher with just teeth alone. There were teeth in the Late Cretaceous–earliest Paleocene that were clearly ungulates (the zhelestids and *Protungulatum*), primates (*Purgatorius*), and carnivore relatives (*Cimolestes*), but most of the placental teeth of the Late Cretaceous were insectivorous mammals, both true Insectivora (related to living moles, shrews, and hedgehogs) or members of unrelated groups that just happened to have the same diet. Now that the new molecular phylogenies are out, they suggest that nearly all the major placental orders differentiated in the Cretaceous, and a bit earlier than the traditional phylogeny suggests. The discrepancy is still being resolved, but the data point in the same direction: the explosive radiation was already underway with the primitive members of the lineages before the nonavian dinosaurs disappeared. Only after the landscape was cleared of large animals did these lineages then diverge in ecology and body size and begin to occupy the newly available niches, ultimately specializing into things as different as bats and whales.

Creationists, of course, don't keep up with or understand any of this, and they misinterpret most of what they do bother to read. Once again, Gish (1995:184) decried the supposed lack of transitional forms among the orders and supported his case with grossly out-of-date quotations. Clearly, he has no idea how much progress we have made in the past 20 years. In some cases, he is simply wrong. He claims (p. 188) that there are no transitional forms for rodents and quotes Romer's 60-year-old textbook as his source. If he had bothered to read anything more recent (published in the 1970s and 1980s, long before his 1995 edition; most recently summarized by Meng et al. [2003] and Meng and Wyss [2005]), he would have learned about the amazing discoveries of the anagalids (fig. 13.13) and mimotonids from the Paleocene beds of China. These fossils are now hailed as a classic primitive linking taxon that apparently gave rise to both rodents and rabbits in Asia during the Paleocene. In addition to the transitional anagalids, the most primitive true rodents and rabbits also appeared in Asia before they spread to other continents in the early Eocene—and the differences between them are so subtle that only specialists can tell them apart. What Gish doesn't dare mention, of course, is that once rodents appear in the record in the Eocene of the rest of the northern continents, they have an extraordinary fossil record, with hundreds of specimens spanning every meter of section in places such as the Big Badlands of South Dakota. As Martin (2004) points out, in many parts of the Cenozoic, you can get hundreds of distinctive rodent teeth in every meter of sediment and use them to date the rocks very precisely. And if Gish had read any *recent* works on rodent paleontology, there are many, many transitional forms between the major groups of rodents. Not only that, but the molecular phylogenies of rodents are now closely reflecting the traditional anatomical phylogenies, so we have a very robust database of their evolution.

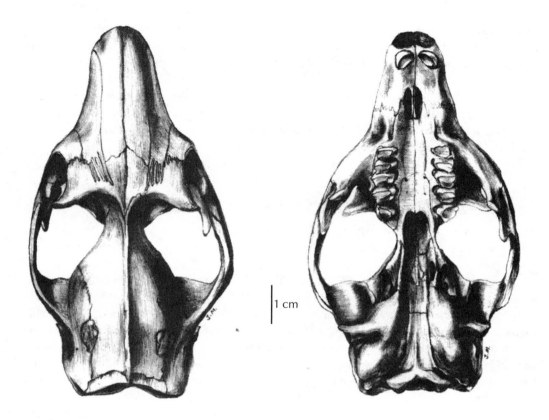

1 cm

FIGURE 13.13. The Paleocene and Eocene fossils of China include anagalids and eurymylids, which are the transitional form linking rabbits and rodents with other mammals. This is the skull of *Rhombomylus*, which has rodent-like and rabbit-like features such as the chisel-shaped incisors, diastema, and cheek teeth, yet it is transitional between both groups. (Courtesy Meng Jin)

In other cases, creationist accusations are simply misleading and unreasonable. Gish (1995) and Davis and Kenyon (2004:102) argue that the fossil record is so good that even bats should fossilize easily, and these creationists harp on the fact that the earliest bats from the Eocene seem to look a lot like modern bats. If they knew anything at all about bats and fossilization processes, they would realize that bats have very delicate skeletons and tiny hollow bones, so they are very rarely fossilized (Simmons and Geisler 1998; Simmons 2005). We have just a handful of bat fossils over the entire Cenozoic, and most of them are just jaws and teeth. There are just a few extraordinarily complete bat specimens with wing membranes from amazing localities like Messel in Germany and the Green River Shale in Wyoming, all stagnant lake deposits that happened to preserve beautiful fossils, but the creationist emphasis on these one or two extraordinary specimens gives the false impression that we should be finding these lucky accidents all the time. If we were fortunate enough to have a similar extraordinary deposit like Liaoning or Messel for the Paleocene, we might find better transitional bat fossils, but more than a century of looking still hasn't produced such a locality. More importantly, the creationists are wrong when they claim that these Eocene bats look just like modern bats. That may be true for an untrained, unobservant amateur, but anyone who knows bat anatomy can tell how primitive these fossils are. For one thing, the earliest bats did not yet have the ear structure necessary for the modern system of echolocation that bats use to catch insects on the wing. Their large skulls and eyes show that they probably hunted by day using sight, not at night using echolocation. In addition, Eocene bats have many other primitive features of the skull, hands, and feet that are not found in any other living bat. They may have had wings, but to someone who actually knows their fossils and mammals, a bat is not *just* a bat!

Lions and Tigers and Bears, Oh My!

A lion's work hours are only when he's hungry; once he's satisfied, the predator and prey live peacefully together.
—Chuck Jones, Animator

Bats and rats usually don't interest too many people (except as pests) nor do most of the other orders of mammals, such as xenarthrans (which have an excellent fossil record with many transitional fossils of ground sloths and huge armored armadillos) or the insectivores (which have an amazing fossil record going back to the Cretaceous). So we will not dwell on these examples further for space reasons, but we will focus on two large mammal groups that do excite people: the hoofed mammals (or ungulates), subject of the next chapter, and the carnivorous mammals.

Lots of people love cats and dogs, and many are pet owners, so carnivores are near and dear to many hearts. Those science documentaries on cable TV and the children's books about prehistoric mammals love to show saber-toothed cats, dire wolves, and cave bears, but there is far more to carnivore evolution than just these glamorous creatures. Carnivorous mammals are usually not as commonly fossilized as their herbivorous prey. Because it usually takes many prey animals to support one predator, the population numbers of carnivores are always small, and their chances of fossilization are further diminished. Nevertheless, they have an excellent fossil record going back through the entire Cenozoic, with far more diversity of forms than can be seen today with modern cats, dogs, bears, and their relatives. We can trace the living order Carnivora back to the most primitive groups, an assemblage of highly primitive carnivorans known as miacids known from the Paleocene of the northern continents (fig. 13.14). All of these creatures were small, archaic weasel-sized creatures that look nothing like their descendant families but had all the hallmarks of the Carnivora. In the next few million years of the Eocene and Oligocene, we soon see the divergence of a number of modern families (fig. 13.15). For example, the dogs appear in the middle Eocene and soon have an incredible evolutionary radiation of dozens of genera and hundreds of species. They evolved into a wide spectrum of forms, from some that were smaller than weasels (*Hesperocyon*) to the huge borophagine dogs, hyena-like forms with crushing teeth for breaking bones—far more diversity than dogs show today (fig. 13.16).

True cats appear in the early Oligocene with *Pseudailurus*, a creature that looked more like a weasel than a cat (fig. 13.17). But by the Miocene, they began to take on their catlike

FIGURE 13.14. The most primitive true carnivorans are known as miacids, which were shaped roughly like weasels or raccoons, but were much more primitive. They were probably ancestral to nearly all living carnivorans. (Drawing by Carl Buell)

characteristics and evolved a variety of forms, including several different acquisitions of saber-like canines. In addition, there was an earlier family of catlike forms known as nimravids, which parallel cat evolution in many ways (including the evolution of several "saber-toothed" forms) but are unrelated to cats.

Bears have a history going back to the early Oligocene as well, but nearly all the early bears were small badger-like forms that then developed into fast-running dog-shaped forms like *Hemicyon*. Only late in their evolution did bears become large and develop teeth for an omnivorous diet. The fossil record of the mustelids (weasels, skunks, otters, badgers, wolverines, and their kin) and of raccoons and their relatives are also quite good (see Baskin 1998a, 1998b), although all the early members of these families are very primitive and would look nothing like their living descendants if you saw them today. Nevertheless, the details of their teeth, skulls, and skeletons are highly distinctive and unmistakable. Considering their rarity, the fossil record of most carnivorans is remarkably good, and

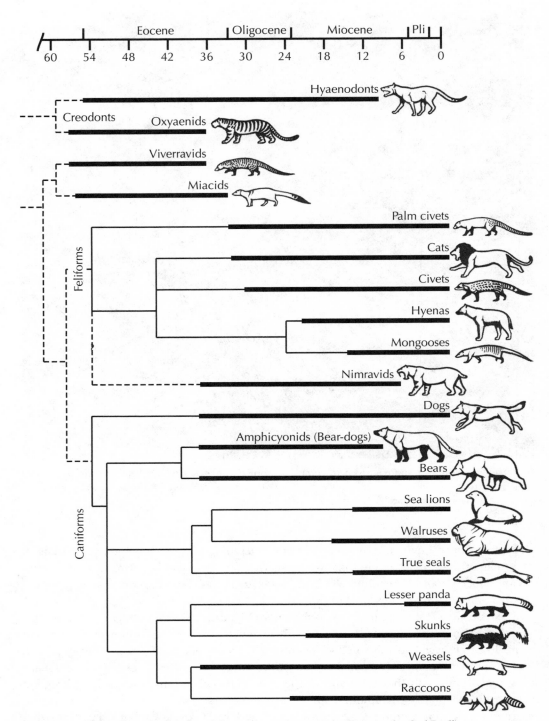

FIGURE 13.15. The evolutionary history of the carnivorous mammals. (Drawing by Carl Buell)

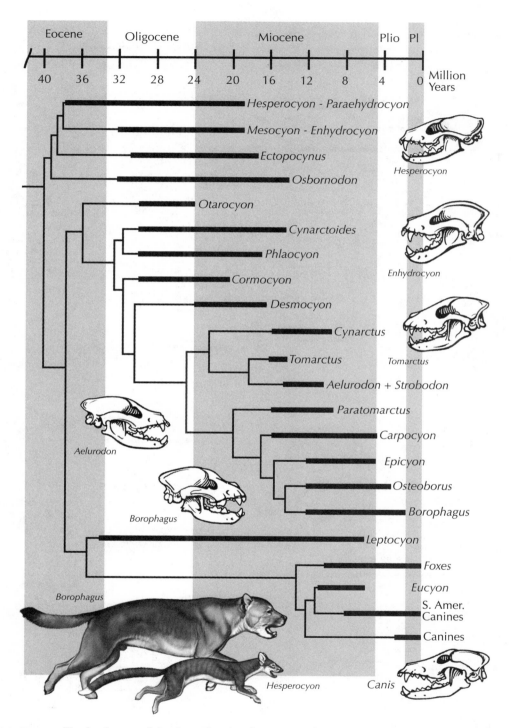

FIGURE 13.16. The family tree of the dogs, showing the variation from weasel-like *Hesperocyon* to the huge hyena-like bone-crushing borophagine dogs. (Based on information supplied by Xiaoming Wang; drawing by Carl Buell)

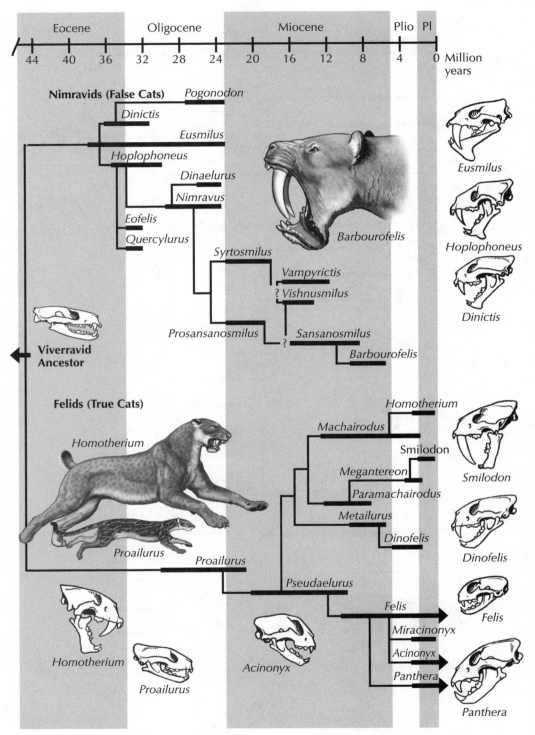

FIGURE 13.17. The family trees of the Felidae, or cat family, and of the Nimravidae, or "false cats," which parallel cats in many ways but are not closely related. (Drawing by Carl Buell)

there is no shortage of diversity of fossils or transitional forms between nearly every fossil group known.

Finally, one of the most amazing of all transitions in mammals is the origin of pinnipeds (seals, sea lions, and walruses). At one time, paleontologists thought that seals were related to weasels and that sea lions and walruses were related to bears, but recent cladistic and molecular phylogenetic analyses (Wyss 1987, 1988) have shown conclusively that all pinnipeds are monophyletic and closely related only to bears (fig. 13.15). And there are beautiful transitional fossils that link them. Oligocene deposits of Europe yield bears known as amphicynodontids, which were terrestrial animals yet had many features that link them to pinnipeds. Lower Miocene beds of California and Oregon yield the enaliarctines, which are the first truly marine relatives of seals and sea lions (Mitchell and Tedford 1973; Barnes 1989; Berta et al. 1989; Berta and Ray 1990). Although they retained many primitive skull features seen in the bearlike amphicynodontids, they also have some specializations of seals and sea lions, including enlarged eyes, an enlarged nasal cavity for regulating the temperature of the blood as they swim, and larger openings for the muscles that control their lips and whiskers. They also have reduced their olfactory lobes of the brain (since the sense of smell is not very important to aquatic mammalian predators) and improved the drainage of blood to their brains as an aid to diving. Their bodies (fig. 13.18) also had rudimentary flippers and streamlined shapes, so they would definitely remind us of the living seals, although they were very primitive looking and their flippers were clearly not as advanced as seen in modern pinnipeds. However, their bodies were still not as fully aquatic as the later seals and sea lions. Instead, they may have had lifestyles not too different from that of the sea otter.

Not long after the enaliarctines, we find the first members of all the living pinniped groups in the late early to middle Miocene, including the first true seals (*Pontophoca*, *Praepusa*, and *Cryptophoca* in the middle Miocene of Europe and *Leptophoca* in the middle Miocene of North America), the first sea lions (*Pithanotaria* in the middle Miocene of the Pacific

FIGURE 13.18. The earliest known relative of the seals, *Enaliarctos*, which was seal-like in many features but did not have the fully specialized skull, nasal region, or ear region of modern seals and sea lions. Their webbed feet and hands were more like the condition in otters and not fully modified into flippers. (Courtesy A. Berta)

Northwest of North America), and the first walruses (the desmatophocines of the early Miocene of North America and *Prototaria* of the early Miocene of Japan).

The best documented of these transitions is the evolution of walruses. The most primitive walruses, such as the middle Miocene *Proneotherium* and *Neotherium* (fig. 13.19A), are barely different from *Enaliarctos*, except that they are larger and more robust, and already begin to show the size and canine tusk differences between males and females that is so characteristic of walruses and not enaliarctines. Slightly later in the middle Miocene, we find *Imagotaria*, which was about the size of the living walrus, with ear bones already suited to underwater hearing. *Imagotaria* was beginning to develop the tusk-like canines and simplified peg-like molars and premolars in its cheek teeth that all walruses have. By the late Miocene and early Pliocene, there were at least eight different kinds of walruses along the Pacific Rim, most of which looked like unusually large sea lions with short to medium-sized tusks. They included *Pontolis*, *Gomphotaria* (fig. 13.19C), and *Dusignathus*, with completely peg-like simplified cheek teeth, large lower tusks, and small upper canine tusks, and *Aivukus* (fig. 13.19B), with slightly larger upper tusks, broad cheek teeth, and a deep lower jaw with small canines. *Aivukus* apparently lived in shallow water as a bottom feeder and crushed its food like modern walruses do. Then, in the latest Miocene and early Pliocene, the walruses spread through the Central American seaway (there was no Panamanian land bridge until the middle Pliocene), up the Atlantic Coast, and on to Europe and the Mediterranean. In the early Pliocene of Europe and the African coast of the Mediterranean, we find fossils of *Alachtherium*, which looked slightly more like a modern walrus, with large tusks, reduced peg-like molars, a deep lower jaw with no canines, and only a few other cheek teeth. In almost every respect, it is an ideal intermediate between the primitive walruses and the modern groups. Finally, the late Miocene and Pliocene also yield an even better transitional form, *Valenictus* (fig. 13.19D), which has long tusks (but still considerably shorter than the tusks of the living walrus, *Odobenus*, fig. 13.19E), and almost no teeth in the lower jaw. These walruses also had the characteristic arching of the palate that we see in living walruses. This arched palate, combined with the action of the tongue, allows living walruses to suck their prey (mostly mollusks) into their mouths, where they then crush the shells and suck out the contents.

If a creationist saw the end-members of this transitional series, they would not guess the connection, until we put all the transitional fossils in between them and demonstrated this dramatic example of land animals adapting to marine life. The walruses, in particular, give us an amazing array of transitional forms from sea lion–like early forms to those with intermediate conditions of the tusks and cheek teeth. Ironically, some of the best specimens

FIGURE 13.19. The evolution of walruses from primitive forms that resembled sea lions. (A) *Proneotherium* ▶ from the early Miocene, which has short canines and relatively primitive teeth but several distinctive features found only in walruses. (B) *Aivukus* from the late Miocene with larger canine tusks and simpler peg-like cheek teeth. (C) The dusignathine walrus *Gomphotaria*, with its big upper and lower tusks. (D) The more advanced walrus *Valenictus*, with upper tusks almost as large as those of the living species, a highly arched palate, and greatly reduced cheek teeth. (E) The living walrus, *Odobenus rosmarus*. (Photo [C] courtesy L. Barnes; [B] after Repenning and Tedford 1975, courtesy U.S. Geological Survey; all others courtesy T. Demére)

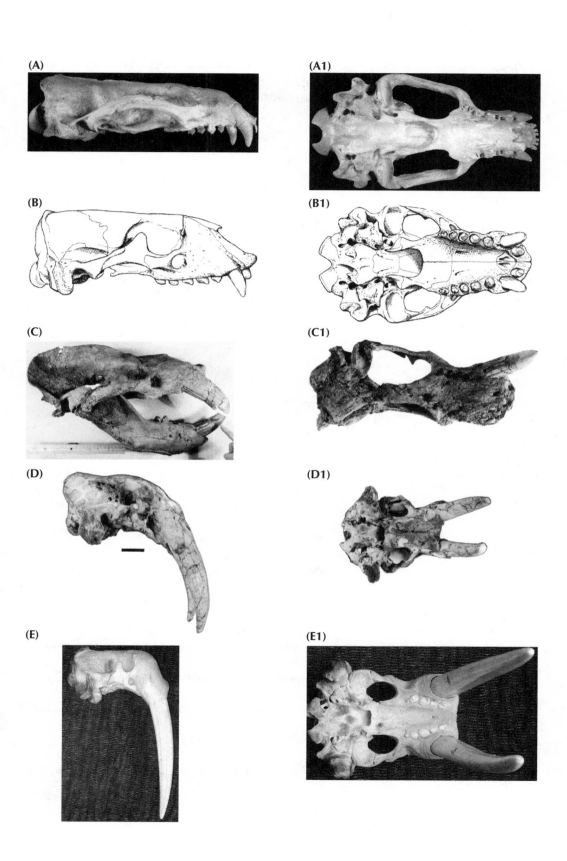

(A)　　　　　　　　　　　(A1)

(B)　　　　　　　　　　　(B1)

(C)　　　　　　　　　　　(C1)

(D)　　　　　　　　　　　(D1)

(E)　　　　　　　　　　　(E1)

of these "transitional" walruses are on display at the San Diego Natural History Museum, only a short drive from the original ICR headquarters. Gazing at them in the display case is one of the most powerful proofs that transitional forms are real. We will see an even better transitional series when we examine whales in chapter 14.

For Further Reading

Benton, M. J., ed. 1988. *The Phylogeny and Classification of the Tetrapods*. Vol. 2, *Mammals*. Oxford, U.K.: Clarendon.

Benton, M. J. 2014. *Vertebrate Palaeontology*. 4th ed. New York: Wiley-Blackwell.

Carroll, R. L. 1988. *Vertebrate Paleontology and Evolution*. New York: Freeman.

Gittleman, J., ed. 1996. *Carnivore Biology, Behavior, and Evolution*. Ithaca, N.Y.: Cornell University Press.

Hopson, J. A. 1994. Synapsid evolution and the radiation of non-eutherian mammals. In *Major Features of Vertebrate Evolution*, ed. D. R. Prothero and R. M. Schoch. Paleontological Society Short Course 7:190–219.

Janis, C., K. M. Scott, and L. L. Jacobs, eds. 1998. *Evolution of Tertiary Mammals of North America*. Vol. 1, *Terrestrial Carnivores, Ungulates and Ungulate-Like Mammals*. New York: Cambridge University Press

Janis, C., G. F. Gunnell, and M. D. Uhen, eds. 2008. *Evolution of Tertiary Mammals of North America*. Vol. 2, *Small Mammals, Xenarthrans, and Marine Mammals*. New York: Cambridge University Press.

Kielan-Jaworowska, Z., R. L. Cifelli, and Z.-X. Luo. 2004. *Mammals from the Age of Dinosaurs: Origins, Evolution, and Structure*. New York: Columbia University Press.

Li, C. K., R. W. Wilson, and M. R. Dawson. 1987. The origin of rodents and lagomorphs. *Current Mammalogy* 1:97–108.

Luckett, W. P., and J.-L. Hartenberger, eds. 1985. *Evolutionary Relationships Among Rodents*. New York: Plenum.

McKenna, M. C., and S. K. Bell. 1997. *Classification of Mammals*. New York: Columbia University Press.

McLoughlin, J. C. 1980. *Synapsida: A New Look into the Origin of Mammals*. New York: Viking.

Novacek, M. J. 1992. Mammalian phylogeny: shaking the tree. *Nature* 356:121–125.

Novacek, M. J. 1994. The radiation of placental mammals. In *Major Features of Vertebrate Evolution*, ed. D. R. Prothero and R. M. Schoch. Paleontological Society Short Course 7:220–237.

Novacek, M. J., and A. R. Wyss. 1986. Higher-level relationships of Recent eutherian orders: morphological evidence. *Cladistics* 2:257–287.

Peters, D. 1991. *From the Beginning: The Story of Human Evolution*. New York: Morrow.

Prothero, D. R. 1994. Mammalian evolution. In *Major Features of Vertebrate Evolution*. ed. D. R. Prothero and R. M. Schoch. Paleontological Society Short Course 7:238–270.

Prothero, D. R. 2006. *After the Dinosaurs: The Age of Mammals*. Bloomington: Indiana University Press.

Prothero, D. R. 2013. *Bringing Fossils to Life: An Introduction to Paleobiology*. 3rd ed. New York: Columbia University Press.

Prothero, D. R. 2016. *The Princeton Field Guide to Prehistoric Mammals*. Princeton, N.J.: Princeton University Press.

Rose, K. D., and J. D. Archibald, eds. 2005. *The Rise of Placental Mammals*. Baltimore, Md.: Johns Hopkins University Press.

Savage, R. J. G., and M. R. Long. 1986. *Mammal Evolution: An Illustrated Guide.* New York: Facts-on-File.

Szalay, F. S., M. J. Novacek, and M. C. McKenna, eds. 1993. *Mammal Phylogeny.* New York: Springer-Verlag.

Turner, A., and M. Anton. 2004. *National Geographic Prehistoric Mammals.* Washington, D.C.: National Geographic Society.

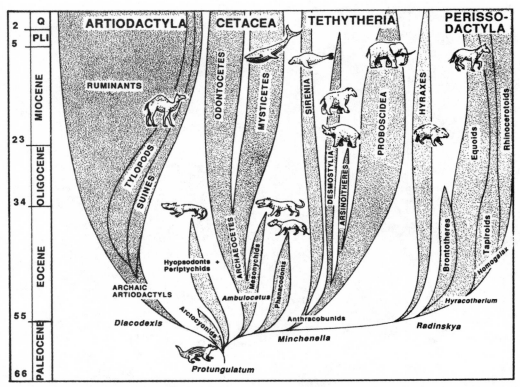

FIGURE 14.1. The evolutionary radiation of the hoofed mammals or ungulates. (Drawing by C. R. Prothero; after Prothero 1994b)

BOSSIES AND BLOWHOLES 14

Thundering Hooves

We live in a world where most of the attention gets grabbed by the carnivores—and mammals are no exception. We have TV shows entitled "Fangs" but none (alas) called "Molars," and the hoofed mammals are often regarded as little more than fodder.
—Christine Janis, 2003

After rodents and bats, the third-largest group of placental mammals is the ungulates, or hoofed mammals. Hoofed mammals make up about a third of the genera of living and extinct mammals, and nearly all large-bodied herbivorous mammals are ungulates. Most people are familiar with the common domesticated animals that provide us with food and services, such as horses, donkeys, cows, sheep, goats, pigs, as well as camels and alpacas, and know of more exotic hoofed mammals in the zoo or wild, such as rhinos, giraffes, antelopes, and hippos. (If elephants are members of Afrotheria, they may not be ungulates, but we will discuss these large hoofed mammals in this chapter anyway).

But the diversity of hoofed mammals of the past is much greater than the limited selection that survives today, with twice as many extinct families and genera. These include a wide array of hornless rhinoceroses, including the indricotheres, which towered over elephants as the largest land mammals that ever lived; bizarre chalicotheres, which were related to tapirs but looked like horses, except that they had claws like a ground sloth and knuckle-walked on their long forelimbs like gorillas; a great variety of North American camels, none of which had humps, but some of which did great impersonations of gazelles and giraffes; a number of extinct giraffes, none with long necks; pigs with horns; elephants without tusks that looked more like tapirs; and last but not least, an incredible array of transitional fossils that show how whales evolved from terrestrial hoofed mammals. All of this is well preserved in the fossil record because most ungulates are large-bodied and heavy-boned animals that tend to fossilize well.

Until recently, most of their history was not well understood or well documented. Part of the problem was logistical: the huge collection of fossil ungulates in the American Museum of Natural History in New York has still not been fully published since it became available for study in the 1970s. This collection occupies a separate wing of the museum, with 10 floors of storage and labs and offices and includes a whole storage floor of mastodonts and mammoths, a whole floor of North American rhinos, a whole floor of North American camels, a whole floor of horses, and three other floors of all the rest of the mammals. I was fortunate to arrive there as a graduate student in 1976 and got to work on many different groups of mammals. Many of these new collections contain cabinet after cabinet of undescribed complete skulls and skeletons of many species that had previously been known only from isolated teeth and jaws. For example, I did a lot of research on horses, peccaries, and camels (which is still ongoing), but my major task was documenting the long history of North American rhinos. This group had not been updated in a century, but I published a comprehensive monograph on the rhinos after more than 20 years of work (Prothero 2005).

Another part of the problem was conceptual: early work on many groups of ungulates focused too often on trying to find primitive ancestors and link them to descendants, ignoring shared derived characters. When cladistics came along, it made the study of ungulates much more rigorous, especially when cladistic analyses were completed on many of the major groups (see Prothero et al. [1986] on rhinos; Prothero et al. [1988] on the ungulates; and the chapters in Prothero and Schoch [1989] on perissodactyls and in Prothero and Foss [2007] on artiodactyls; and Prothero [2016] for the most recent information). Part of the problem had been a paraphyletic "wastebasket" group, the "Condylarthra," which had been used to cluster any archaic ungulates that were not members of the living orders. With the aid of cladistics, this wastebasket has been broken up, and anyone who still uses the archaic and obsolete term "condylarth" (such as creationists) reveals their ignorance of the current state of research.

We can trace ungulates back as far as 85 million years ago in the early Late Cretaceous to fossils known as zhelestids from Uzbekistan (Archibald 1996). By the latest Cretaceous (67 million years ago), there were primitive ungulates known as *Protungulatum* that had many distinctive features, such as larger body size than any other latest Cretaceous mammal, more rounded cusps on the teeth, and a distinctive ankle region. After the extinction of the nonavian dinosaurs, the ungulates quickly dominated the habitats on the ground beneath the trees and radiated very rapidly in the Paleocene, so they are by far the most common fossils in Paleocene strata from the Bighorn Basin of Wyoming or the San Juan Basin of New Mexico. Most of these archaic ungulates (once called condylarths) are members of extinct groups that look nothing like the living hoofed mammals and were only distantly related to them (fig. 14.1). By the late Paleocene, we find the first fossils that clearly represent the origins of the major living orders of ungulates. These include *Radinskya* from the Paleocene of Mongolia, the primitive relative of the odd-toed *perissodactyls*, or horses, rhinos, and tapirs; *Phosphatherium*, *Eritherium*, *Daouitherium*, and *Minchenella* from Africa and Asia, the earliest members of the elephant-mastodont clan; *Diacodexis* and *Dichobune*, the earliest members of the even-toed hoofed mammals, or *artiodactyls*; and the mesonychids, which are distantly related to the whales. By the early Eocene, all of these ungulate groups were evolving rapidly into many different families and genera, most of which are now extinct. This story is so detailed and interesting that we can review only a few highlights in a chapter like this, but I recommend you take a look at my new book, *The Princeton Field Guide to Prehistoric Mammals* (published by Princeton University Press in 2016).

Tiny Horses, Hornless Rhinoceroses, and Thunder Beasts

The geologic record of the ancestry of the horse is one of the classic examples of evolution.
—William Diller Matthew, "The Evolution of the Horse: A Record and Its Interpretation"

Let us start with the most familiar example of mammalian evolution, the origin of the horse (figs. 14.2 and 14.3). This is one of the oldest cases used to support evolution in the fossil record and still one of the best—and therefore the most distorted and misrepresented by the creationists. Shortly after Darwin's book was published, fossil horses in Europe were connected into an apparent evolutionary series by Thomas Henry Huxley, French paleontologist Albert Gaudry, and Russian paleontologist Vladimir Kowalewsky. Then, in 1876, Huxley made a trip to North America, where he visited the amazing collections of fossil horses at Yale University amassed by crews working for the pioneering paleontologist Othniel C. Marsh. Huxley soon realized that the few European fossil horses were occasional immigrants to the Old World, not a continuous lineage that evolved in Europe, and that most of horse evolution had taken place in North America. Marsh pulled out drawer after drawer of specimens for Huxley, completely documenting the stages in the transition from tiny four- and three-toed early Eocene horses up to the modern *Equus*. Huxley was so amazed that he threw out his planned lecture notes and used Marsh's specimens for his featured lecture instead. This early research on horses continued until it culminated in the famous book on horse evolution published by Matthew (1926), quoted in the epigraph above (fig. 14.2).

In general terms, these century-old diagrams are still valid. Horses did start as tiny beagle-sized animals with four toes on their front feet and three on their hind feet, low-crowned teeth for eating soft leaves, and relatively small brains and short snouts. These early Eocene horses have long been known as *Eohippus* (but that name is invalid for most of them) and *Hyracotherium* (but Hooker [1989] showed that *Hyracotherium* is a member of a native European group known as palaeotheres, not a true horse). Froehlich (2002) analyzed the North American fossils in detail and found that the old name *Eohippus* is only applicable to one of the species, *E. angustidens*. Instead, many of these early Eocene horses belong to

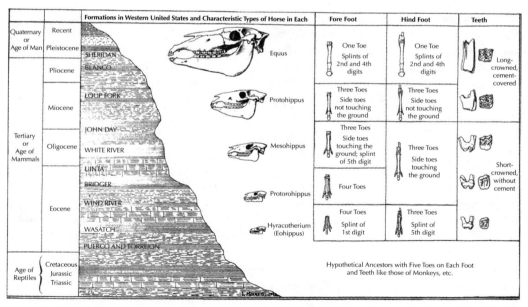

FIGURE 14.2. The evolution of horses as it was portrayed a century ago when there were relatively few fossils. The overall trend through time is clear: larger size, longer limbs, reduction of side toes, development of a longer snout and larger brain, and especially the development of higher-crowned cheek teeth for eating gritty grasses. However, a century of further collecting has shown that horse evolution is a more complicated bushy branching tree, rather than this oversimplified linear sequence. (After Matthew 1926)

Protorohippus, while others are assigned to a variety of genera, including previously proposed names such as *Xenicohippus*, *Systemodon*, and *Pliolophus*, as well as new genera such as *Sifrhippus*, *Minippus*, and *Arenahippus*. The old days when all early Eocene horses could be lumped into one genus (whether *Eohippus* or *Hyracotherium*) are long gone!

From these early Eocene horses, there is a general size increase through Eocene forms like *Orohippus* and *Epihippus*, culminating in the German shepherd–sized horse *Mesohippus* and *Miohippus* from the late Eocene and Oligocene (Prothero and Shubin 1989). These horses had three robust toes on their hands and feet, slightly higher-crowned teeth, and many other more horselike features of the snout and skull. By the early Miocene, horses began an explosive radiation (fig. 14.3) into multiple lineages. Some of these Miocene horses retained low-crowned teeth for leaf eating (the anchitherines), but most evolved higher-crowned teeth for eating gritty grasses and also developed longer legs and toes (with reduced side toes) to run fast across the grasslands. Finally, most of these diverse horse lineages went extinct in the late Miocene, leaving only the lineage leading to the modern genus *Equus* to flourish in the Pliocene and Pleistocene.

The biggest difference between the old pattern of horse evolution (fig. 14.2) and our modern one (fig. 14.3) is its bushiness. Back in the late 1800s and early 1900s, there were only a few horse fossils known at each level, and it was easy to imagine one continuous lineage evolving through time in a "ladder-like" fashion. But tens of thousands of additional horse fossils have been collected since the 1920s. Just as we have seen in every other instance, horse evolution is a bushy, branching pattern, with multiple lineages coexisting at the same time (fig. 14.3). Gingerich (1980, 1989) and Froehlich (2002) document multiple species of horses coexisting in the early Eocene. Prothero and Shubin (1989) found that three different species of *Mesohippus* and two of *Miohippus* coexisted in some upper Eocene beds in Wyoming. Some Miocene quarries in Nebraska, such as Railroad Quarry A in the Valentine Formation of north-central Nebraska, yield as many as 12 different contemporary species of horses. Even today, the genus *Equus* is very speciose, with not only domesticated horses and their ancestral stock, Przewalski's horse, but three species of zebras, and several more species of wild asses and onagers. Thus the general trend of the classic diagram shown in figure 14.2 is factually correct, but a gross oversimplification that does not capture the bushy pattern that we now recognize.

Naturally, this example has been presented so often that creationists feel obligated to attack and distort it and defuse its powerful impact. Their lies about the horse evolution story never actually involve looking at real specimens or doing their own research. They are just quotations out of context that distort what the author is really saying or quotations from really old sources that are no longer describing the current state of our knowledge. For example, Gish (1995:189–197) devoted nine pages to the topic, all of which are highly misleading and dishonest. His most common tactic is quotes out of context that have authors pointing out that the old simplistic linear horse evolution story (fig. 14.2) is no longer valid—without giving the rest of the quote in which the author points out that the fossil record is now much *better*, and the horse phylogeny is very bushy! In other cases, he quotes authors as saying that we have few gradual transitions between species—but as we showed in chapter 4, under the model of punctuated equilibrium, we don't expect gradualism in horses. Nevertheless, by looking at the features of each successive species, we can see the evolutionary trends in the group just fine. Remember, just because they don't show gradual continuous change doesn't mean that they show no evolution! In fact, some paleontologists argue that they *can* show

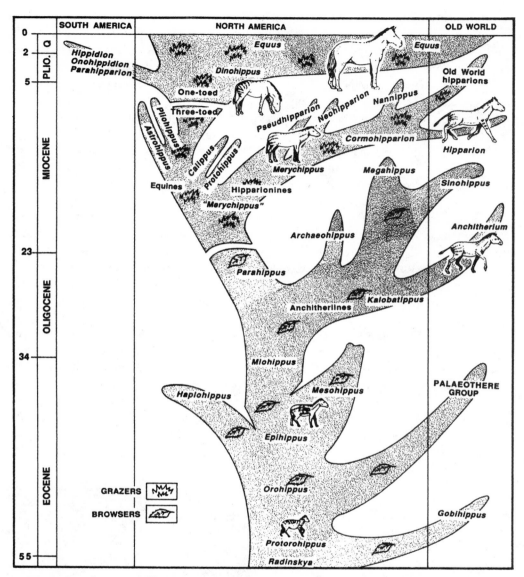

FIGURE 14.3. A modern view of horse evolution, emphasizing the bushy branching nature of their history, as many more fossils have been found and new species named. However, the overall trends toward higher-crowned teeth (shown by the symbols for browsing leaves or grazing grasses), larger size, longer limbs, and reduction of side toes are still true. (Drawing by C. R. Prothero; after Prothero 1994b)

gradual change between horse species, one example being Phil Gingerich (1980, 1989), with his excellent detailed record of early Eocene horses from the Bighorn Basin of Wyoming. Similarly, when Neil Shubin and I (Prothero and Shubin 1989) looked closely at the horses from the Eocene and Oligocene deposits in the Big Badlands and related areas, we found that what had been interpreted as a gradual transformation from *Mesohippus* to *Miohippus* made more sense as a bushy pattern of evolution. Nonetheless, the transformation from one species to another is very subtle. MacFadden (1984, 1992) has documented the subtle distinctions between different species of Miocene horses, which form a nice series of species transformations, yet Gish (1995) misquotes him to indicate that horses didn't evolve at all! Again and again, Gish falsely claims there are no transitional forms between the species, but if he had bothered to read the original literature closely, or better yet, look at the real specimens, he would have been overwhelmed by the continuous gradation in fossil species from the Eocene to *Equus*. Finally, Gish falsely claims (based on out-of-context quotations from outdated books written 50 years ago) that there are no transitional forms linking horses to anything else. But those transitional forms have been known for a long time. They include the archaic ungulates known as phenacodontids (documented by Radinsky 1966, 1969), and the Paleocene Mongolian creature *Radinskya* (fig. 14.4), which is such a good link between early perissodactyls and their relatives among arsinoitheres and elephants that McKenna et al. (1989) had a hard time deciding what group to put it in.

Intelligent design creationist books also mention horse evolution but do not discuss it in any detail. Davis and Kenyon (2004:96) deny the existence of the evolutionary sequence of horses, yet make no further mention of it anywhere else in their book. Wells (2000:195–207) discusses how the ideas of horse evolution have moved away from the old linear "straight-line" notions to the modern complex phylogeny, but nowhere does he dispute the reality of horse evolution. Instead, his convoluted argument seems to suggest that if we change our notions from a simplistic linear model to a more complex bushy model, we are denying that horse evolution occurred! Of course our notions have changed—we have more fossils and more data. We would be bad scientists if our ideas *didn't* change in the face of new data. Sarfati (2002) makes the absurd claim that "the other animals in the sequence show hardly any more variation between them than *within* horses today" (133). Clearly, he has never looked at the actual specimens, because no one would mistake the tiny collie-sized Eocene horses, nor *Mesohippus*, which was the size of a Great Dane, for any living horse (not even the tiniest pony). Not only are they dramatically smaller than all living horses, but they have a

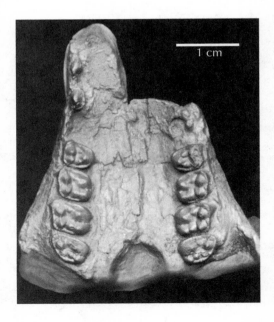

FIGURE 14.4. The Paleocene Chinese fossil *Radinskya*, which is a primitive relative of the perissodactyls, shows that they originated in Asia in the Paleocene. The differences between this skull and the earliest horses and rhinos are very subtle. (Photo courtesy M. C. McKenna)

whole range of primitive anatomical features (three functional digits on their hands and feet, shorter snout, much lower crowned teeth, and many other features) that clearly distinguish them from any living horse, no matter how small.

In fact, if creationists had bothered to spend any time at all looking at real fossils, they would have been amazed by how subtle the transition is from phenacodontids and *Radinskya* to the early perissodactyls. Even more surprising, the earliest Eocene horses, rhinos, and tapirs are also very hard to tell apart—yet modern horses look nothing like the living tapirs or rhinos today. This fact struck me when I was working on my undergraduate research project on early Eocene mammals from the Bighorn Basin of Wyoming. Although the literature on the subject was clear, it was a major challenge trying to tell the earliest horse teeth from the teeth of *Homogalax*, the earliest member of the rhino-tapir lineage. They are virtually identical in size and in cusp-by-cusp detail (fig. 14.5), except that *Homogalax* tends to have slightly better connections of the crests between the cusps. Anyone without a sharp eye would miss this difference completely, and the same is true of the skulls and skeletons. All of the early perissodactyls (horses, rhinos, tapirs, and brontotheres) look so similar when they begin their evolution that only a trained eye can tell them apart. Yet we can trace the evolution of each of these distinct lineages through time, and they soon begin to look very different, so that by the late Eocene, they are dramatically distinct in size and body shape, and even a kid could tell them apart. This is one of the best examples of how we can document the origin of many modern distinct lineages back to ancestors that converge to the point of being virtually indistinguishable.

But if the evolution of horses is not convincing enough, let's look at my favorite group, the rhinoceroses. They have just as long and dense and detailed a fossil record as horses, yet they have received almost no attention because their systematics was a mess for decades and nothing could be concluded until the valid species were determined using the new collections (Prothero 2005). Once that was done, however, we again see a highly bushy, branching family tree of rhinos (fig. 14.6) in North America (and a similar pattern in Eurasia), with many different families, species, and genera spanning almost 50 million years. The earliest relative of rhinos was the early Eocene form known as *Homogalax*, which also gave rise to tapirs, and yet *Homogalax* is virtually identical to the early Eocene horses (fig. 14.5). By the middle Eocene, we see the split between the tapiroid lineage and the lineages that lead to the three main families of rhinocerotoids. Unlike horses, which evolved mainly in North America with occasional emigrations to Eurasia, rhinos evolved on both hemispheres and immigrated back and forth, so their family tree is much more bushy and dominated by sudden immigration events than that of the horse (fig. 14.6). Although most of the species are distinct, we can still see evolutionary trends, particularly in the front of the snout (fig. 14.7A), where the primitive forms have many incisors and small canines, and as rhinos evolve, they lose most of the incisors and develop sharp, short tusks between their remaining incisors. In addition, their cheek teeth (fig. 14.7B) also show evolutionary trends, such as the modification of the primitive premolars into molar-like crests that resemble the Greek letter "π." My rhino monograph (Prothero 2005) documented many other changes, both gradual and punctuated, such as the size changes in many lineages, and the gradual development in horns within the genus *Diceratherium*. Like horses, rhinos also get larger and more specialized throughout their evolution. They started out with four toes on the front foot, and reduced it down to three by the middle Eocene—but unlike horses, they remained three-toed even today and never became highly specialized one-toed runners like living horses.

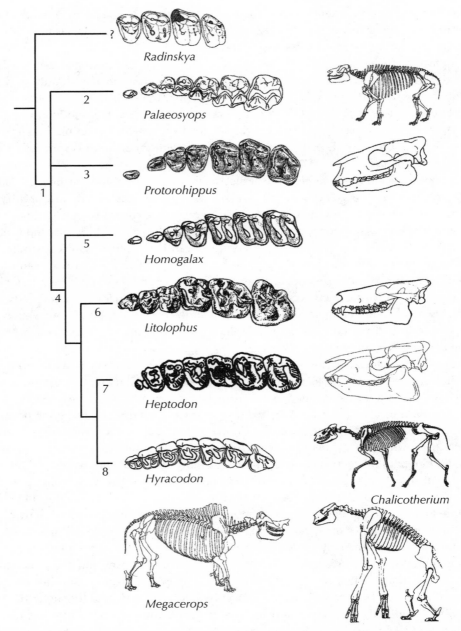

FIGURE 14.5. The evolutionary radiation of perissodactyls, showing the major branches of the horses, rhinos, tapirs, chalicotheres, bronthotheres (*Megacerops*), and other extinct groups. As can be seen from the crown views of the upper left cheek teeth, the details of the crests and cusps are extremely similar between *Radinskya*, the early brontothere *Palaeosyops*, the primitive horse *Protorohippus* (long called "*Hyracotherium*"), the primitive moropomorph *Homogalax*, the chalicothere *Litolophus*, the tapiroid *Heptodon*, and the primitive rhinoceros *Hyracodon*. Shown next to the upper cheek teeth are typical skulls of horses, tapirs, and rhinos, emphasizing how similar they all looked in the early stages of perissodactyl evolution. The numbered branching points are as follows: 1, Perissodactyla; 2, Titanotheriomorpha; 3, Hippomorpha; 4, Moropomorpha; 5, Isectolophidae; 6, Chalicotherioidea; 7, Tapiroidea; 8, Rhinocerotoidea. (Phylogeny after Prothero and Schoch 1989; redrawn by Carl Buell)

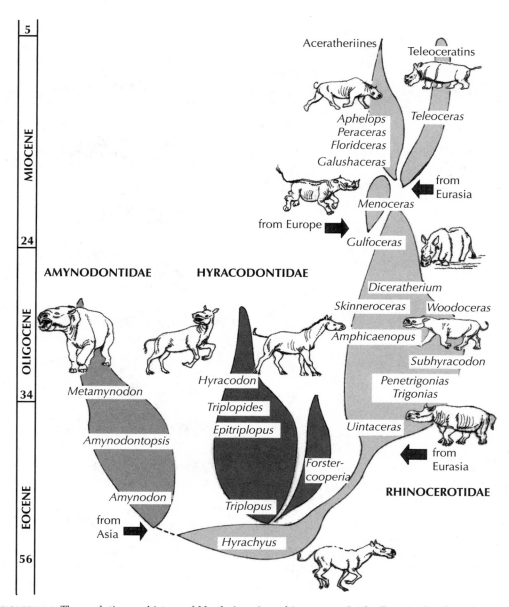

FIGURE 14.6. The evolutionary history of North American rhinoceroses. In the Eocene, they branched into three families, the hippo-like amynodonts, the long-legged running hyracodonts, and the living family Rhinocerotidae. During their evolution, they varied not only in body size and limb and skeletal proportions but also in the number and position of horns (or lack of horns), the details of their teeth, and many other features. (Drawing by C. R. Prothero; after Prothero 2005)

It comes as a shock to many people that most fossil rhinos were hornless. Modern rhino horns are made of tightly compacted hairs and have no bony support (unlike the horns of cattle or antelopes), so they seldom fossilize. We can usually determine the presence of a horn by a patch of spongy roughened bone on the snout or forehead that served as the attachment point. On this basis, rhinos were hornless during most of their evolution, and

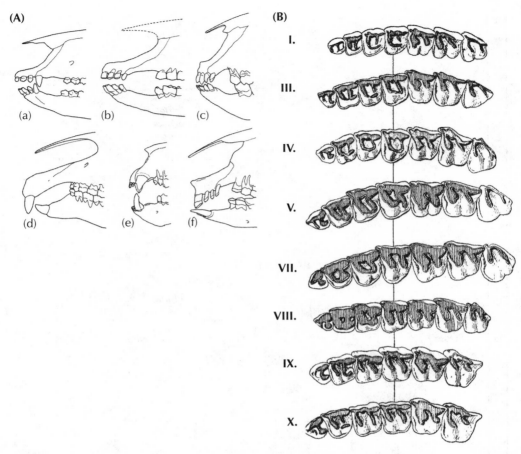

FIGURE 14.7. The different rhinocerotoid families can be distinguished largely through the modifications of their teeth. (A) Primitive forms such as *Hyrachyus* (a) have the tapir-like condition, while hyracodonts like *Ardynia* (b) and *Hyracodon* (c) have reduced their canines and modified their incisors. The gigantic indricothere *Paraceratherium* (d) has lost all its front teeth except for huge conical incisor tusks. The amynodont *Metamynodon* (e) has reduced its incisors and developed heavy chisel-like upper and lower canines. The true rhinoceros *Trigonias* (f), on the other hand, has almost all of its incisors and canines, with a chisel-like upper incisor and a tusk-like lower incisor. (Drawing by Carl Buell). (B) Evolutionary transformation of the cheek teeth of *Hyracodon*, showing the gradual transformation of the premolars into more molarized crests. (After Prothero 2005)

then in the Oligocene (about 28 million years ago) they independently evolved two different types of paired horns on the tip of the nose (*Diceratherium* and *Menoceras*). After these lineages disappeared, there was a long period dominated by hornless rhinos, except for a single nose horn in the *Teleoceras* lineage. Asian rhinos also had a single nose horn, while the two African rhino genera, the black rhino and white rhino, have two horns in tandem. The elephant-sized Eurasian ice age elasmotherine rhinos had a huge horn (1.5 meters or 5 feet long) on their foreheads, but none on their noses!

Closely related to the rhinos are the tapirs, which look vaguely piglike, except that they have an odd number of toes and a flexible snout or proboscis. Most people have seen them only in zoos (if at all) because all the living species are restricted to the tropics, including Southeast Asia (the Malayan tapir) and three species in Central and South America. But tapirs were widespread in North America and Eurasia through most of the past 50 million years, although their teeth are adapted for chopping leaves, so they have always been restricted to the more forested regions. Tapirs, too, have an impressive fossil record of their evolution (fig. 14.8). The early more primitive forms like *Homogalax* mentioned earlier (fig. 14.5) can barely be distinguished from the earliest horses and rhinos. The only clue is that the cross crests on their molars are slightly more developed (foreshadowing the strong crests on later tapir teeth), but otherwise their skulls and skeletons look almost identical to the earliest rhinos and horses. However, through the rest of the Eocene they become more and more specialized, and more and more like living tapirs. Even by the middle Eocene, their teeth

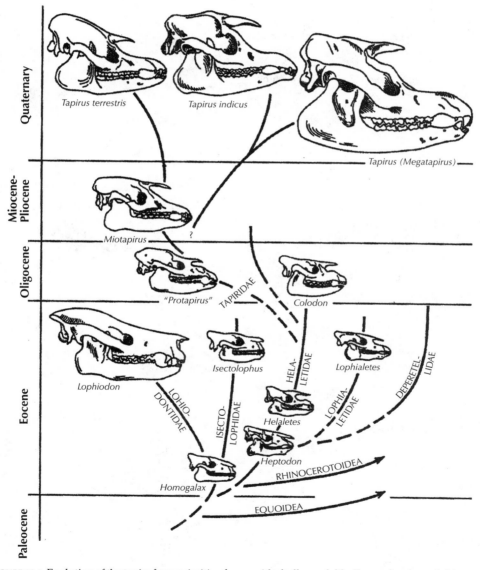

FIGURE 14.8. Evolution of the tapirs from primitive forms with skulls much like Eocene horses and rhinoceroses, through progressively more specialized forms that have a deeper retraction of the nasal notch, indicating a larger proboscis. (Modified from Prothero and Schoch 2002)

have more strongly developed cross crests. My friend Matt Colbert described a new species of tapir, *Hesperaletes*, from the Eocene deposits of San Diego (not far from the old headquarters of the ICR) that is the oldest fossil to show evidence of a nasal notch on the skull for attachment of the proboscis. Through the past 40 million years, tapir skulls show a deeper and deeper retraction of the nasal notch, suggesting a larger and better-developed proboscis, until we reach the condition in living tapirs, which have the largest proboscis of all. During this time their molars become more and more highly specialized, until they are simple cross crests ideally suited for chopping leaves.

Finally, we should mention one more example from the perissodactyls: the brontotheres, also known as the titanotheres (figs. 14.9 and 14.10). At the climax of their evolution in

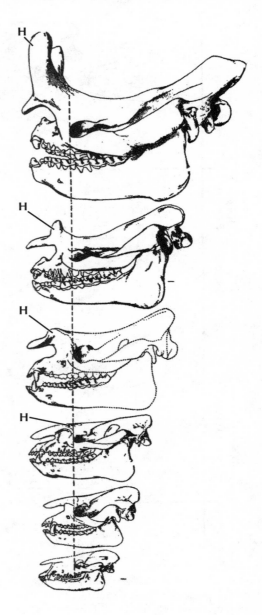

FIGURE 14.9. Conventional linear view of brontothere evolution through the Eocene from primitive forms like *Palaeosyops* that are barely distinguishable from contemporary horses (fig. 14.5) through larger and larger forms that eventually developed two blunt horns (H) on their noses. (After Osborn 1929)

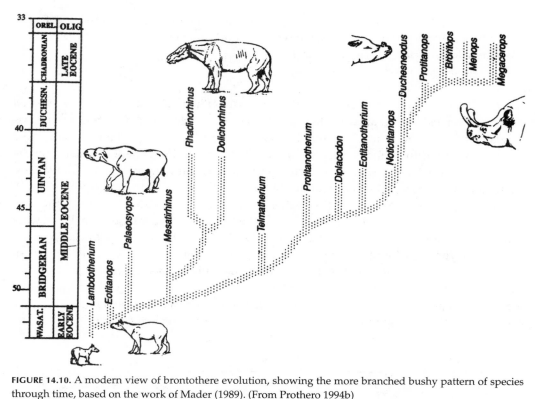

FIGURE 14.10. A modern view of brontothere evolution, showing the more branched bushy pattern of species through time, based on the work of Mader (1989). (From Prothero 1994b)

the latest Eocene, these elephant-sized beasts sported a pair of blunt bony battering rams on their noses, yet they evolved from beagle-sized unspecialized early Eocene ancestors (*Palaeosyops*) that are difficult to distinguish from the earliest Eocene horses and tapir-rhino ancestors. A century ago, the famous paleontologist Henry Fairfield Osborn spent years studying them and published a huge two-volume work (Osborn 1929) that was chock-full of bad taxonomy (even by the standards of his day). It was further confused by his unorthodox ideas about *orthogenesis*, or straight-line evolution going out of control in one direction without the restraint of natural selection. Unfortunately, this outdated example (fig. 14.9) continues to be published in textbooks today, even though it falsely portrays their evolution as a single straight lineage. In fact, brontothere evolution was very bushy as well (fig. 14.10), with multiple lineages coexisting in the middle and late Eocene, after which the entire group died out. Lest the creationists attempt to misquote me, brontotheres *do show evolutionary changes in size and horn development through time*, but it is in the context of many different species branching out, not a single lineage marching to their ultimate extinction as Osborn once thought.

Kingdom of Cloven Hooves

Nevertheless, these shall ye not eat, of them that chew the cud or of them that divide the cloven hoof; as the camel and the hare, and the coney; for they chew the cud, but divide not the hoof, therefore they are unclean unto you. And the swine, because it divideth the hoof, yet cheweth not the cud, it is unclean unto you; ye shall not eat of their flesh, nor touch their dead carcass.

—Deuteronomy 14:7–8

After the Perissodactyla, the second great living order of hoofed mammals is the even-toed hoofed mammals, or Artiodactyla. They are "even toed" or "cloven hoofed" because the axis of symmetry of the foot runs between the third and fourth toes, and so they usually have either two toes or four. Today, they are the most diverse and abundant ungulates on the planet, with over 190 living species, including pigs, peccaries, hippos, camels and llamas, deer, pronghorns, giraffes, sheep, goats, cattle, and dozens of species of antelope (fig. 14.11). Nearly every domesticated animal we eat is an artiodactyl (pigs, sheep, goats, cattle, and deer), and they provide us with all of our milk (whether from a cow, goat, or camel) and wool (from either sheep or alpacas). Almost every large herbivore you might see in East Africa is an artiodactyl except for zebras, rhinos, and elephants. Artiodactyls are indeed a modern success story, yet they acquired that dominance gradually as the odd-toed perissodactyls (especially horses, rhinos, tapirs, and brontotheres) dominated in the Eocene and gradually were displaced, while artiodactyls (especially the ruminants, such as camels, sheep, goats, and cattle) with their superior mode of digestion came to dominate the earth.

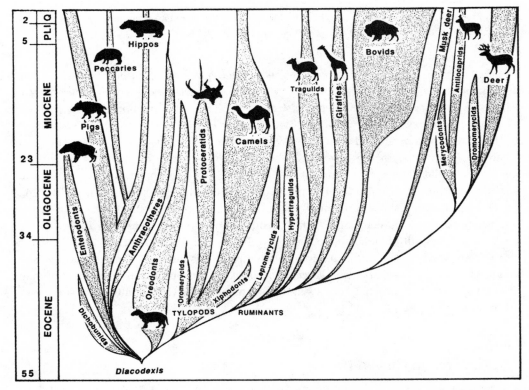

FIGURE 14.11. Evolutionary history of the even-toed hoofed mammals, or artiodactyls. (Drawing by C. R. Prothero; from Prothero 1994b)

FIGURE 14.12. The earliest even-toed hoofed mammals, or artiodactyls, such as *Diacodexis* and *Dichobune*, were small primitive forms that looked nothing like their living descendants, the pigs, hippos, camels, deer, giraffes, pronghorns, cattle, sheep, antelopes, and goats. Nevertheless, they had the characteristic teeth found in all primitive artiodactyls and distinctive features of the skull and ankle region that mark them as ancestors of this great order of mammals. (Drawing by Carl Buell)

Yet once again, a look at the first known artiodactyls from the early Eocene reveals a form that no one would connect with a cow, giraffe, or camel (fig. 14.12). *Diacodexis* and *Dichobune* were tiny creatures, about the size of a rabbit, with long delicate limbs and simple primitive teeth and skulls. Some of them had very long hind limbs and were apparently well adapted to hopping and leaping as well. Both anatomical cladograms of the ungulates (Prothero et al. 1988) and the molecular phylogenies (Murphy et al. 2001a, 2001b) place their branching point at the very base of the ungulate radiation (fig. 14.1). Ken Rose (1987) has suggested that the Paleocene archaic ungulate *Chriacus* looks very much like a possible transitional form between the earliest ungulates and artiodactyls. Once artiodactyls appeared in the early Eocene, they became very stereotyped and distinctive in many features. Their simple cusped teeth soon evolved into half-moon-shaped crests, giving them their signature *selenodont* teeth. Their legs and feet were long and delicate from the very beginning and soon began to lengthen the central toes and reduce the side toes (but in a different way than horses). All artiodactyls have a distinctive double-pulley bone in their ankle known as the astragalus, which made their legs very efficient for front-to-back running motions (but prevented them from rotating their foot or hind leg out of the front-to-back plane).

By the middle Eocene, artiodactyls were undergoing an explosive radiation with many different families, most of which are now extinct (fig. 14.11). However, the primitive relatives of pigs were present, as were the first camels and the first ruminants. In the Oligocene, another explosive evolutionary radiation occurred, this time with mostly living families, such as the camels and early ruminants, and by the Miocene, the pronghorns, giraffes, antelopes, and cattle were beginning to diversify as well. Some of these families have extraordinary fossil records, and we can trace their lineages back to the Eocene, then look forward to

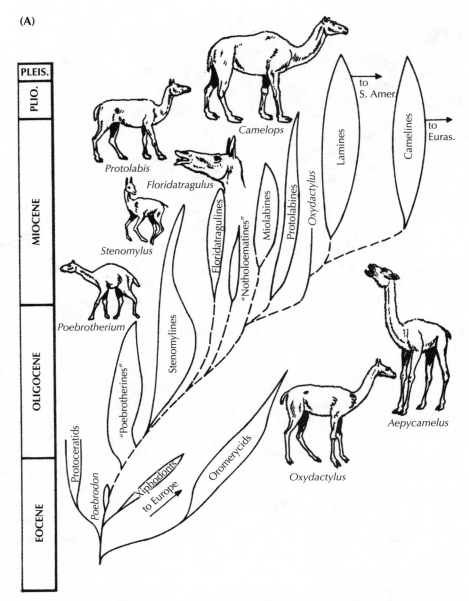

FIGURE 14.13. The evolution of camels in North America. (A) The family tree of camels, showing the great diversity of forms, from small primitive deerlike creatures to the gazelle-like stenomylines, the short-legged protolabines and miolabines, the long-legged long-necked "giraffe camels," and the modern humpless South American camels (alpaca, llama, vicuña, guanaco), which are more typical of the whole family. Only the living African dromedary and two-humped Asian Bactrian camels have humps. (Drawing by C. R. Prothero; after Prothero 1994b) (B) Evolutionary trends within the camels, from the tiny oromerycid *Protylopus* through the Oligocene camel *Poebrotherium* through more advanced *Procamelus*. Although their history is not a straight line of evolution but a bushy branched pattern, there are trends toward larger body size, loss of the front teeth, longer snouts and larger eyes, longer legs and toes (reducing to just two toes fused together), and higher-crowned cheek teeth. (After Scott 1913)

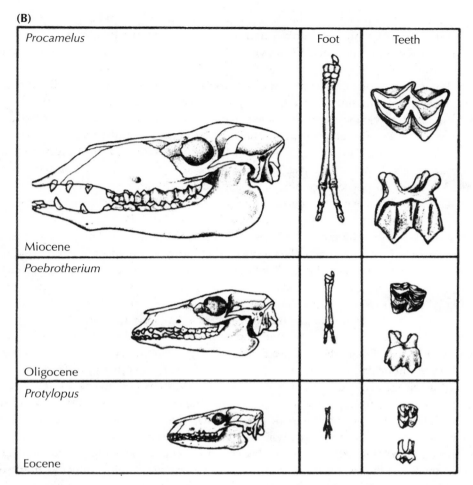

(B)

Procamelus	Foot	Teeth
Miocene		
Poebrotherium		
Oligocene		
Protylopus		
Eocene		

FIGURE 14.13. (*Continued*)

how they evolved into a wide diversity of forms in the Oligocene and Miocene. For space reasons, we'll look at just two examples, the camels and the giraffes.

Most people are surprised to learn that extinct camels did not have humps and that the camel family evolved in isolation in North America (fig. 14.13A). They only escaped this continent in the late Cenozoic when they reached South America 3 million years ago to evolve into llamas, guanacos, and vicuñas, and Eurasia about 7 million years ago, where they evolved into the African dromedary and the Asian Bactrian camels. After all this success, they vanished from their ancestral North American homeland at the end of the last ice age 10,000 years ago. Fossil camels are also surprising in their amazing array of ecological types, far exceeding the limited forms we see today (Honey et al. 1998). The earliest camels were tiny rabbit-sized creatures (*Poebrodon*) that are known from isolated teeth and jaws from the late middle Eocene of Utah, Texas, and California. But by the late Eocene and early Oligocene they had evolved into sheep-sized creatures known as *Poebrotherium* (Prothero 1996), which are common in the Big Badlands of South Dakota. *Poebrotherium* has all the hallmarks of a typical early camel: very high-crowned selenodont teeth, long limbs that were nearly fused into a cannon bone, and the signature features of the skull and skeleton as well (fig. 14.13B). Yet its proportions looked more like those of an antelope or a gazelle, and it apparently had no hump either. In the late Oligocene and early Miocene, camels underwent an explosive evolutionary radiation (fig. 14.13A) into relatively short-limbed varieties (protolabines and miolabines); tiny delicate gazelle-like forms with extraordinarily high-crowned teeth (stenomylines); long-legged, long-necked forms that looked much like the modern guanaco or vicuña (aepycamelines); and even a group that evolved long necks and performed the roles of treetop browsers that giraffes occupied in the Old World. Some of these late Miocene and Pliocene "giraffe-camels" were huge as well, with appropriate names like *Gigantocamelus* and *Titanotylopus*. Then, after spreading to Eurasia and South America in the late Miocene and Pliocene, camels dropped in diversity during the ice ages, and only a few species were left when they vanished from North America 10,000 years ago.

For our final example from the artiodactyls, let us consider the giraffe. As we have been pointing out all along, the modern representatives of many mammalian families are very atypical of most of the members of the family during its evolution. Most fossil rhinos did not have horns, most fossil camels did not have humps, and most fossil giraffes did not have long necks. Extinct giraffids (Solounias 2007) sported a wide diversity of horn shapes and body sizes (fig. 14.14). Most fossil giraffes looked more like the short-necked okapi (fig. 14.14B), a shy white-and-brown-striped denizen of the African rain forests, and the only other living giraffid. Others, like *Sivatherium*, got to be huge and heavy with moose-like palm-shaped horns. In the late Miocene, we finally see the lineage that leads to the modern *Giraffa*. Although the fossil teeth are common enough, complete skeletons with the neck vertebrae are rare, so we can see how the lineage evolved, but didn't yet have fossils to show how the neck got longer. The oldest known fossils of the genus *Giraffa jumae* from the late Miocene of Africa already seems to have a long neck, so we need to seek fossils from earlier in the Miocene. Nikos Solounias (1999) has shown the mechanisms by which the neck lengthens, and the prevailing ideas have to be revised. Giraffes lengthen each vertebra somewhat but actually add an extra vertebra to the neck, then shift the last neck vertebra to the shoulder region. This is why the giraffe neck begins *behind* the forelimbs and why giraffes have their distinctive posture with the legs out in front and the neck balanced almost over the center of the body, rather than sticking forward all the time as in most mammals. Finally, Solounias has just described a description of a classic transitional form (fig. 14.15): a giraffe fossil with an intermediate-length neck, longer than that of the okapi and the other extinct forms but shorter than that of the living giraffes. For so many years, people have speculated about how giraffes got their long necks, and now we finally have the fossils to show exactly how it happened! Once again, the fossil record has yielded a transitional form that the creationists claimed could never exist.

Walking Whales

These dogmatists, who by verbal trickery can make white black, and black white, will never be convinced of anything, but *Ambulocetus* is the very animal that they proclaimed impossible in theory. . . . I cannot imagine a better tale for popular presentation of science or a more satisfying, and intellectually based political victory over lingering creationist opposition.

—Stephen Jay Gould, "Hooking Leviathan by Its Past"

(A)

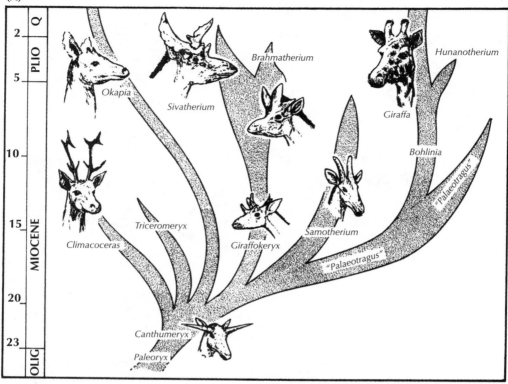

(B)

FIGURE 14.14. Evolution of the giraffe family. (A) The modern okapi is more typical of the group, with its short neck and relatively short horns or "ossicones." Some fossil giraffids, however, had very unusual branching and flaring cranial appendages. (Drawing by C. R. Prothero; after Prothero 1994b) Only the lineage of the modern giraffid evolved a long neck. (B) The modern okapi, much more typical of the giraffe family than its long-necked cousin. (Photo by the author)

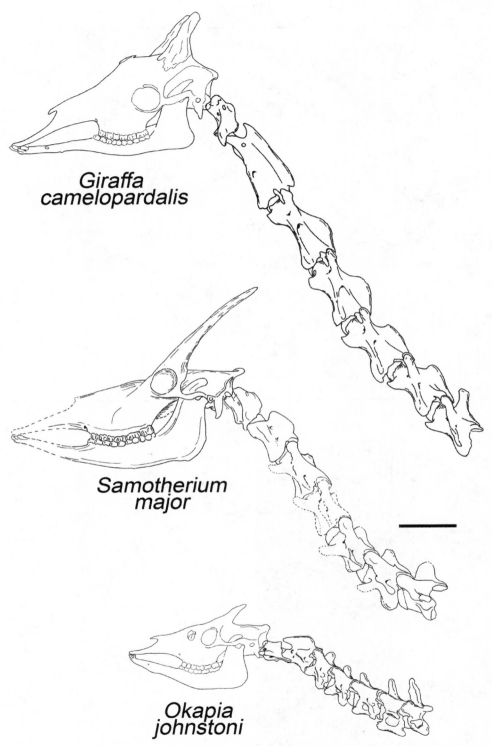

Giraffa
camelopardalis

Samotherium
major

Okapia
johnstoni

FIGURE 14.15. Neck vertebrae of a recently discovered fossil giraffid, *Samotherium major*, that are intermediate in length between those of primitive giraffids and the modern long-necked species. This amazing discovery is a true "missing link" between okapis (bottom) and the long-necked modern species (top). (Drawing courtesy N. Solounias)

People are startled to learn that most fossil rhinos didn't have horns, most fossil camels didn't have humps, and most fossil giraffes had short necks. But they are even more surprised to learn that whales are related to hoofed mammals and are descended from a group of carnivorous hoofed mammals. In debates, creationists love to exploit this public ignorance of the fossil record and zoology by putting up a slide of "Bossie" the cow and "Blowhole" the whale and a ridiculous cartoon of an intermediate between a cow and a whale. But when we said that whales are descendants of ungulates, we did not say "cows." Apparently, when creationists hear the words "hoofed mammal," cows are the only kind they can think of. Actually, hippos would be better models for a modern relative of whales, and they are not nearly so different from whales (both are large and aquatic).

Ever since people realized that whales and dolphins were mammals, they have speculated about how they might have evolved from land-dwelling mammals and from which group of mammals they originated. By the 1830s and 1840s, specimens of huge primitive whales known as archaeocetes (fig. 14.16) were being discovered in middle Eocene beds of Alabama, but these specimens were fully aquatic, with flippers and tail flukes and a sinuous 24-meter-long (80 foot) body. Clearly, the origin of whales must have occurred before the middle Eocene, but nothing was known of their fossil record prior to that time. In 1966, Leigh van Valen and others had shown that the skulls and teeth of primitive whales looked very much like the predatory archaic hoofed mammals known as mesonychids. Even though mesonychids were land mammals with hooves, there were many similarities in the skull and skeleton (especially the large, serrated triangular bladelike teeth) that suggested close relationship with archaeocete whales. Yet for over a century, there were no transitional fossils known between mesonychids and archaeocetes.

Until very recently, paleontologists were comfortable with the idea that whales were related to mesonychids, and the fossil evidence seemed to bear this out. Then, in the late 1990s, molecular studies showed that among living mammals, the artiodactyls (and particularly the hippos) were the nearest relatives of the whales. This wasn't too surprising because artiodactyls and whales are very closely related on the ungulate cladogram (fig. 14.1), although we always thought they were sister taxa, not that whales were nested *within* artiodactyls (Prothero et al. 1988). But in 2001, two independent groups of scientists (Gingerich et al. 2001; Thewissen et al. 2001) found specimens of early whales that preserved the ankle region (fig. 14.17). Amazing as it seems, these fossils clearly showed that early whales had feet bearing ankles with the characteristic double-pulley astragalus, the signature feature of the whole order Artiodactyla. Since then, we've rethought the evidence, and now most scientists would agree that whales are a group that evolved from the hippo-pig lineage within artiodactyls and that mesonychids are the distant relatives of both whales and artiodactyls (Geisler and Uhen 2005).

The breakthrough in our understanding of whale ancestry occurred when scientists began to collect fossils in the lower Eocene beds of Pakistan. In 1983, Phil Gingerich and colleagues described *Pakicetus*, based on a skull with an archaeocete braincase but lacking ears that were capable of echolocation and with teeth intermediate between those of mesonychids and archaeocetes (fig. 14.16). *Pakicetus* came from river sediments bordering shallow seaways, suggesting that it might have been a semiaquatic predator that waded in rivers part-time to find food. The skeleton of *Pakicetus* is still quite wolflike, with long slender limbs and a tail, so it still resembles a mesonychid in most features. The chemistry of its bones showed it lived in freshwater.

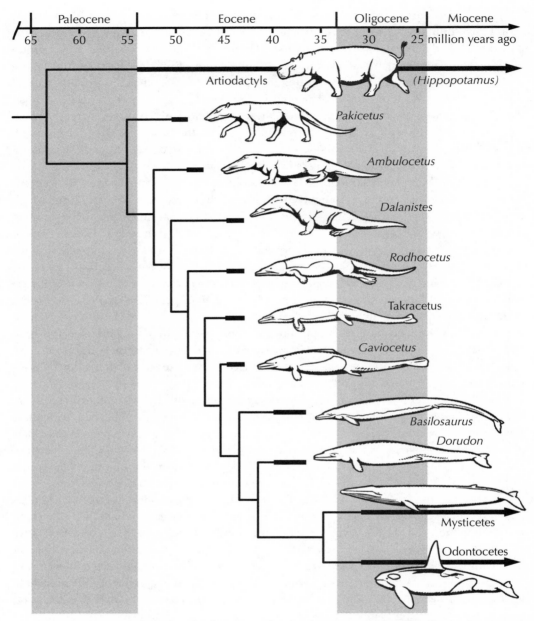

FIGURE 14.16. Evolution of whales from land creatures, showing the many transitional fossils now documented from the Eocene beds of Africa and Pakistan. (Drawing by Carl Buell)

The next development occurred a few years later, when Gingerich et al. (1990) described new specimens of the archaeocete *Basilosaurus* from the middle and upper Eocene deposits of Egypt (in a locality just west of the Pyramids). Although it was like other archaeocetes in being fully aquatic, these new specimens had something never previously preserved: the hind limbs. In most whales, there are no external hind limbs, but the remnants of the hip and thigh bones are buried in muscles along the spine halfway down the body (fig. 4.9). These specimens, however, had tiny hind limbs (about as large as a human arm on a 24-meter-long

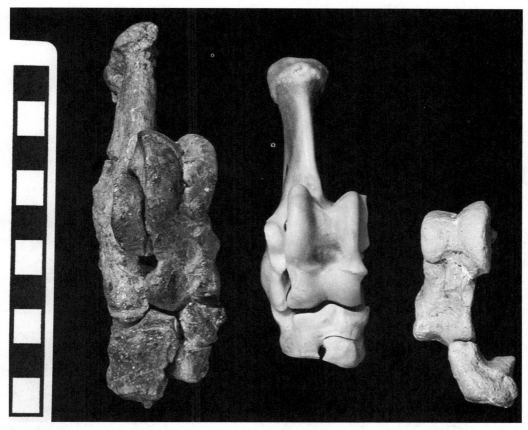

FIGURE 14.17. Ankle bones of middle Eocene whale *Rodhocetus balochistanensis* (left) and *Artiocetus clavis* (right) from Pakistan, compared with those of the pronghorn *Antilocapra americana* (center). Note double-pulley astragalus characteristic of mammals of the order Artiodactyla. (Photo courtesy P. D. Gingerich, University of Michigan Museum of Paleontology)

body!), which clearly did not function for locomotion. Like the vestigial hind limbs of modern whales buried inside the body, these tiny limbs were functionless relics of the day when "whales" did walk on land. Since this discovery, other archaeocetes such as *Takracetus* and *Gaviocetus* have been found to retain vestigial hind limbs.

The most important discovery occurred when Thewissen et al. (1994) discovered and described *Ambulocetus natans*, whose name means "walking swimming whale." Found in middle Eocene marine beds of Pakistan, it was about the size of a sea lion (fig. 14.18), with functional flippers on both its forelimb and huge hind limb (which still had vestigial hooves as well). Its skull and teeth, however, were still like those of mesonychids. On the basis of its highly flexible vertebrae, Thewissen et al. (1994) suggested that *Ambulocetus* swam with an up-and-down flexure of its body, similar to the swimming motion of an otter rather than paddling with its feet like a penguin or seal or wriggling side to side like a fish. This is a precursor to the up-and-down motion of a whale's tail flukes as it swims through the water.

Further discoveries (mostly in the middle Eocene of Pakistan) followed one after another. *Dalanistes*, for example, had fully functional front and hind limbs with webbed feet and a long tail but was much more whalelike with a longer snout. *Rodhocetus* was even more like a dolphin, yet still retained functional hind limbs. *Indohyus*, on the other hand, provides

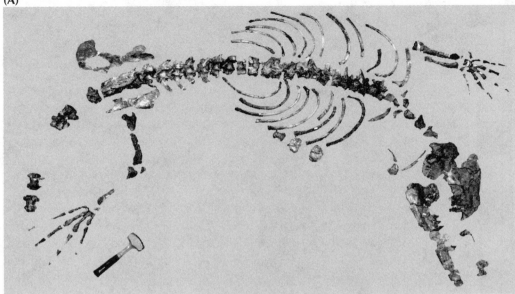

FIGURE 14.18. *Ambulocetus natans*, the primitive whale from the Eocene of Pakistan that still retains a mesonychid-like head, large functional webbed hands and feet, and a semiaquatic mode of life. (A) Photograph of the nearly complete skeleton laid out in anatomical position. (Photo courtesy J. G. M. Thewissen). (B) Reconstruction of *Ambulocetus* lunging out of the water to capture another Eocene mammal. (Reconstruction by Carl Buell)

a link between *Pakicetus* and the primitive land relatives of whales and hippos known as anthracotheres. As the years go by, more and more transitional whales are being discovered, so that by now the amazing transformation from land mammals to whale is one of the best examples of evolutionary transitions in the fossil record (fig. 14.16). This may not make creationists happy, but the fossils cannot be denied.

Creationists have been flummoxed by all this new evidence. The ID creationist textbook *Of Pandas and People: The Central Question of Biological Origins* (Davis and Kenyon 2004:101–102) claims "there are no transitional fossils linking land mammals to whales." They could not be more wrong. This false statement is carried over from their 1989 edition into their 2004 edition, yet the 1980s and 1990s yielded an amazing array of transitional whale fossils that clearly link terrestrial land mammals to full-fledged aquatic whales. These fossils have been well documented in many television shows and websites, in popular books such as Carl Zimmer's (1998) *At the Water's Edge: Macroevolution and the Transformation of Life*, and in high-profile scientific journals such as *Science* and *Nature*, so there is no excuse for creationist ignorance or denial of these fossils. Davis and Kenyon (2004:101, figs. 4–5) illustrate two extremes of the whale evolutionary sequence (the terrestrial mesonychids and the aquatic archaeocetes) but falsely state that there are no transitional forms between them. Sarfati (2002:135–141) snipes at the fact that a few of these spectacular whale fossils are not 100 percent complete and suggests that we cannot draw any conclusions from fossils that are 80 percent complete! He also falls back on the false idea that if whales did not form a clear series of ancestors and descendants, then they do not show an evolutionary transition. But as figure 14.16 shows, one could not ask for a more clear-cut series of intermediate forms from fully terrestrial to fully whalelike fossils.

Every time Duane Gish got into another debate, his opponent brought up *Ambulocetus* and the amazing sequence of transitional whale fossils and showed any open-minded individual how whales evolved. Gish (1995:199–208) blustered on for nine pages about these new discoveries, but his discussion is very confused and self-contradictory. He never addressed the main evidence but sniped at minor quibbles between specialists. After he has mentioned numerous red herrings and blown smoke screens on irrelevant points, he makes the following revealing comment (Gish 1995:203): "Confused? So are we." Nowhere does he address the obvious intermediate anatomy of *Ambulocetus*, *Dalanistes*, *Rodhocetus*, and all the rest of these new fossils. His verbal gyrations basically boil down to this: if it is a modern whale without hind limbs it is really a whale, but if it is a transitional form with a whale's head but intermediate forelimbs and hind limbs (for either walking or paddling), it cannot be a whale, but some unknown fossil! Essentially, he dodges the problem by defining whales in his mind so they cannot have the possibility of intermediate forms. That is not intellectually or scientifically honest and shows the complete bankruptcy of his illogical thinking. And the final clinching evidence is the fact that living whales *do* have hind legs—they are merely vestiges of the hip bone and thigh bone, usually buried deep in their muscles and not visible on the surface (fig. 4.9). Nevertheless, this is final proof (if all the molecules and fossils were not already enough) that whales indeed are descended from four-legged land mammals.

Dumbo and the Mermaid

There is no creature among all the Beasts of the world which hath so great and ample demonstration of the power and wisdom of almighty God as the Elephant.
—Edward Topsell, *The Historie of Foure-Footed Beastes*

Now that we have seen that most of the popular large hoofed mammals—horses, rhinos, camels, giraffes, and whales—have excellent fossil records that document transitional forms going all the way back to the Cretaceous, we need to look at one more group: the elephants and their relatives. Elephants, too, have an excellent fossil record in the late Oligocene and more recent rocks, because mastodonts left Africa about 18 million years ago and migrated among all the northern continents (fig. 14.18). Unfortunately, we are somewhat handicapped because most of their early evolution took place in Africa, and we have a relatively poor fossil record in Africa before the early Oligocene. Nevertheless, we can trace their lineage back from the modern Asian and African elephants and their extinct relatives, the mammoths and mastodonts, through more primitive lineages with a wide variety of tusks and different lengths of trunks. Some (the anancines) had two huge long straight tusks protruding from their skulls, while others (the stegotetrabelodonts) had four long straight tusks; others (the deinotheres) had two tusks that curled down from the lower jaw, and still others (the amebelodonts) had their lower tusks flattened into large shovel-like blades. Going back farther into the early Oligocene, the famous Fayûm beds of Egypt (source of the archaeocete whales with tiny hind limbs) also produce very primitive, small mastodonts with short jaws and even shorter tusks, known as *Palaeomastodon* and *Phiomia*. In the early Oligocene, the various lineages of proboscideans (elephants, mammoths, and mastodonts) are very primitive and hard to tell apart, typical of the early stages of an evolutionary radiation (figs. 14.19 and 14.20). These primitive forms can be traced back to the ultimate transitional fossil, *Moeritherium*, from the late Eocene of Egypt. Superficially, it looked more like a tapir or a pygmy hippo than an elephant and probably only had a short proboscis, not a long trunk. But a close look at the skull shows that it had very short tusks in the upper and lower jaws, the teeth of a primitive mastodont (*not* those of a tapir or hippo), and the details of the ear region and other part of the skull (such as the condition of the jugal bones in the zygomatic arch) are unique to the Proboscidea as well.

All of these fossils have been known for decades, but in the last few years, paleontologists have found even older and better transitional forms. There is the 1984 discovery of an even more primitive proboscidean, *Numidotherium*, from the early Eocene of Algeria (Mahboubi et al. 1984). Although the specimen is very incomplete, it already had the high forehead, the retracted nasal opening (indicating a short proboscis), short upper tusks, mastodont-like teeth, and the lower front jaw is beginning to develop a broad scoop, a diagnostic feature of mastodonts. It was only a meter tall (3 feet) at the shoulder, smaller even than *Moeritherium*, yet it already had the limb characteristics found in later, larger mastodonts. In 1996, Gheerbrant and others reported the discovery of an even earlier proboscidean, *Phosphatherium*, from the late Paleocene of Morocco. The fossil consists only of a partial skull (typical of the poor preservation of mammal fossils in the Paleocene worldwide), but the teeth already show the distinctive mastodont pattern at the very beginning of proboscidean evolution. Thus we now have fossils to trace modern elephants continuously back through many different transitional forms to forms that are almost 60 million years old, and that brings us almost to the time when all the hoofed mammal lineages diverged. Other fossils from the Paleocene of North Africa, such as *Daouitherium* and *Eritherium*, are even more primitive, although known only from teeth and jaws.

But proboscideans are not the only members of this clade. In his groundbreaking 1975 paper on the cladistic classification of mammals, Malcolm McKenna suggested that the closest living relatives of the elephants was the sirenians or "sea cows," better known to us as the manatees and dugongs. Although sirenians are aquatic forms with flippers and a broad tail fluke and no hind limbs that look superficially nothing like elephants, all the evidence shows

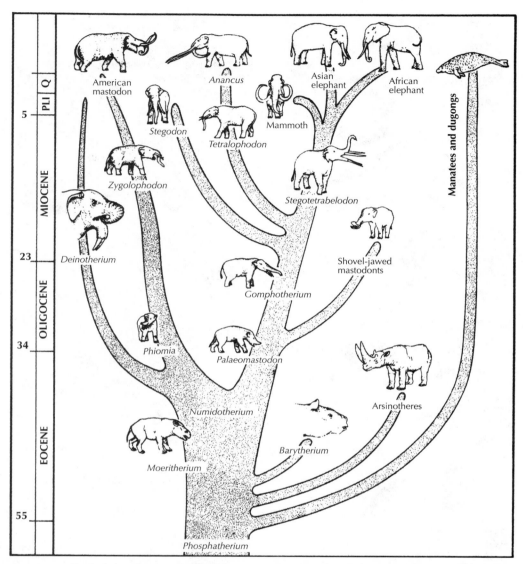

FIGURE 14.19. Evolutionary history of the elephants and their kin (Proboscidea), starting with pygmy hippo-like forms like *Moeritherium* with no trunk or tusks, through mastodonts with short trunks and tusks, and concluding with the huge mammoths and the two living species. Early in their history, the other tethytheres branched off from the Proboscidea. These include the manatees, order Sirenia, the extinct desmostylians, and the extinct horned arsinotheres. (Drawing by C. R. Prothero; from Prothero 1994b)

their close relationship. There are many details of the skull and jaw, and even of the teeth, that closely resemble the early mastodonts. In addition, both living species of sirenian have the same unique mode of tooth replacement that elephants have. Instead of pushing out baby teeth from below, as most mammals do, elephants and sirenians have horizontal tooth replacement. Their tooth row consists of a long "conveyer belt" of molars, with new teeth erupting from the back of the jaw, then pushing the rest of the tooth row forward until the older worn teeth are dropped off the front edge of the mouth. This unique condition occurs

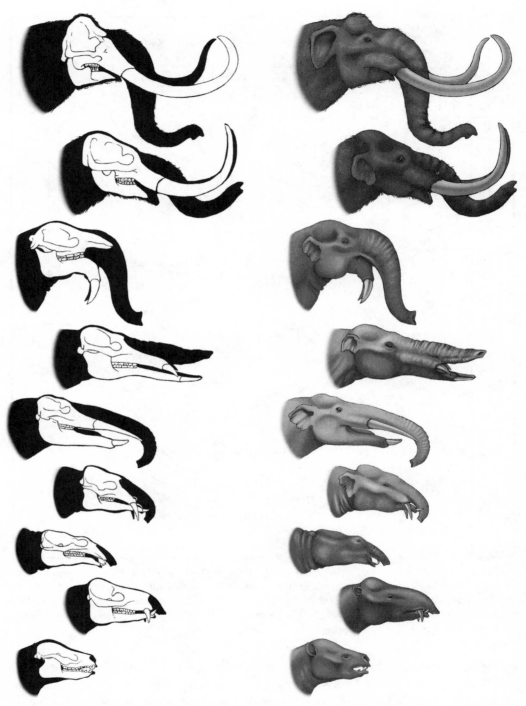

FIGURE 14.20. Details of the evolution of the skull, tusks, and trunk of proboscideans, from the pygmy hippo-like *Moeritherium* through mastodonts with longer tusks and trunks to mammoths. (Drawing by M. P. Williams).

in no other group of mammals and is strong evidence for a clade of Proboscidea plus Sirenia that McKenna (1975) called the "Tethytheria."

Since McKenna's original daring hypothesis, the Tethytheria has been supported by many additional anatomical analyses, so the relationship of manatees and elephants is one of the best established in all of science. Every molecular system that has been examined, from the proteins to the mitochondrial DNA to the nuclear DNA always clusters these two groups together, so there is a complete convergence of evidence. In addition to the sirenians, several other previously mysterious groups of fossil mammals now appear to cluster within the Tethytheria. They include the huge two-horned elephant-sized beasts from the Oligocene of Egypt known as arsinoitheres (fig. 14.19). These had once been a complete zoological mystery, placed in their own order Embrithopoda for lack of a better place. But McKenna and Manning (1977) linked them to the Paleocene Mongolian mystery fossil known as *Phenacolophus*, and since then other arsinoitheres have been found in Eocene beds of Turkey and Romania. Yet another mysterious group was the desmostylians, a hippo-like marine mammal that is found only in the Oligocene and Miocene of the North Pacific. These too placed in their own order for lack of a better hypothesis, until Domning, Ray, and McKenna (1986) described very primitive fossils of desmostylians called *Behemotops* and showed that they had the characteristic tethythere configuration of the jaws and teeth. Once again, the fossil record has yielded a transitional form that links a previously isolated group with other groups of mammals.

Finally, let's return to the manatees, peacefully sleeping in the shallow warm waters of the tropics and munching sea grasses. According to some historians, the legend of the mermaids may have come from sailors who saw manatees floating upright, feeding their babies at their paired breasts (a configuration also found in humans and elephants) and possibly with seaweed draped over them that resembled hair, and imagined that they were mermaids. Close up, of course, they are so plug-ugly that they could never be mistaken for beautiful half-women/half-fish, but never underestimate what months at sea can do for homesick and horny sailors! But this myth, and the legend of the sirens trying to lure Odysseus's sailors to their doom with their beauty and seductive songs, is the basis for the name of the order, Sirenia.

When we look at manatees up close, they have many remarkable specializations. Their skulls exhibit many unique features, especially in the way the upper bones are modified into a snout. They have horizontal tooth replacement, and some also have short tusks as well. Their ribs are unique among mammals in that they are extremely dense and heavy (*pachyostotic*). These act as diving ballast and help keep the manatee floating at the proper depth. Last but not least, the front limbs are modified into a flipper (different from the detailed bone configuration found in whale or ichthyosaur flippers), the hind limbs have vanished completely, and the tail is a broad flat horizontal fluke like that found in whales. Its shape is rounded in manatees, but with pointed lobes in dugongs.

Sirenian fossils are well known, although they consist mostly of their distinctively dense and heavy rib fragments, plus a few decent skulls that show their evolution (Domning 1981, 1982). But in 2001, another remarkable transitional form was discovered that clearly catches the sirenians in the act of evolving from land mammals. Known as *Pezosiren portelli* (literally, "Portell's walking sirenian") (fig. 14.21), it is a nearly complete skeleton from the Eocene of Jamaica that was described by Daryl Domning (2001). The skull is much like many other primitive sirenians, with all the hallmarks in the skull bones and teeth, and the ribs are thick and heavy, showing that it too was mostly aquatic. But instead of flippers it has four perfectly good walking limbs, with strong shoulder girdles, hip bones, and even well-developed

FIGURE 14.21. The mounted skeleton of *Pezosiren portelli*, the manatee with feet rather than flippers, next to Daryl Domning, who described and named it. (Photo courtesy Raymond L. Bernor)

hands and feet! One could not imagine a better transitional form: a creature with all the skull and skeletal features of manatees, yet it still has the ability to walk on land.

In many respects, *Pezosiren* is comparable to *Ambulocetus*, which is a beautiful example of a walking whale making the transition to aquatic life, and to the enaliarctines, which are the transition fossils linking terrestrial bears to seals and other pinnipeds. Just a few years ago, we had no transitional forms to show how terrestrial ancestors of marine mammals like sirenians, whales, and seals went back to the sea and became aquatic, and now we have excellent fossil transitions for all three groups! The creationist websites have tried to discredit *Pezosiren*, but reading their discussions just shows how laughably incompetent they are in anatomy or paleontology. Their basic argument boils down to the idea that because it was terrestrial, it couldn't be a sirenian! They fail to notice all the uniquely sirenian features of the skull (particularly in the snout and teeth, with their horizontal tooth replacement) and especially the heavy ribs for ballast, showing that this creature was hippo-like with both terrestrial and aquatic features—just as we would expect for a form making the transition between terrestrial and aquatic lifestyles. To a creationist, it is either in the fully aquatic sirenian "kind" or it's some other sort of terrestrial mammal, but their conceptual blinders make it impossible for them to recognize a fossil that is perfectly intermediate between two of their "created kinds." *Pezosiren* is the stuff of creationist nightmares, because you couldn't ask for a better transitional fossil: half manatee, half walking land mammal.

Thus we have seen that the fossil record of hoofed mammals is *full* of transitional forms, showing how nearly all the familiar large ungulates (horses, rhinos, giraffes, elephants, and so on) evolved and how two groups of marine mammals (whales and sirenians) evolved from land ancestors. Even more examples are given in my new book, *The Princeton Field Guide to*

Prehistoric Mammals. In the previous chapter, we saw a plethora of transitional forms among the other groups of placental mammals, and if we had more space, we could provide them for nearly every group. But it's time to get to the real $64,000 question that interests us the most: what about us humans? That will be the subject of our chapter 15.

For Further Reading

Benton, M. J., ed. 1988. *The Phylogeny and Classification of the Tetrapods*. Vol. 2, *Mammals*. Oxford, U.K.: Clarendon.

Benton, M. J. 2014. *Vertebrate Palaeontology*. 4th ed. New York: Wiley-Blackwell.

Carroll, R. L. 1988. *Vertebrate Paleontology and Evolution*. New York: Freeman.

Janis, C., K. M. Scott, and L. L. Jacobs, eds. 1998. *Evolution of Tertiary Mammals of North America*. Vol. 1, *Terrestrial Carnivores, Ungulates and Ungulate-like Mammals*. New York: Cambridge University Press.

Janis, C., G. F. Gunnell, and M. D. Uhen, eds. 2008. *Evolution of Tertiary Mammals of North America*. Vol. 2, *Small Mammals, Xenarthrans, and Marine Mammals*. New York: Cambridge University Press.

MacFadden, B. J. 1992. *Fossil Horses*. New York: Cambridge University Press.

McKenna, M. C., and S. K. Bell. 1997. *Classification of Mammals*. New York:Columbia University Press.

Novacek, M. J. 1992. Mammalian phylogeny: shaking the tree. *Nature* 356: 121–125.

Novacek, M. J. 1994. The radiation of placental mammals. In *Major Features of Vertebrate Evolution*, ed. D. R. Prothero and R. M. Schoch. Paleontological Society Short Course 7:220–237.

Novacek, M. J., and A. R. Wyss. 1986. Higher-level relationships of Recent eutherian orders: morphological evidence. *Cladistics* 2:257–287.

Prothero, D. R. 1994. Mammalian evolution. In *Major Features of Vertebrate Evolution*, ed. D. R. Prothero and R. M. Schoch. Paleontological Society Short Course 7:238–270.

Prothero, D. R. 2005. *The Evolution of North American Rhinoceroses*. New York: Cambridge University Press.

Prothero, D. R. 2006. *After the Dinosaurs: The Age of Mammals*. Bloomington: Indiana University Press.

Prothero, D. R. 2013. *Bringing Fossils to Life: An Introduction to Paleobiology*. 3rd ed. New York: Columbia University Press.

Prothero, D. R. 2016. *The Princeton Field Guide to Prehistoric Mammals*. Princeton, N.J.: Princeton University Press.

Prothero, D. R., and S. Foss, eds. 2007. *The Evolution of Artiodactyls*. Baltimore, Md.: Johns Hopkins University Press.

Prothero, D. R., and R. M. Schoch, eds. 1989. *The Evolution of Perissodactyls*. New York: Oxford University Press.

Prothero, D. R., and R. M. Schoch. 2002. *Horns, Tusks, and Flippers: The Evolution of Hoofed Mammals*. Baltimore, Md.: Johns Hopkins University Press.

Rose, K. D., and J. D. Archibald, eds. 2005. *The Rise of Placental Mammals*. Baltimore, Md.: Johns Hopkins University Press.

Savage, R. J. G., and M. R. Long. 1986. *Mammal Evolution: An Illustrated Guide*. New York: Facts-on-File.

Szalay, F. S., M. J. Novacek, and M. C. McKenna, eds. 1993. *Mammal Phylogeny*. New York: Springer-Verlag.

Thewissen, J. G. M., ed. 1998. *The Emergence of Whales: Evolutionary Patterns in the Origin of Cetacea*. New York: Plenum.

Turner, A., and M. Anton. 2004. *National Geographic Prehistoric Mammals*. Washington, D.C.: National Geographic Society.

FIGURE 15.1. Satirical cartoon from Darwin's time, posing the question that most British were struggling with: Are we an ape's reflection? (By permission of the Trustees of the British Museum)

THE APE'S REFLECTION? 15

The Only Transition That Really Matters

We must, however, acknowledge, as it seems to me, that man with all his noble quali-
ties, still bears in his bodily frame the indelible stamp of his lowly origin.
—Charles Darwin, *The Descent of Man*

Throughout the second half of this book, we have documented example after example
of transitional forms in everything from microfossils to mollusks to mammals. We could continue to do this for hundreds of pages more, but it wouldn't really make a difference to the
creationists or to those who are confused and misled by them. The only transition that really
matters is, of course, the evolution of humans. Many creationists readily concede many of
the examples we have just discussed as evolution as just variation within "created kinds,"
although we have documented many examples of macroevolutionary changes that exceed
anyone's definitions of "kinds." But to many people, those are just stories about critters.
They only care about humans and whether or not we are specially created in God's image
or "just another ape."

The idea that we might be related to apes was shocking when it was first proposed after
On the Origin of the Species was published in 1859. Darwin deliberately avoided the subject
in his already controversial book and finally dealt with it near the end of his life with *The
Descent of Man and Selection in Relation to Sex* (1871). Thomas Henry Huxley, however, was
not afraid of offending the sensibilities of the Victorians and boldly published *Zoological
Evidences of Man's Place in Nature* in 1863, with explicit diagrams showing the detailed skeletal similarities between humans and the great apes. Nevertheless, because of the strong
religious beliefs of the nineteenth century, most people still refused to accept the idea. They
were horrified to look into the mirror and see an image of themselves as an ape (fig. 15.1).

As the years have passed, however, the gulf between humans and the rest of the apes has
narrowed considerably. Instead of the old "screaming hooting monkey" stereotype, we have
discovered just how similar the apes are to humans. Decades of field research by pioneering anthropologists like Jane Goodall with the chimpanzees and the late Dian Fossey with
the mountain gorillas have demystified these majestic creatures and surprised us with their
amazing behavioral similarities to humans. Both chimpanzees and gorillas can learn sign
language, communicate in simple sentences, and make and use simple tools. Their societies
are very sophisticated compared with those of any other animal and show us many insights
into the complexities of human societies as well. Over a century of research by hundreds of
anthropologists has documented more and more connections between apes and humans. In
nearly every other westernized country, polls show that a majority of educated people no
longer object to the idea that humans and apes are related or at least have come to terms with
the fact that humans are part of the animal kingdom and a part of nature as well, not above it.

Yet the idea is still offensive to many in the United States, where many polls show that a majority of Americans still do not accept that we are related to the apes. Nearly all of this is driven by strong religious beliefs, of course, plus common misconceptions about apes (not yet defused by all the documentary footage of amazing chimpanzees on television), and especially by a determined campaign of creationist misinformation. Still, it is surprising that more than 150 years after Darwin's book was published, we still cannot come to terms with overwhelming evidence from both biology and the fossil record. My colleagues in the anthropology departments all over the country face this all the time. More than any other group of scientists, they are under attack. They have had to waste a lot of time undoing creationist mischief and clarifying ape and human evolution for the general public when they could be doing real research, discovering something new and useful instead.

Whatever one's personal religious beliefs about humans and their "specialness," as scientists we must stick to objective evidence and testable hypotheses. Many different religious beliefs have held many different concepts of humanity and its relation to God or the gods, but that cannot concern us here in a work of science. As discussed in chapter 1, science cannot and should not deal with the supernatural or with untestable hypotheses. Science cannot and should not deal with issues of the soul or other concepts that are important in the religious perspective but cannot be dealt with in a scientific fashion. This is not to say that the soul does not exist (science cannot decide one way or the other) or that humans don't have some special element of God in them (again, not a scientific question). As long as we are talking about scientific evidence for any hypothesis (whether it is the evolution of rhinos or of humans), we have to stick by the rules of science and exclude supernatural hypotheses because they cannot be tested in a scientific manner. Of course, everyone is entitled to their own personal beliefs, but they are not entitled to impose them on others or to call their ideas scientific when they are clearly not.

And that is where we run into the most outrageous lies and distortions broadcast by the creationists. They are pretty bad at misleading people about the evolution of other animals, but when it comes to discussing the human fossil record, they hit rock bottom. To a creationist, every human fossil has to be discredited somehow because it goes against their very innermost beliefs to acknowledge the existence of these fossils. The hominin fossil record has improved enormously over the past several decades. We can spread out the impressive array of hominin fossils that connect us to the apes and to all of the rest of the animal kingdom (fig. 15.2) just to get a sense of the quality of the human evolutionary record. If these were any other fossils other than hominins, most people would be duly impressed and agree that the case was well established. Simply because they are our relatives, the stakes are much higher, and ideas and specimens are attacked that much more vigorously by creationists.

Unfortunately, some anthropologists are unduly naïve and create all sorts of opportunities for misunderstanding and for creationist distortions. The study of human fossils is one of the most crowded and contentious of any scientific field I've ever seen. Although the human fossil record is now quite impressive and includes thousands of specimens (fig. 15.2), there are also thousands of physical anthropologists who must "publish or perish," and who need to make a career somehow. Most of the best fossils are typically studied by those who have the funding and the access to the key sites in Africa and elsewhere, so the rest of the profession has to make their careers whatever way they can. Consequently, every idea and every specimen in hominin paleontology is challenged and restudied and reinterpreted

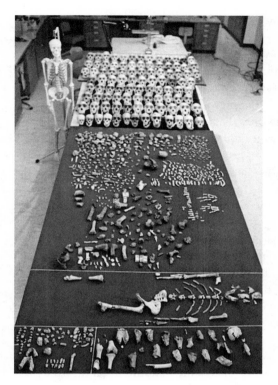

FIGURE 15.2. The hominid fossil record is now becoming very complete, contrary to creationist falsehoods. This table displays the huge collection of fossils of *Australopithecus afarensis* discovered in Ethiopia and described by Don Johanson and Tim White. In the foreground is the partial skeleton of "Lucy," the oldest nearly complete hominid skeleton known. In the background is a selection of modern chimpanzee skulls for comparison, and a skeleton of *Homo sapiens*. If we were to lay out all the thousands of fossils of hominids in the cabinets of the Kenya National Museum, it would be even more impressive. (Photo courtesy D. Johanson, Institute of Human Origins)

many times, and any creationist who wants to mine for quotes can find some person (no matter how unqualified he or she might be) to say something that out of context seems to deny the validity of a given fossil.

Let us not forget that anthropologists are humans and make mistakes, too. Like all people who are trying to find out new information, they sometimes get ahead of their data or their interpretations are colored by their expectations. Roger Lewin's (1987) excellent book *Bones of Contention: Controversies in the Search for Human Origins* shows this side of anthropology very nicely, and I highly recommend it for anyone who wants to see what the profession is really like. Sometimes those biases can lead to serious mistakes. A good example is how a clever forger duped the British anthropological establishment with the Piltdown hoax, which fit their prejudices that humans evolved large brains first, and seemed to put Raymond Dart's *Australopithecus africanus* out of the picture. In another case, Henry Fairfield Osborn got excited about a curious tooth that looked very like a hominin and published a premature report on "Nebraska Man" before he was corrected by other scientists.

The key point to all of this is that *science is self-correcting*. Individual scientists can make mistakes or be misled by their biases, but there are so many other hard-boiled skeptical scientific critics out there that mistakes are soon caught and corrected. The Piltdown forgery was eventually exposed when new scientific methods came along that could show its fraudulent nature, and it was already suspected as a fake before then because it was not consistent with the emerging fossil record from Africa.

"Nebraska Man" was not accepted by any other paleontologist except Osborn and was quickly corrected when better specimens appeared that showed it was the misleadingly worn tooth of a fossil peccary or javelina known as *Prosthennops*. (I'm currently working on a book on the fossil peccaries of North America, so I've gotten to know these fossils well). It is *not* a "pig" as creationists claim. Typical of their ignorance of most biological topics, creationists don't know enough zoology to know that pigs and peccaries are separate families that evolved on different continents; they are only distantly related. In addition, Osborn's mistake was not as ludicrous as the creationists try to make it appear. The tooth itself is very badly worn and apparently rotated in the socket, giving it a peculiar appearance. In addition, peccaries and pigs both have omnivorous diets and their teeth have squared crowns with low rounded cusps, *very* similar to another group of omnivores, the primates. I've surprised creationists on more than one occasion by putting the teeth of apes and peccaries side by side, and they can't tell the difference.

A quick look at some of the creationist pamphlets and books shows just how misleading and dishonest their presentations are. Typical of the genre is the little pamphlet *Big Daddy*, published by creationist Jack Chick. It is an insidious cartoon for easily swayed young minds, with a professor proclaiming the facts about human evolution, then being "corrected" by a polite young Christian who shatters the professor's ideas about science. The garish yellow centerfold features a "march of hominins" that we discussed in chapter 5, perpetuating the misconception that human evolution is a single linear sequence, not a branching bush of many species. And each example in the cartoon picks just one hominin fossil and attempts to discredit it. One by one, they distort the hominin fossil record or just plain make it up.

For example, they say of "Peking Man" that "all evidence has disappeared." Not true! The original specimens of "Peking Man" (the *Homo erectus* specimens from Zhoukoudian cave near Beijing, not a distinct species) were lost when the Chinese and Americans were fleeing with the fossils during the Japanese invasion in 1939. But many good casts were made of the original material, and many more new and better specimens have been found in subsequent excavations. Creationists point to Piltdown man, because this forgery was accepted for a while. Of course, it was scientists (*not* creationists) who eventually discovered the forgery. "Nebraska Man," as we outlined already, was the mistake of one scientist and was corrected within a year—and it is *not* the tooth of a pig! Creationists claim that Neanderthal man is just based on an arthritic skeleton. Not true! One famous specimen of Neanderthal was indeed arthritic, and its malformities influenced some early reconstructions, but there are now dozens of normal, undiseased specimens, and they clearly show that Neanderthals represent a distinct species that is *not* modern *Homo sapiens*. They were much more robust and heavy limbed than we are, with a brain larger than ours but with a distinctive flatter skull with large brow ridges, protruding face, no chin, and a bulge in the back of the skull (see fig. 15.6). The mysterious "New Guinea Man" is something that only appears in creationist publications. No legitimate anthropologist has ever made the claim that they are anything but modern *Homo sapiens*. Finally, Cro-Magnon was *always* considered to be modern *Homo sapiens*, so the creationist attempt to suggest otherwise is deceptive and misleading.

More importantly, these eight examples are used to discredit the entire human fossil record, yet every other human species that they cannot discredit (of which there are now dozens) is conveniently not mentioned—an obvious attempt at trickery and distortion.

This brief summary of how they managed to get every single example in their attack on hominin fossils *completely wrong* is representative of their tactics and of the abysmal level of their understanding of human fossils. Any minor mistake, or any account that seems to discredit a fossil, is good enough to be perpetuated over and over again. These same examples pop up in nearly every other creationist publication and website as well, often cribbed word for word, complete with all the same mistakes and misspellings.

There is no point in wasting more space in this chapter to correct every single creationist lie about the hominin fossil record. The example given above is very typical, and I have not seen anything more sophisticated in any of their other publications. Just as is the case with all the other creationist misstatements we have already discussed, their tactics usually include quoting out of context, quoting old outdated sources that don't reflect the modern knowledge of the fossils, or quoting scientists who were considered cranks even when they were active (such as Solly Zuckerman, Gish's personal favorite). Creationists don't do any legitimate peer-reviewed anthropological research themselves nor do they bother to actually work with the fossils or even learn the basics of anthropology and human anatomy to see what the fossils really look like. They just do book reports and pull quotes out of context; that's as far as their scientific curiosity goes. Once they've found what they think is a damaging quote, either they don't bother to find out the context or they deliberately mislead the reader about the entire context of the quotation. This may be a sneaky way to confuse people and win them to their cause, but it is dishonest, unethical, and unscientific.

So let's stop wasting our time with the distorted creationist view and briefly review what the hominin fossil record *does* show.

The Truth About Human Fossils

It is . . . probable that Africa was formerly inhabited by extinct apes closely allied to the gorilla and chimpanzee, and as these two species are now man's nearest allies, it is somewhat more probable that our early progenitors lived on the African continent than elsewhere.

—Charles Darwin, *The Descent of Man*

When Darwin published *On the Origin of Species* in 1859, there were still no good hominin fossils for him to point to, and this was still true when he wrote *The Descent of Man* in 1871. Although the first Neanderthal specimen was known, it was usually misinterpreted as the skeleton of a diseased Cossack who had died in a cave and didn't figure in the early ideas about human evolution. The first genuine hominin fossil that was truly different from us was Eugène Dubois's famous "Java Man" specimens of *Homo erectus*, originally described as *Pithecanthropus erectus* in 1896. As outlined by Swisher et al. (2000), the specimens were controversial and misinterpreted for many years because they were so incomplete (just a skull cap, a thigh bone, and a few other fragments) and did not fit the biases of anthropology at the time. On top of that, Dubois's own paranoid behavior made his ideas harder to accept. But Raymond Dart's 1924 description of the South African skull known as the "Taung Child," *Australopithecus africanus*, was the first good fossil hominin species that was not a member of the genus *Homo*. With its discovery almost 100 years ago, it should have clinched the case that humans had evolved and did so in Africa. Nevertheless, the hominin fossil record (especially of our earlier ancestors, who all lived in Africa) was still quite poor for several decades after 1924 because most anthropologists focused on very young European material and could not accept Darwin's insight that our closest ape relatives lived in Africa, and therefore we probably originated there, too.

Contrary to what creationists say, the fossil record of humans is no longer as poor as it was even 40 years ago. Decades of hard work in the field by hundreds of scientists has turned up thousands of hominin fossils (fig. 15.2), including a few good skeletons and many good skulls that show clearly how humans have evolved over 7 million years. This avalanche of new discoveries every year has occurred despite the handicap that hominin fossils are delicate and rare, and only one or two are found for every hundred fossil pig or fossil horse specimen found in the same beds in eastern Africa. A tour through the bomb-proof hominin vault in the Kenya National Museum in Nairobi is a revelation: a whole room full of fossils that document our evolution, and whose existence the creationists must deny. Many other museums in Africa, Europe, and Asia have similar large collections of our early ancestors, so there are lots of fossils to work with now (and even more ideas about how to interpret them, which is a good thing in science). Fortunately, there are now books such as Eric Delson's (1985) *Ancestors: The Hard Evidence*, which was once part of a traveling museum exhibit. These books and exhibits allowed the public to see these precious fossils up close for the first time and realize that the creationist view of the human fossils record is a lie. Since then, anthropologists have been much better at displaying and publicizing the quality of the hominin record and demonstrating that they do not have any skeletons hiding in their closets. Now there are a number of beautifully illustrated books with excellent full-color photographs of the actual specimens published (listed at the end of the chapter), and there is no longer any excuse for someone who wants to see what the fossils really look like.

The entire story of human evolution is too long and detailed to be discussed in a single chapter, so we will just touch upon the highlights. The short version is this: dozens of human species and genera are now known, forming a very bushy family tree that spans almost 7 million years of human evolution, mostly in Africa (fig. 15.3). The exact details of how all these fossils should be named or how they are interrelated is always controversial because many of the specimens are incomplete, and anthropologists are famous for being argumentative and contentious. But no matter how the arguments swing from year to year, the amazing quality of the hominin fossil record is an objective fact, not someone's interpretation or guesswork.

First, let us place humans in a broader context. We are members of the order Primates, the group that includes not only ourselves and the great apes but also the Old World monkeys (Cercopithecidae), New World monkeys (Cebidae), and lemurs, lorises, bush babies, pottos, and many other archaic primates still alive today (fig. 15.4). We can trace the fossil record of most of these lineages back to the Cretaceous and Paleocene primate *Purgatorius*, whose name has interesting and ironic religious implications. (It was so named from Purgatory Hill in the Hell Creek Formation of Montana, where it was found). In the early Cenozoic, the globe was much warmer and more densely vegetated, and there was a huge diversity of different archaic primates distantly related to lemurs and other "prosimians," including the plesiadapids, adapids, omomyids, and many other groups (see Beard 2004, for an outstanding account). If you collect fossils in Paleocene or early Eocene beds in places such as the Bighorn Basin of Wyoming, primates are one of the most common fossils, and enormous

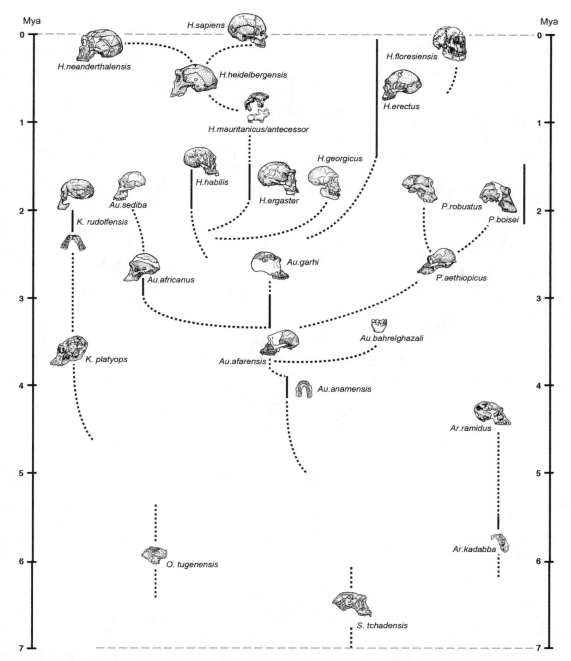

FIGURE 15.3. The current family tree of hominids, showing the extremely bushy branching pattern of human evolution, with many different fossil species at the same time at some time intervals. Sketches of the best specimens of each species are shown next to their time ranges. (Courtesy I. Tattersall)

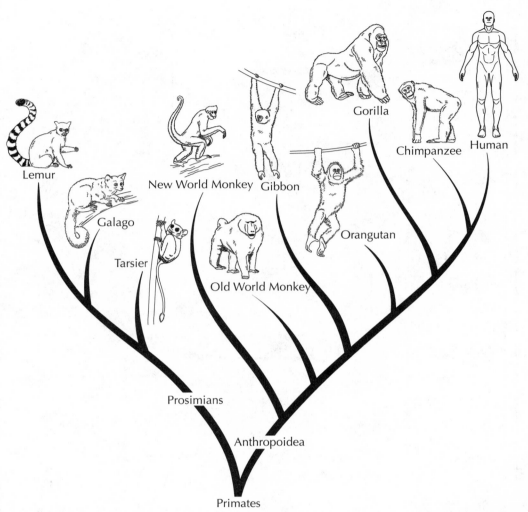

FIGURE 15.4. Family tree of the major groups of living primates, showing the close relationships between the great apes and humans and the more distant relationships of the Old World monkeys, New World monkeys, and lemurs, lorises, and bush babies.

collections of their jaws and teeth are now stored in museums around the world. But as world climate became cooler and drier in the Oligocene and the forests vanished, primates became scarce, too. They vanished from North America and Europe, where they had once flourished, and became restricted to Southeast Asia and Africa. During the Oligocene, one group of primates, the Platyrrhini or New World monkeys, managed to make the crossing of the South Atlantic, where they radiated into the great diversity of prehensile-tailed monkeys including spider monkeys, colobuses, and howler monkeys, marmosets, and the like. Most primate evolution was confined to Africa, where the Old World monkeys (baboons, macaques, rhesus monkeys, and their relatives) flourished, alongside the earliest members of the ape lineage (*Aegyptopithecus*, *Propliopithecus*, *Apidium*, among others), which are documented from the Oligocene Fayûm beds of Egypt.

In the Miocene, apes were more diverse and successful than the monkeys and occasionally were found in Europe as well as their African homeland. By the middle Miocene, primitive members of the orangutan lineage (*Sivapithecus*) are known from deposits dated at 12 million years old in Pakistan, giving us some of the first ape fossils that belong to modern lineages. Unfortunately, we have lots of fossil apes but no clear-cut fossils of the chimpanzee or gorilla clan just yet, probably because both of these apes have always lived in forests where there is scant chance of fossilization. Most of the fossil apes declined and died out by the end of the Miocene, and Old World monkeys came to dominate the primate adaptive zones ever since then.

The oldest specimen that can be truly described as a member of our own family was discovered and described only a few years ago. Nicknamed "Toumai" by its discoverers, its formal scientific name is *Sahelanthropus tchadensis*. The best specimen is a nearly complete skull (fig. 15.5A) from rocks about 6–7 million years in age from the sub-Saharan Sahel region of Chad (Brunet et al. 2002). Although the skull is very chimp-like with its small size, small brain, and large brow ridges, it had remarkably human-like features, with a flattened face, reduced canine teeth, enlarged cheek teeth with heavy crown wear, and an upright posture at the very beginning of human evolution. Just slightly younger is *Ororrin tugenensis*, from the upper Miocene Lukeino Formation in the Tugen Hills in Kenya, dated between 5.72 and 5.88 million years ago. *Ororrin* is known mainly from fragmentary remains, but the teeth have the thick enamel typical of early hominins, and the thigh bones and shin bones clearly show that it walked upright. Slightly younger still are the remains of *Ardipithecus ramidus kadabba*, found in Ethiopian rocks dated between 5.2 and 5.8 million years ago. These consist of a number of fragmentary fossils, but the foot bones show that hominins used the "toe off" manner of upright walking as early as 5.2 million years ago. Thus our human lineage was well established by the latest Miocene and fully upright in posture, even though our brains were still small and primitive and our body size not much different than that of contemporary apes.

The Pliocene saw an even greater diversity of hominins (fig. 15.3), with a number of archaic species overlapping in time with the radiation of more advanced hominins. Archaic relics of the Miocene included *Ardipithecus ramidus ramidus*, found in Ethiopia in 1992 from rocks 4.4 million years in age, which had human-like reduced canine teeth and a U-shaped lower jaw (instead of the V-shaped lower jaw of the apes). Nearly complete skeletal material of this species has now been discovered and reported by Tim White and his colleagues, making it the oldest known fossil hominin skeleton. Rocks in Kenya about 3.5 million years in age yield primitive forms like *Kenyapithecus platyops*. By 4.2 million years ago, however, the first members of the advanced genus *Australopithecus*, the most diverse member of our family in the Pliocene, are also found. The oldest of these fossils is *Australopithecus anamensis* from rocks near Lake Turkana in Kenya ranging from 3.9 to 4.2 million years in age. These creatures were fully bipedal, as shown not only by their bones but also by hominin trackways near Laetoli, Tanzania. The most famous of these early australopithecines is *A. afarensis* (from rocks 3.0–3.4 million years ago near Hadar, Ethiopia), better known as "Lucy" by its discoverers Don Johanson and Tim White (figs. 15.2 and 15.6A). When it was discovered in the 1970s, *Australopithecus afarensis* was the first early hominin to clearly show a bipedal posture (based on the knee joint and pelvic bones) but was not as upright as later hominins. These were still small creatures (about 1 meter or 3 feet tall) with small brains, and very ape-like in having large canine teeth and a large overhung jaw.

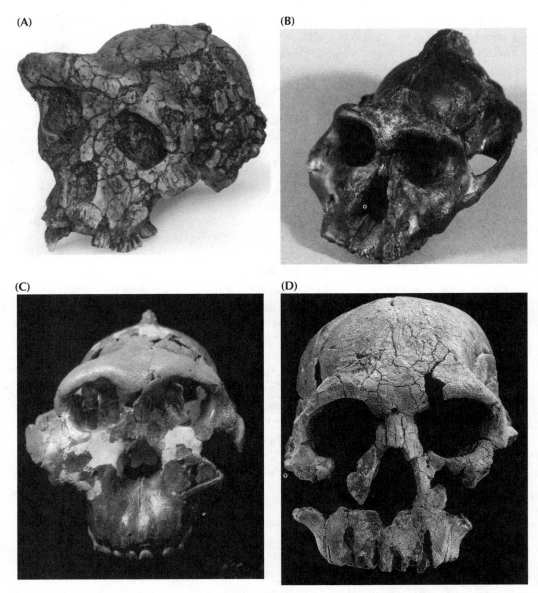

FIGURE 15.5. Some of the better known fossil hominid skulls from Africa. (A) The oldest known hominid, *Sahelanthropus tchadensis*, nicknamed "Toumai," from beds 6–7 million years old in Chad. (From Brunet et al. 2002; by permission of the Nature Publishing Group) (B) The famous "Black Skull," KNM-WT 17000, a specimen of the very primitive robust australopithecine, *Paranthropus aethiopicus*, discovered in beds 2.5 million years in age on the west shore of Lake Turkana. (Photo courtesy A. Walker) (C) Mary Leakey's "Nutcracker Man," the super-robust *Paranthropus boisei*, originally named "*Zinjanthropus*" *boisei*, OH5. It was recovered from 1.8-million-year-old rocks in Olduvai Gorge, Tanzania. (Photo courtesy National Museums of Kenya) (D) Richard Leakey's most famous discovery, the best skull of a primitive member of our own genus, *Homo rudolfensis* (originally referred to as *Homo habilis*), KNM-ER 1470, from beds dated at 1.88 million years old on the eastern shore of Lake Turkana. (Photo courtesy National Museums of Kenya)

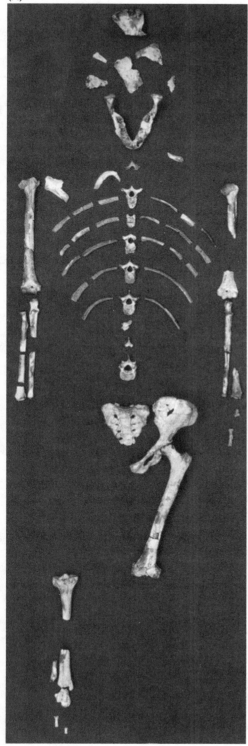

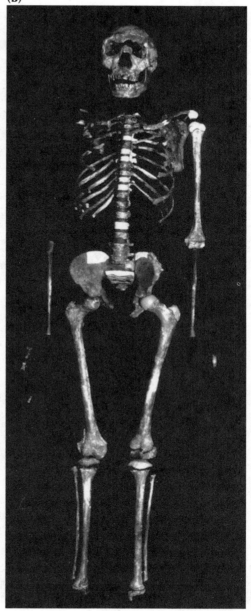

FIGURE 15.6. Two of the oldest known relatively complete hominid skeletons. (A) The skeleton known as "Lucy," the nearly complete specimen of *Australopithecus afarensis* from rocks 3.4 million years old in Ethiopia. (Courtesy D. Johanson, Institute of Human Origins) (B) The famous "Nariokotome Boy," KNM-WT 15000, a nearly complete skeleton of *Homo ergaster* (formerly referred to *Homo erectus*), from rocks 1.6 million years old on the western shores of Lake Turkana, Kenya. (Courtesy A. Walker)

By the late Pliocene, hominins had become very diverse in Africa (fig. 15.3). These included not only the primitive forms *Australopithecus garhi* (dated at 2.6 million years) and *A. bahrelghazali* (dated at 3.4 million years) but also the best-studied australopithecine, *Australopithecus africanus*. Originally described by Raymond Dart in 1924 based on a juvenile skull (the "Taung Baby"), for decades the Eurocentric anthropology community refused to accept it as ancestral to humans. But as more South African caves yielded better specimens to paleontologists like Robert Broom (especially the adult female skull known as "Mrs. Ples"), it became clear that *Australopithecus africanus* was a bipedal, small-brained African hominin, not an ape. This went contrary to all the accepted notions, which postulated that human evolution was driven by brain size, that bipedalism was secondary, and that it had occurred in Europe or Asia. The Piltdown forgery was deliberately set up to reinforce this bias, but by the 1950s, when Piltdown was exposed as a fraud, the evidence from *Australopithecus africanus* became undeniable. *Australopithecus africanus* was a rather small, gracile creature, with a dainty jaw, small cheek teeth, no skull crest, and a brain only 450 cc in volume. On the basis of its gracile and very human-like features, *Australopithecus africanus* is also the best candidate for ancestry of our own genus *Homo*.

In addition to *Australopithecus africanus*, the late Pliocene of Africa also yields a number of highly robust hominins. For a long time, they were lumped into a very broad concept of the genus *Australopithecus*, either as distinct species or even dismissed as robust males of *Australopithecus africanus*. In recent years, however, paleoanthropologists have come to regard them as a separate robust lineage, now placed in the genus *Paranthropus*. The oldest of these is the curious "Black Skull," discovered in 1975 by Alan Walker and his crew on the shores of West Lake Turkana, Kenya, from rocks about 2.5 million years in age (fig. 15.5B). Although its brain is small, and it would have had a small body as well, the skull is robust with large skull crest and massive molars and an advanced dish-shaped face. Currently, scientific opinion places the Black Skull as the most primitive member of *Paranthropus*, *P. aethiopicus*. It was followed by the most robust of all hominins, *P. boisei*, from rocks in East Africa ranging from 2.2 to 1.2 million years in age (fig. 15.5C). This fossil was nicknamed "Nutcracker Man" for its huge thick-enameled molars, robust jaws, wide flaring cheekbones, and strong crest on the top of its head, indicating a diet of nut or seed or bone cracking. Originally found by Mary Leakey at Olduvai Gorge in 1959, it was named *"Zinjanthropus boisei"* by Louis Leakey, who made his reputation from it. The rocks of South Africa between 1.6 and 1.9 million years in age yield the type species of *Paranthropus*, *P. robustus*. These too had massive jaws, large molars, and large skull crests but were not as robust as *P. boisei*. *Paranthropus robustus* lived side by side in the same South African caves as *A. africanus*. It was not only more robust but also larger than that species as well, with some individuals weighing as much as 150 pounds.

Finally, the latest Pliocene saw the first members of our own genus *Homo*, which are easily distinguished from contemporary *Australopithecus* and *Paranthropus* by a larger brain size, flatter face, no skull crest, reduced brow ridges, smaller cheek teeth, and reduced canine teeth. The first of these to be described was *Homo habilis* ("handy man"), discovered in the 1960s by Louis and Mary Leakey in Olduvai Gorge, Tanzania, from beds about 1.75 million years in age. Originally, all of the early *Homo* specimens were shoehorned into the species *H. habilis*, but now paleoanthropologists recognize that this material is too diverse to belong to one species, so several are now recognized. These include the very modern-looking skull (fig. 15.5D) of *H. rudolfensis* (from beds ranging from 1.9 to 2.4 million years in age), which made Richard Leakey's reputation, and the very advanced but short-lived *Homo ergaster* (fig. 15.6B), from beds 1.6–1.8 million years in age. These species are known not only from bones but also from their primitive tools, choppers and hand axes of the "Olduwan culture."

Many of these archaic Pliocene taxa persisted into the early Pleistocene (as recently as 1.6 million years ago), including *Paranthropus robustus* and *P. boisei, Homo ergaster*, and *H. habilis* (fig. 15.3). The best-known fossil of *H. ergaster* is a nearly complete skeleton found on the shores of West Lake Turkana in 1984, and known as "Nariokotome Boy" (fig. 15.6B), which would have been 2 meters tall if fully grown.

By 1.9 million years ago, however, a new species had appeared: *Homo erectus* (fig. 15.7). This creature was not only bipedal and erect (as its species name implies) but also almost as large in body size as we are. Some individuals reached 190 centimeters (6 feet) in height. Its brain capacity was about 1 liter, only slightly less than ours. Like earlier species of *Homo*, they made crude choppers and hand axes ("Acheulian culture" tools) and certainly knew how to make and use fire and how to construct stone and wooden dwellings and small villages. Originally, *Homo erectus* was confined to Africa, where all of our other ancestors had long lived. By around 1.8 million years ago, we have evidence that *H. erectus* finally escaped the African homeland, for specimens from Indonesia (originally described as "*Pithecanthropus*

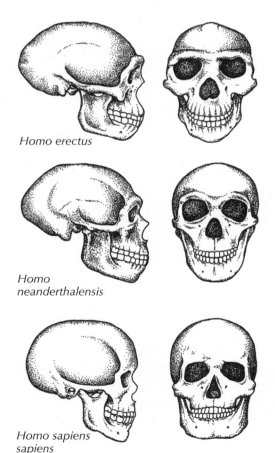

Homo erectus

Homo neanderthalensis

Homo sapiens sapiens

FIGURE 15.7. Comparison of the skull of *Homo erectus* (including specimens known as "Java Man" and "Peking Man"), Neanderthal, and modern *Homo sapiens. Homo erectus* had a brain capacity about half of ours, much heavier brow ridges, and a more protruding face. Yet it had upright posture and an essentially modern skeleton; it was clearly a member of our genus, and not "just an ape" as creationists allege. On the other hand, it is clearly not a modern human, but a transitional form between australopithecines and ourselves. (Drawing by Carl Buell)

erectus" or "Java Man") have been dated at that age (Swisher et al. 2000), and specimens are also known from elsewhere in Eurasia, such as Romania, that are almost as old. By about 500,000 years ago, we have abundant fossils of *H. erectus* in many parts of Eurasia, including the famous specimens from the Chinese caves at Zhokoudian, known as "Peking Man" and dated as old as 460,000 years ago. The latest dating suggests that *H. erectus* may have persisted as recently as 27,000 years ago, outlasting even the Neanderthals and overlapping with modern *H. sapiens* (Swisher et al. 2000). *Homo erectus* was thus the first member of our family to live outside Africa, and it roamed through the entire Old World (except Australia and the glaciated regions). *Homo erectus* was not only the first widespread hominin species but also one of the most successful and long-lived species, spanning more than 1.8 million years in duration between 1.9 and 0.03 million years ago. During most of that long time, it was the only species of *Homo* on the planet and changed very little in brain size or body proportions. If longevity is a measure of success, then it could be argued that it was even more successful than we are.

By about 300,000 years ago, another species was established in western Europe and the Near East: the Neanderthals. These were the first fossil humans to be discovered, although they were originally dismissed as the remains of diseased Cossacks that had died in caves. The first complete descriptions of skeletons were based on a specimen that suffered from old age and disease, so for decades Neanderthals were thought to be stoop-shouldered and primitive, the classic stereotypical grunting "cavemen." Modern research (Stringer and Gamble 1993) has shown that Neanderthals were very different from this stereotype. Although their skulls are distinct from ours in having a protruding face, large brow ridges, no chin, and a flatter skull that sticks out in the back (fig. 15.8), they had a slightly larger brain capacity than we do, and they practiced a complex culture. The famous discoveries at Shanidar Cave in Iraq showed that Neanderthals buried their dead with elaborate religious rituals and even flowers, suggesting that they had at least some kind of religious beliefs and possibly belief in an afterlife. Their bones (and presumably bodies) were robust and muscular and slightly shorter than the average modern human, but they also lived exclusively in the cold climates of the glacial margin of Europe and the Middle East, where their stocky build (similar to a modern Inuit or Laplander) would be an advantage. Their tool kits and culture were also more complex, with Mousterian hand axes, spearheads, arrowheads, and other complex devices, as well as bone and wooden tools. Some of these tools show complex working and simple carving, so they were artistic as no hominin before had ever been.

For decades, anthropologists treated Neanderthals as a subspecies of *Homo sapiens*, but recent work suggests that they were a distinct species. The best evidence of this comes from the Skhul and Qafzeh caves in Israel, where layers bearing Neanderthal remains are interbedded and alternate with layers containing early modern humans. In addition, Neanderthals appeared later than the earliest archaic *Homo sapiens*, so they could not be our ancestors but rather an extinct European side branch. Recently their DNA has been sequenced and they are clearly not *Homo sapiens*, but now there is some evidence that all modern humans have a bit of Neanderthal DNA in them, so there must have been some interbreeding between the two species. DNA analysis of just two tiny bones from Denisova Cave in Siberia has revealed the existence of yet another species of humans, the Denisovans, who are poorly known at this time.

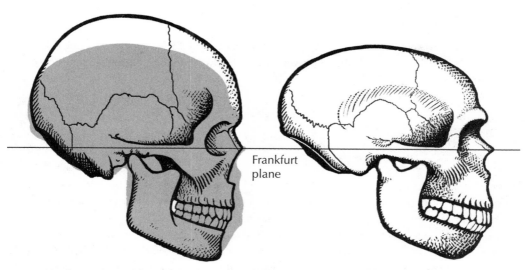

FIGURE 15.8. Comparison of the profiles of a Neanderthal and a modern *Homo sapiens*. Although Neanderthals had brains slightly larger than ours, their skulls were shaped very differently, with a much flatter cranium that protruded in the back, prominent brow ridges, a bulging face, and no real chin. Their skeletons were also distinctive, with much more heavy robust bones than any living human. Contrary to creationist claims that they are just deformed modern humans, they are very distinctive and clearly not a member of our species. (Drawing by Carl Buell)

Finally, we see the first fossil skulls and skeletons that look almost indistinguishable from our own species. Some of these "archaic *Homo sapiens*" are known from deposits in Africa dating as far back as 500,000 years ago. About 90,000 years ago, skulls from Africa (such as Klasies Mouth Cave in South Africa) are almost completely modern in appearance and are universally regarded as *Homo sapiens sapiens* (our species and subspecies). Like *Homo erectus*, early *Homo sapiens* spent most of its history in Africa, and finally migrated to Eurasia about 45,000 years ago. There it came into contact with Neanderthals, and for about 9,000 years they coexisted. Mysteriously, Neanderthals vanished 36,000 years ago. Whether they were wiped out by *Homo sapiens* or by some other cause is highly controversial. Whatever happened, modern *Homo sapiens* soon took over the entire Old World, developing complex cultures (the "Cro-Magnon people") that produced the famous cave paintings of Europe and many kinds of weapons and tools.

This brief review of the hominin fossil record hardly does justice to the richness and quality of the specimens or to the incredible amount of anatomical detail that has been deciphered. If it all seems like too much to absorb, just gaze at the faces of the skulls in the Fig. 15.5. They look vaguely like modern human skulls, but they definitely show the progression from more primitive hominins that creationists want to call "mere apes" (even though they were all completely bipedal and had many other human characteristics) up through forms that creationists want to call "modern humans" (even though they had many distinctive anatomical features, like those found in Neanderthals, that make them a distinct species). Anyone without advanced anthropological training and without biases or prejudices can glance at these fossils and see the hallmarks of human ancestry.

The Third Chimpanzee

The next time you visit a zoo, make a point of walking by the ape cages. Imagine that the apes had lost most of their hair, and imagine a cage nearby holding some unfortunate people who had no clothes and couldn't speak but were otherwise normal. Now try guessing how similar those apes are to us in their genes. For instance, would you guess that a chimpanzee shares 10 percent, 50 percent, or 99 percent of its genetic program with humans?

—Jared Diamond, *The Third Chimpanzee*

If the fossil record of human evolution was not proof enough, the clinching evidence is found in every cell in your body. When molecular biologists first began comparing the molecules from apes, monkeys, and humans back in the 1960s, they made a startling discovery: we are extremely similar to chimps and gorillas at the molecular level. The first evidence of this came when Vince Sarich's famous experiments showed by the relatively crude method of comparing immune responses that there was little difference between the genes of chimps and humans. Then the DNA-DNA hybridization method came along and got better results still. This technique took the DNA of chimps and humans and separated the strands in a solution by warming it up. As the solution cooled down, the individual unpaired strands of DNA link up again, with some combining a strand of chimp and a strand of human DNA. When you heat up that hybrid DNA again, it unzips and separates. The more genes the two strands have in common and the more similar they are, the harder it is for them to unzip, so that the temperature of separation is directly proportional to the number of genes in common.

When these results were first published by Sibley and Ahlquist in 1984, they caused a shock in the scientific community. *It turns out that human and chimpanzee DNA is 97.6 percent identical!* Less than 3 percent of our DNA is different from that of a chimpanzee. Likewise, our DNA shares about 96 percent of its genes with gorillas, and it is 94.7 percent identical with that of gibbons, 91.1 percent identical with that of rhesus monkeys (an Old World monkey, of family Cercopithecidae), 84.2 percent identical with that of the capuchin monkey (a New World monkey, of family Cebidae), but only 58 percent identical with that of a primitive "prosimian" such as a galago. Not only did the sequence of similarity match exactly with the branching sequence of primates (fig. 15.4), which made sense, but the real shock was just how similar our genes are to those of most of the great apes. Molecular biologists pointed out that the genetic similarity between humans and chimps was closer than between any other two species they had studied, such as two closely related species of rats or two kinds of frogs. Since these experiments, the actual sequence of the mitochondrial DNA and the nuclear DNA have been determined for both chimps and humans, and the results are the same (within a fraction of a percent). There's no two ways about it. The chimps are our "kissing cousins" among the animal kingdom, just a few genes removed from being fully human.

Jared Diamond puts it all in perspective with an interesting analogy. Suppose you were a molecular biologist from another planet and you were given just DNA samples of humans and the two chimpanzee species (the common chimp, *Pan troglodytes*, and the pygmy chimp or bonobo, *Pan paniscus*), plus a sampling of DNA from the primates and other groups of animals. You sequence the genes, plot the results, and conclude that humans are just a third species of chimpanzee, genetically closer than just about any other three species of animals on the planet. For example, the genetic similarity between lions and tigers is only 95 percent, yet they can even interbreed in zoos. Without seeing the differences in our bodies and behavior, the conclusion is unavoidable: we are very closely related.

What do we make of this? Although there are many different ideas suggesting what is going on here, the basic idea is that the 1–2 percent of the genome that differentiates us from chimps must be the regulatory genes that turn on and turn off the structural genes (which make up most of the 97.6 percent that is the same). We have the genes for most parts of the ape body, and the monkey body too, and every once in a while there is a genetic mistake or atavism, and humans express the long-repressed genes that we still carry to make a tail (fig. 15.9).

(A)

(B)

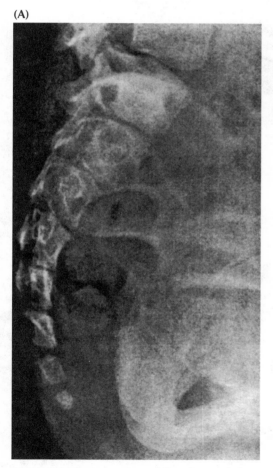

FIGURE 15.9. Every once in a while, a human is born with an atavistic tail, a throwback to our evolutionary past, when the regulation that normally shuts down our genes for tails fails to operate. The human tail comes complete with fully developed vertebrae, muscles, and other features of animal tails. (A) X-ray of a human with well-developed tail vertebrae. (B) Image of two humans with fully developed tails. (From Bar-Maor et al. 1980; used with permission of the *Journal of Bone and Joint Surgery*)

(A)   **(B)**

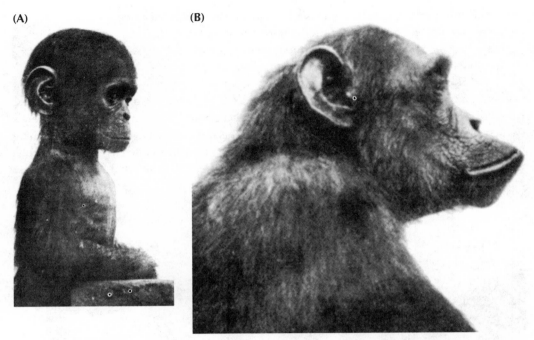

FIGURE 15.10. (A) Juvenile chimpanzees have many characteristics of the skull found in adult humans, including upright posture, relatively large brain, small brow ridges, and a less protruding snout. (B) As they grow into adult chimps, these features all become more apelike. Since the 1920s, many anthropologists have argued that much of what makes us human is retention of juvenile ape characteristics into adulthood (*neoteny*). (From Naef 1926)

In fact, since the 1920s, many biologists and anthropologists have argued that much of what differentiates us from the chimpanzee is neotenic retentions of juvenile ape characters. If you look at a juvenile chimpanzee (fig. 15.10), its skull is much like that of a human, with a large brain, small brow ridges, short snout, and upright posture. Then, during development to an adult, the chimpanzee develops the larger snout with long canines, big brow ridges, and forward slouching posture of the head. If regulatory genes tweak our embryonic development a tiny bit, we can make most of the characteristics that mark us as human just by becoming juvenile apes that reach sexual maturity without every truly growing up.

You may have read some of the fascinating works of Aldous Huxley, the famous novelist and author of the dystopian classic *Brave New World* (a high school reading list favorite). He was also the brother of the famous evolutionary biologist Julian Huxley, and both were grandsons of Darwin's "bulldog," Thomas Henry Huxley. Aldous knew these ideas about human neoteny very well because of his brother's influence. In 1939, he published a novel entitled *After Many a Summer Dies the Swan*. The theme of the novel is immortality, and how humans are always striving to find a way to extend their lives beyond what nature intended. The main character is a millionaire (modeled after William Randolph Hearst, whom Huxley met when he was a Hollywood screenwriter in the 1920s) named Jo Stoyte, who is attempting to live forever by hiring a classic "mad scientist," Dr. Obispo, to do research on extending his life. Dr. Obispo discovers that the third earl of Gonister in England had lived several centuries without any signs of aging, apparently by ingesting carp guts. Archival records showed that he had fathered children when he was over 100 years old. (This is a science fiction novel, remember!). Dr. Obispo seduces the millionaire's mistress (modeled on Hearst's real mistress, actress Marion Davies), and the millionaire accidentally kills the scientist's assistant in a jealous rage. He and Dr. Obispo have to flee the law, so they run to England and try to find out what happened to the third earl of Gonister. Finally, they break into his castle, and in the basement they find him, still alive and over 300 years old—all grown up to become an adult ape. I've spoiled the punch line for an excellent novel, but it's worth reading for the richness of detail and the amazing ironies that Huxley was so great in capturing.

In recent years, even more startling breakthroughs have occurred in molecular genetics. For centuries, the scientific community was deeply racist, and treated non-white people as inferior, or even a different species from white people. But the molecules paint a completely different picture. When you put the DNA of all of the human "races" in the mix with the DNA of Neanderthals and most of the living species of chimps and gorillas, a surprising result emerges. As shown in figure 15.11, the genetic differences among *all* the human "races" are tiny. All humans are far more genetically similar to one another than the populations of West African chimps are to one another, and the same is true of other populations of chimps and gorillas. As anthropologists have been saying for years, human "races" are genetically

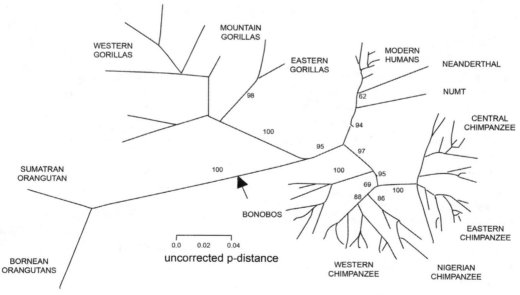

FIGURE 15.11. Molecular phylogeny of apes and humans, showing their genetic distance from one another based on mitochondrial DNA. All human "races" are much more similar to one another than two populations of gorillas or chimpanzees are to each other. (Modified from Pascal Gagneux et al., "Mitochondiral Sequences Show Diverse Evolutionary Histories of African Hominoids," *Proceedings of the National Academy of Sciences USA* 93 [1999], fig 1b: ©1999, National Academy of Sciences USA)

meaningless, and their basis is only a tiny part of our genome. In fact, evidence shows that most of the "racial differences" such as skin color and the shape of the eyes are a very recent change in human evolution, occurring sometime after nearly all modern human lineages emerged from Africa about 70,000 years ago. This is an important thing to think about when we find issues of race coming up in society.

Some people may find it unsettling that we are 97.6 percent genetically identical to the chimpanzee, but no amount of creationist propaganda can change the truth found in every cell in your body. Instead, it is better to accept that the evidence of science has shown us just how much an integral part of nature we really are; we are part of creation, and we cannot treat the rest of creation as something to be abused. So the next time you see a chimpanzee in a zoo, have pity on your close relative and think hard about the implications of that handful of genes that differentiate you from it.

As Smith and Sullivan (2007:100) put it:

> Do people come from monkeys? Not at all. We do share a common ancestor with chimpanzees, and before them, with the group that became monkeys. But to say we come from monkeys is simply wrong, and evolution has never claimed it. . . . Unless we want to live in a web of lies, we can't pick and choose what to believe, not when the raw data of genetics, fossil studies, and anatomical studies are laid before us.

Anyone is free to argue whether they like being a primate or whether they like being related to chimpanzees—but that's not the question. The question is whether or not we're descended from monkeys, and the evidence is in: we're not, but we are related to them.

For Further Reading

Beard, K. C. 2004. *The Hunt for the Dawn Monkey: Unearthing the Origin of Monkeys, Apes, and Humans.* Berkeley: University of California Press.

Conroy, G. C. 1990. *Primate Evolution.* New York: Norton.

Delson, E. C. 1985. *Ancestors: The Hard Evidence.* New York: Liss.

Diamond, J. 1992. *The Third Chimpanzee: The Evolution and Future of the Human Animal.* New York: HarperCollins.

Johanson, D., and B. Edgar. 1996. *From Lucy to Language.* New York: Simon & Schuster.

Lewin, R. 1987. *Bones of Contention: Controversies in the Search for Human Origins.* Chicago: University of Chicago Press.

Lewin, R. 1988. *In the Age of Mankind: A Smithsonian Book on Human Evolution.* Washington, D.C.: Smithsonian Institution Press.

Lewin, R. 1998. *Principles of Human Evolution: A Core Textbook.* New York: Blackwell.

Marks, J. 2002. *What It Means to be 98 Percent Chimpanzee.* Berkeley: University of California Press.

Pääbo, S. 2014. *Neandertal Man: In Search of Lost Genomes.* New York: Basic.

Prothero, D. R. 2016. *The Princeton Field Guide to Prehistoric Mammals.* Princeton, N.J.: Princeton University Press.

Sibley, C. G., and J. E. Ahlquist. 1984. The phylogeny of hominoid primates, as indicated by DNA-DNA hybridization. *Journal of Molecular Evolution* 20:2–15.

Stringer, C., and C. Gamble. 1993. *In Search of Neanderthals: Solving the Puzzle of Human Origins.* London: Thames and Hudson.

Swisher, C. C., III, G. H. Curtis, and R. Lewin. 2000. *Java Man*. New York: Scribner.

Tattersall, I. 1993. *The Human Odyssey: Four Million Years of Human Evolution*. Upper Saddle River, N.J.: Prentice Hall.

Tattersall, I. 2015. *The Strange Case of the Rickety Cossack and Other Cautionary Tales from Human Evolution*. New York: St. Martin's.

Tattersall, I., and J. Schwartz. 2000. *Extinct Humans*. New York: Westview.

FIGURE 16.1. Creationist attacks on science are not just attempts to replace evolutionary biology with religious dogma but to replace other sciences with pseudoscience: astronomy with astrology; neurology with phrenology; physics with magic; and chemistry with alchemy. (Cartoon from the Philadelphia Inquirer; copyright Universal Press Syndicate)

WHY DOES IT MATTER? 16

Deceit in the Name of the Lord

It is ironic that several of these individuals, who so staunchly and proudly touted their religious convictions in public, would time and again lie to cover their tracks and disguise the real purpose behind the ID Policy.
—Judge John Jones, *Kitzmiller et al. vs. Dover Area School District*

Lying lips are an abomination to the Lord.
—Proverbs 12:22

If you tell a lie big enough and keep repeating it, people will eventually come to believe it.
—Joseph Goebbels, Nazi propaganda minister

As discussed in chapter 2, creationism is not about science, but about political power and dictating the agenda for schools and textbooks now and eventually exerting control over society. Creationists play by whatever rules (dirty or otherwise) they need to in order to win. I have tried to document how they routinely distort or deny the evidence, quote out of context, and do many other dishonest and unethical things—all in the name of pushing their crusade. I was raised in a Christian church and learned Bible verses every Sunday, so it appalls me to see how unethically these supposedly "Christian" men and women act in their battle against their perceived foes. It makes you wonder whether they have second thoughts about violating the word and spirit of many parts of the scripture with their lies and deceptions.

How do they reconcile this un-Christian behavior with their Christian beliefs? Psychologists have long shown that humans are very good at self-deception and trying to convince themselves of anything that they fervently want to believe. Given a strong belief system, humans can convince themselves that black is white or ignore obvious evidence and focus only on what they want to see, missing the forest for the trees.

Psychologists have found that despite our hopes, humans are not rational creatures after all. Instead, we have what Michael Shermer calls "believing brains." We are all built to hold certain core beliefs or worldviews about ourselves, and anything that conflicts with this core belief will be rejected, dismissed, or simply ignored. Psychologists call this *reduction of cognitive dissonance*. Our brains are actually highly compartmentalized, with different ideas held in different parts of the brain, creating a dissonance or conflict in our minds. Thus, we are constantly struggling to reconcile or justify these conflicting beliefs. For example, one part of our brain may want to believe we are moral, but another part remembers when we told a white lie or broke the speed limit. Then our brain does all it can to rationalize and justify these conflicts, trying to make us feel less conflicted and remorseful about these inconsistencies between our beliefs about ourselves and our behavior. As George Orwell put it, "We are all capable of believing things which we know to be untrue, and then, when we are finally proved wrong, impudently twisting the facts as to show that we were right. Intellectually, it is possible to carry on this process for an indefinite time; the only check on it is that sooner or later a false belief bumps up against a solid reality."

All humans live with these conflicts most of the time, but in some cases the conflict is extreme. Creationists' entire sense of self and worldview are tied up in their literalistic belief in the Bible, and anything that challenges or conflicts with it *must* be false, no matter how strong the evidence. This explains the incredible mental gymnastics and twisted logic and outright denial of the facts that are right in front of their noses. And that core belief is extremely powerful in the mind of a creationist. Not only does it define who they are, but they are even more concerned about salvation and going to hell. Mountains of evidence about evolution will not shake them or make them pay attention, when they believe eternal torture in the underworld is the alternative.

This kind of dogmatic, inflexible religious belief explains many aspects of their bizarre behavior. Apparently, to the creationists, lying and deception are lesser sins than accepting evolution, and they are willing to sacrifice their integrity in their crusade against what they believe to be the source of all evils in the world. Their intellectual blinders are so strong that they see only what they want to see and read only what they want to read in a quotation, all in the name of their religious beliefs. To creationists, a literal belief in the Bible is essential to their religious salvation, and everything else (including science) must be sacrificed so their souls can go to heaven. A famous quote by Judge Braswell Deen (Pierce 1981:82) of Georgia says it all: "This monkey mythology of Darwin is the cause of permissiveness, promiscuity, pills, prophylactics, perversions, pregnancies, abortions, pornography, pollution, and proliferation of crimes of all types."

British reporter Bruno Maddox (2007:29) described the creationists' attempt at reduction of cognitive dissonance in his *Discover* magazine column, "Blinded by Science," focusing on his visit to the Answers in Genesis creationism museum in Kentucky. He wrote,

> I find myself reminded of F. Scott Fitzgerald's proposition in *The Crack-Up*, that "the test of a first-rate intelligence is the ability to hold two opposed ideas in mind at the same time and still retain the ability to function." Fitzgerald's first-rate mind, of course, eventually stopped retaining its ability to function, and watching [creationist Jason] Lisle try to reconcile the cutting edge of modern planetary physics with the offhand assertions of a religious tract written thousands of years ago by an unknown assortment of bearded semi-cave dwellers, I found myself wondering how long the poor chap has.

> For the record, I have even less patience now with the creationist agenda than I did going in, because I now suspect that they don't really believe the falsehoods with which they are trying to flood the world. But at the same time I got the clear impression that they don't have any choice. I thought I was going to meet people who love God and therefore hate science. What I found instead were people who love God but who have at least a pretty serious crush on science as well, and thus find themselves in the Fitzgeraldian nightmare of waking up every day and trying to believe in both. They will—they must—spend their lives, and brains, trying to think of ways that patently false ideas can be made to seem, if not actually true, at least not quite so patently false. . . .

Not to overdo the Fitzgerald, but I shall think of [the creationists] often, as day after day they beat on, boats against the current of truth, borne back ceaselessly into being just completely, utterly wrong.

One of the chief mechanisms that humans use to hang on to their core beliefs and reduce cognitive dissonance is called *confirmation bias,* or remembering the successes and forgetting the failures. "Psychics," fortune-tellers, and other con artists take advantage of this human weakness when they do a "cold reading," trying to "predict the future" for an unsuspecting victim. If you listen carefully to what they say, you'll realize that they throw out a lot of random vague guesses until they get a "hit," then they follow the victim's body language and verbal cues to refine their guesses and amaze the victim. If you keep a tally, you'll see that most of their guesses are wrong—but the victim only remembers the successful "hits" and comes away amazed that the "psychic" knew so much.

The same is true of most horoscopes and astrology. Most of the statements are vague platitudes that tend to be true of most people, and even a detailed personal horoscope will be full of misses—but most listeners will only be impressed by the "hits." Confirmation bias (also called "cherry-picking" the best data or examples) explains how many creationists can ignore the conflicts of different verses in the Bible (discussed in chapter 2) or read a passage about evolution and "quote mine" just a tiny piece of the text that seems to fit their beliefs. The power of confirmation bias in the brain completely blocks the overall context of the evidence or the statements in the text, and only a tiny out-of-context quotation registers in their cognition as important—because it seems to agree with their existing biases. This is why quote mining, as we have seen throughout this book, seems so bizarre and frustrating to most of us—but makes perfect sense to creationists, whose filters and biases can only allow certain ideas that fit their belief system.

Another factor is at work here as well: *tribalism.* We are all products of our backgrounds, especially our families and communities, and we learn and accept whatever our families and peers teach us. Adopting these ideas is essential to our sense of belonging and acceptance by our families and communities, and going against their beliefs is very difficult for most people. Most people don't want to be the "black sheep" but just want to fit in and belong. Rebelling or breaking from family or community is truly terrifying to to almost anyone. People from small towns in rural America know this phenomenon well. The first question people will ask you after "What is your name?" is "What church do you attend?" Your church membership in a small town defines your place in that society, and if you're not a church-going Christian, you can expect harsh judgment from the rest of the small community. With powerful incentives like this, it is no wonder that most creationists will never listen to evidence that threatens their core beliefs and their membership in their churches and communities.

In addition, people often don't realize just how intellectually isolated people in evangelical and fundamentalist churches are. In his brilliant 2008 book *The Great Derangement: A Terrifying True Story of War, Politics, and Religion,* reporter Matt Taibbi describes the experience of going "undercover" and immersing himself as a spy within a fundamentalist church in Texas. He attends all their services throughout the week and goes to their weekend "retreats" and special seminars about creationism. Although he hides his true beliefs well and doesn't give away his reactions to their bizarre notions, he confesses in his writing how he is shocked and sickened and truly amazed by the kind of illogical behavior these people exhibit. What is more revealing is their social and intellectual isolation from the rest of the world. Despite the availability of 24/7 news channels and unlimited Internet access, these people only read and listen to what their church approves of and are made to feel very guilty if they expose themselves to any media that might challenge their beliefs.

The scientific and educational community has despaired for years about how to make the evidence for evolution plainer or the message clearer to the creationists so they can learn the truth about evolution. If Taibbi is right, however, none of it makes a difference. Hard-core creationists are deeply embedded in this belief system and isolated from any outside information, and no improvements in the quality or quantity of evidence or how we communicate it will reach them. When they are exposed to evidence, they will ignore it or dismiss it or try to rationalize it away. Most of the time they will avoid being exposed to challenging ideas in the first place. We've seen this over and over throughout this book. A classic example is an interview posted on YouTube between Richard Dawkins and creationist Wendy Wright (just Google those two names and it will be the first hit). For most of us, it's excruciating to watch. Over and over again, Dawkins patiently and gently shows examples of transitional fossils to the creationist, and over and over again she just repeats the mantra, "There are no transitional fossils." No amount of evidence could ever reach a person so thoroughly indoctrinated and brainwashed by creationism.

Instead of trying to reach people who are deep in the creationist cult, our efforts are better spent reaching people who may have been raised as fundamentalists but are beginning to open their minds and listen to evidence that was once forbidden to them. Based on the hundreds of letters and emails I received and on the reviews of the first edition of this book on Amazon.com, such people were the audience that this book reached most effectively. The hard-core creationist community never even bothered to attack or take notice of this book when it first appeared in 2007, and they ignored it for years afterward. But those people who were uncertain about what to think and were fence-sitting between fundamentalism and science were reached effectively, and they responded very positively.

Bit by bit, this will be the way that we can slowly erode the creationist hold on people: by reaching those who are beginning to open their minds and question what they were taught. Thanks to the Internet and books like this, it's much easier than ever to do so. When I grew up in the 1960s, there were no books disputing or debunking creationism or explaining evolution clearly, let alone books that challenged religion. I was searching hard for such resources as I began to doubt my Presbyterian background and look for answers in my community. It's still that way in many small towns in America today. But thanks to the Internet, even the most isolated person can now find all the evidence needed with just a few clicks of a mouse.

Although many rank-and-file creationists may sincerely believe these ideas because their leaders tell them to, and they've never heard of or read other sources, many of the leaders themselves may be exploiting creationism for power and money. We've all heard the accounts of hypocritical evangelists who turned out to have feet of clay, from Jimmy Swaggart to Jim Bakker to the recent case of the Rev. Ted Haggard. The creationist leaders also have their share of crooks. The most outrageous is Kent Hovind, who called himself "Dr. Dino" (even though he has no experience in dinosaurs or paleontology and his "Ph.D." was bought from an online diploma mill) and even built a Dinosaur Adventure Land theme park in Pensacola, Florida, to promote creationist ideas. He just finished doing a 10-year sentence in the Federal Correctional Institution in Florence, Colorado, for cheating and lying—to the IRS. Their creationist ministries give men like these lots of money and an enormous influence over a lot of people, and it's no surprise that they abuse the privilege.

What often shatters the beliefs of creationists, however, is not the misdeeds of their leaders, but the fact that the leading creationist authors and debaters keep repeating deceptive and discredited arguments, even after they have been shown to be lies. This is often the critical moment that causes a crisis of faith that turns people away from creationism. Saddest of all, however, is how these fanatical "Christians" treat members of their own flock when they stray from the strict creationist dogma when finally confronted by the evidence for evolution. The story of oil geologist Glenn Morton (http://www.oldearth.org/whyileft.htm), who was abused and harassed by his fellow "flood geologists" when his own discoveries led him away from creationism, is heart-rending. Likewise, these stories show that the "brotherly Christian love" of some fanatical creationists is a sham (see www.talkorigins.org/origins/postmonth/jan03; www.talkorigins.org/origins/postmonth/nov02.html).

As we have seen throughout this book, these supposed "Christians" will twist the facts and refute logic to salvage their unshakable belief that a literal interpretation of the Bible is final truth. To them, nothing else matters. Their religion is foremost in their lives, and if the scientific world didn't intrude, they wouldn't pay much attention to science at all. But evolution, cosmology, geology, anthropology, and many other sciences *do* impinge on their beliefs, so they try to attack science by pretending to be scientists and using scientific language wherever possible. Unlike real scientists, however, creationists do not follow the first rule of science that we discussed in chapter 1: a scientific hypothesis must be testable and falsifiable, and scientists must be willing to give up their cherished ideas if the data show they are wrong. Thomas Henry Huxley said it best: "Sit down before a fact as a little child, be prepared to give up every preconceived notion, follow humbly wherever and to whatever abysses nature leads, or you shall learn nothing."

Christian evolutionary biologist Ken Miller of Brown University, who has beaten all the ICR's debaters many times, provides an insight. In his book, *Finding Darwin's God* (1999:172), he describes an encounter with Henry Morris, whom he had just beaten in debate the night before. He asked Morris about his discredited positions, and said "Do you actually believe all this stuff?" Morris replied, "You don't realize what is at stake. In a question of such importance, scientific data aren't the ultimate authority. Scripture tells us what the right conclusion is. And if science, momentarily, doesn't agree with it, then we have to keep working until we get the right answer. But I have no doubts as to what that answer will be" (Miller 1999:173).

Morris may be sincere and believe he is serving God. But he could not have provided a more chilling indictment of the narrow self-righteousness, fanaticism, and antiscientific attitude that is creationism.

Why Should We Care?

We've arranged a global civilization in which the most critical elements profoundly depend on science and technology. We have also arranged things so that almost no one understands science and technology. This is a prescription for disaster.
—Carl Sagan, *The Demon-Haunted World*

An educated citizenry is the only safe repository for democratic values.
—Thomas Jefferson

> In a democracy, it is very important that the public have a basic understanding of science so that they can control the way that science and technology increasingly affect our lives.
> —Stephen Hawking

> Scientific literacy may likely determine whether or not democratic society will survive into the 21st century.
> —Leon M. Lederman, Nobel laureate

So what if the creationist extremists have weird beliefs? Why should we care? How does it make any difference in our own lives? There are so many crazy religious cults in America who believe weird things, and thanks to our Constitution, they all have a right to believe anything they like. If we leave them alone, won't the problem go away?

The short answer to those questions is no; we cannot just ignore creationism. Unlike most religious extremists who are harmless and somewhat amusing (or suicidal, like the Jim Jones cult in Guyana, or the Heaven's Gate cult), the creationists are not planning to leave the rest of us alone. They are fanatics on a crusade to overcome all those who oppose them. The reasons for resisting them are very clear and straightforward.

1. Creationism is a narrow sectarian religious belief and cannot be taught in public schools without violating the Constitution.

In chapter 2, we detailed the history of creationism and showed how every single court has found that their ideas are the religious beliefs of a specific sectarian group. No matter how they disguise their ideas as "creation science" or "intelligent design," the fact remains that they cannot be allowed to push their narrow sectarian beliefs in preference to those of other religions without violating the First Amendment of the Constitution concerning the separation of church and state.

As we saw in chapter 2, creationists have been more and more clever in disguising their religious tracks each time they are defeated in court. Right after the 2005 *Dover* decision, intelligent design (ID) creationism died, and the creationists tried even more subtle strategies. During the past decade, many states have seen creationists propose (and some states adopt) bills allowing creationism in new forms. One form is "teach the controversy," where the teacher gives equal time to creationism, and then "let the kids decide." In other cases, these laws have allowed teachers to introduce "evidence against evolution" (provided by creationists but not mentioning creationism) to science classes. No other topic in science is so targeted except climate change, another inconvenient truth that fundamentalists deny. Still other approaches have been proposing "academic freedom" bills that allow teachers to say anything, including creationism, without anyone stopping them for teaching pseudoscience in science classes. Creationists are always creative in their strategies to sneak religion into public school science classrooms, and they have unlimited time, energy, and resources to do so. Most scientists don't have the time or energy to fight them, because they must pursue their scientific careers and focus on real research, not on fighting political battles with religious extremists.

Nick Matzke (2016) published a clever study, where he used the language and key phrases of many different creationist policies and bills submitted to different state legislatures. He inserted them into a software program that deciphers the evolutionary tree of real organisms. From this he showed that you could trace the ancestry of various creationist bills and documents as they copied each other, then were modified (evolved) into different strains of creationist bills. Almost all versions of creationist laws are copycats of one another, all modified and evolving through time to avoid the problems of the separation of church and state and trying to hide their religious motivations deeper and deeper. But the religious roots of any creationist law or policy are easy to trace.

Why not let them have "equal time" or "teach the controversy," as some (including former president George W. Bush) advocate? Our culture is fond of equal time and fairness, so this sounds OK to lots of people. But science is not about popularity contests but about what scientists have discovered about the real world. There is no time or valid reason to teach outdated and discredited ideas from the past like astrology, the flat earth, the geocentric universe, or creationism. In addition, if we allowed them equal time, we would open the door of public school science classes to any or all religious beliefs that wanted equal time. For example, there are a significant number of extreme creationists (including an entire Flat Earth Society) who point out that the Bible teaches that the earth is flat and believe that all those NASA photos of the earth from space are hoaxes. Should they be allowed equal time in science classes too? Their beliefs are just as sincerely held as those of the more polished and scientific-sounding ID creationists, yet they still fail the fundamental test of science.

The conclusions of the creationists are determined in advance and not subject to testing and falsification, so no matter how much they call it "creation science," it is not science. If we allow creation science, do we also flat-earthers to teach their ridiculous ideas? Do we allow astrology instead of astronomy, or parapsychology and phrenology (reading bumps on the head) instead of psychology, or replace chemistry with alchemy, or physics with magic? (See fig. 16.1.) All of these nonscientific and pseudoscientific notions are believed by at least some of American society, but that doesn't entitle them to equal time in a science classroom.

Science teaching in this country is difficult enough with the incredible crowding of the curriculum and the time lost to standardized testing. Students have short attention spans and are distracted by video games and television, so most science teachers barely have time to cover the basics, let alone take time to discuss an unscientific religious belief just because some noisy minority wants it.

2. The attack on evolution is really an attack on all of science.

If they could, creationists would abolish the teaching of many fields of science—not just evolutionary biology but also geology, paleontology, astronomy, anthropology, and any other field that does not conform to a literal reading of Genesis. More importantly, their attempts to introduce supernaturalism and unscientific ideas to science classes undermine the very foundations on which science is based. This is not just an intellectual issue either. If we reverted to flood geology, we would never find any oil, and our economy would be in a shambles. If we followed creationist astronomy, none of our space program (and the benefits it provides) would be possible.

3. Creationists are threatening, harassing, and intimidating our public schools, universities, and museums.

The creationists are not content just to preach to their followers. They insist on forcing their views on everyone else, even though it is unconstitutional to do so in public schools and museums. Because the courts have turned them down every time, they resort to pressure tactics on school boards and especially on state textbook-adoption boards. Consequently, most high school biology textbooks are still shamefully weak on the topic of evolution, and the majority of high school biology teachers and classes still avoid the topic or teach it in a watered-down version so as not to offend one or two kids who have fundamentalist parents. Many textbooks are forced to avoid the dreaded "E" word altogether and use euphemisms like "organic change through time" just to avoid terrible fights with the school boards and in classrooms.

Lately, the creationists have adopted even more aggressive tactics, including filling young kids with lies about the fossil record and coaching them to talk back, sass, and disrespect their teachers. The foremost example of this is Ken Ham, the head of the huge Answers in Genesis organization, with 160 employees and an annual budget of over $150 million. He tours the country indoctrinating young children to believe that biologists, paleontologists, and geologists are liars, that no transitional fossils exist, that dinosaurs lived with humans, and that the earth is only 6,000 years old. Even more frightening is the way he coaches the kids to challenge their teachers and disrupt classroom activities. His song and dance is chillingly recorded in Alexandra Pelosi's HBO documentary *Friends of God*, about the radical evangelicals who want to take over political power in the United States. An article by Stephanie Simon in the *Los Angeles Times* on February 27, 2006, describes it this way:

> Evangelist Ken Ham smiled at the 2,300 elementary students packed into pews, their faces rapt. With puppets and cartoons, he was showing them how to reject geology, paleontology and evolutionary biology as a sinister tangle of lies.
>
> If a teacher mentions evolution, or the Big Bang, or an era when dinosaurs ruled Earth, Ham said, "You put your hand up and you say, 'Excuse me, were you there?' Can you remember that?"
>
> The children roared their assent.
>
> "Sometimes people will answer, 'No, but you weren't there either,' " Ham told them. "Then you say, 'No, I wasn't, but I know someone who was, and I have his book about the history of the world.' "
>
> He waved his Bible in the air.
>
> "Who's the only one who's always been there?" Ham asked. "God!" the boys and girls shouted.
>
> "Who's the only one who knows everything?"
>
> "God!"
>
> "So who should you always trust, God or the scientists?" The children answered with a thundering: "God!"
>
> A former high school biology teacher, Ham travels the U.S. training kids as young as 5 to challenge scientific orthodoxy. He doesn't engage in the political and legal fights that have erupted over the teaching of evolution. His strategy is more subtle: He aims to give people who trust the biblical account of creation the confidence to defend their views—aggressively.
>
> He urges students to offer creationist critiques of their textbooks, parents to take on science museum docents, professionals to raise the subject with colleagues. If Ham does his job well, his acolytes will ask enough questions—and spout enough arguments—to shake the evolution theory of Charles Darwin.

"We're going to arm you with Christian Patriot missiles," Ham, 54, recently told 1,200 adults gathered at Calvary Temple in northern New Jersey. It was Friday night, the kickoff of a weekend conference sponsored by Ham's global ministry, Answers in Genesis.

To a burst of applause, Ham exhorted: "Get out and change the world!"

Over the past two decades, "creation evangelism" has become a booming industry. Several hundred independent speakers promote biblical creation at churches, colleges, private schools, Rotary clubs. They lead tours to the Grand Canyon or museums to study the world through a creationist lens.

They churn out home-schooling material. A geology text devotes a chapter to Noah's flood; an astronomy book quotes Genesis on the origins of the universe; a science unit for second-graders features daily "evolution stumpers" that teach children to argue against the theory that is a cornerstone of modern science.

But the creationist political pressure, propaganda, and lies are not restricted to public schools. In many smaller colleges (especially community colleges and those that focus primarily on teaching and do not have a strong research emphasis), the professors are just as intimidated by creationist bullies who are eager to disrupt class and trash the professor's reputation on the course evaluation forms. Yet this is in a college setting, where the faculty is supposed to have intellectual freedom and the protection of tenure, and the system is not run by highly politicized local school boards. It should not be so—but it is.

The saddest commentary of all is how creationists have repeatedly used lawsuits and political pressure to intimidate and threaten museums into advocating their particular religious viewpoint (in violation of the Constitution). Failing that, they try to remove any mention of evolution, geology, or astronomy. The Smithsonian Institution, as the largest federally funded science museum in the United States, has been repeatedly attacked in this way. Creationists have pressured sympathetic right-wing politicians to investigate and bully this esteemed institution into removing their displays on paleontology and evolution. Fortunately, they have been rebuffed so far, but they keep trying. Even sadder, a story in the British newspaper the *Telegraph* (dated December 8, 2006) reported that evangelicals in Kenya are trying to force the National Museums of Kenya, repository of most of the important hominin fossils discussed in chapter 15, to remove their displays because they don't want to be exposed to the fact that we evolved in Africa. As the outraged Richard Leakey, the museum's director and famous paleoanthropologist, put it, "The collection it holds is one of Kenya's very few global claims to fame and it must be forthright in defending its right to be at the forefront of this branch of science."

It is one thing when creationists try to intimidate public schools and prevent their children from hearing stuff they don't want to believe. But museums? If they don't want to be exposed to science, they don't have to visit museums! They have no right to force a museum to remove displays that contradict their religious beliefs. Given the political instability in Africa, it is a frightening thought that such a fanatic religious group might take power, or riot in the streets, and attempt to break into the museum and destroy this fantastic and irreplaceable evidence of our own evolution whose very existence they don't want to acknowledge.

The attempt by Kenyan evangelicals to suppress the evidence of human evolution is reminiscent of one of the last scenes in the classic 1968 movie (and Pierre Boulle novel) *Planet of the Apes*. Taylor, the marooned human astronaut (played, ironically, by the conservative icon Charlton Heston), has fled into the Forbidden Zone, where the sympathetic ape, Dr. Cornelius, has shown him an archeological excavation that proves humans lived on the planet before apes became dominant. Once they are all captured by the gorilla storm troopers, the head ape, Dr. Zaius, orders the cave dynamited and the evidence destroyed. As he says in the film, the evidence might shake their belief systems and therefore it is too dangerous.

With the creationist threats to schools, universities, and museums, are we in danger of letting dogmatic religious authorities destroy the evidence of our own evolution because it challenges their belief systems? If the fundamentalists continue to expand their political power, are we in for another Inquisition, with the religious fanatics suppressing and destroying books and evidence, and harassing anyone who doesn't agree with them? Many countries in the world ruled by fundamentalist regimes (especially in fundamentalist Muslim countries, such as Afghanistan under the Taliban or the extremist mullahs ruling Iran) already showed that this is possible.

4. Thanks in part to creationists, the American public is appallingly illiterate in basic science.

Every study and survey that has been done for decades shows consistently that Americans are among the most scientifically illiterate of all westernized nations. Carl Sagan (1996) estimated that 95 percent of Americans were scientifically illiterate, and most could not give the correct scientific answers to the simplest questions such as: What is a molecule? What is a cell? What is DNA? Less than 50 percent knew that the earth goes around the sun once a year, and a small percentage thought the sun goes around the earth. And thanks to creationism, only 35 percent thought that the Big Bang theory was correct, and 48 percent agreed with creationists and the *Flintstones* that humans lived with dinosaurs. The embarrassing list goes on and on, making the United States the laughingstock of the educated world. Every survey that has been conducted in recent years ranks American science literacy near the bottom of the 40 westernized nations that were compared. Countries like Japan, China, the Netherlands, Canada, Sweden, Switzerland, and Germany nearly always rank at the top, while America usually ranks with several underdeveloped nations with a fraction of our wealth and nowhere near the money available to spend on education (fig. 16.2).

There have been lots of arguments about why Americans are so scientifically illiterate, but the evidence is pretty clear. Of all the developed nations of the world, only the United States has a significant creationist influence in politics. There is no significant creationist influence pressuring lawmakers in Canada or in any other European or Asian country with a developed economy. As figs. 16.2 and 16.3 show, the key predictor of science illiteracy is any question that is influenced by creationism, whether it deals with the Big Bang, the age of the earth, or evolution directly. No other variable is so predictive of science illiteracy in a westernized country.

It wasn't always this bad. During the Sputnik scare and space race of the late 1950s and early 1960s, Americans were shocked to discover how far they had fallen behind and brought back rigorous and engaging science education—only to see it languish as creationism has eaten away at the textbooks and the demands of standardized testing have pushed the time in the curriculum toward subjects covered in the test, leaving science (and physical education and art and music and many other subjects) out in the cold.

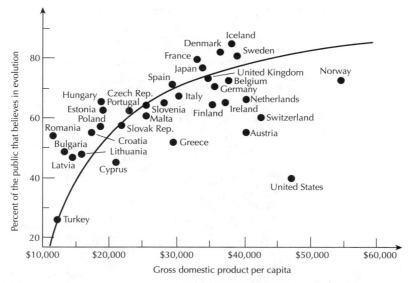

FIGURE 16.2. Plot comparing acceptance of evolution (measured by acceptance of ideas that "human beings, as we know them, evolved from earlier species of animals") versus national wealth (as measured by GDP per capita). Northern European countries are at the top of the graph, followed down the curve by southern European countries, and then by former Soviet Bloc countries of Eastern Europe. The United States is the sole outlier, with science literacy on the level of Turkey, but more spending per student than any country except Norway. (Modified from Prothero 2013a).

5. America has fallen behind many other nations in technological and scientific supremacy, which threatens the economic future of us all.

Thanks to the scientific illiteracy of our general population and the hostile environment for science that creationism is promoting, we are falling behind many other nations in the one area where we used to excel, science and technology. Study after study has documented a "brain drain" of scientists going to other countries with less anti-intellectualism and more favorable climates for science, especially in fields that are opposed strenuously by fundamentalists (like stem-cell research or cloning). America can no longer compete to make the cheapest or best electronics, toys, cars, or most anything else, as China, Korea, India, Singapore, Indonesia, and many other nations have taken those tasks away from us in corporate cutbacks and outsourcing. For many years, we could brag that we won a lion's share of the Nobel Prizes in science, but that dominance is now coming to an end as well. If we can't compete with other nations in manufacturing and commerce, and we give away our advantages in science and technology, what kind of world are we leaving for our children? What does this imply for our national security when we farm out not only blue-collar but also white-collar jobs and then are slaves to other countries that are doing better science and technology as well?

6. Denial of evolution is not just bad science, but it threatens our health and well-being.

As discussed in chapter 3, evolution keeps happening all the time, whether creationists want to believe it or not. Yet if we deny the fact that evolution is happening in viruses and bacteria and in other pathogens and pests, it only makes the problem worse when they

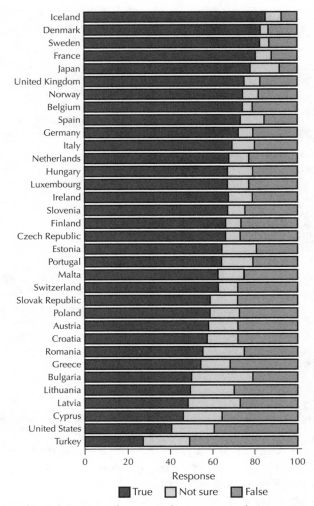

FIGURE 16.3. Percentage of population in each country that accepts evolution as true (dark bar to left) and regards it as false (solid bar to right); the undecided are indicated by the light bar in the middle. Note that nearly every developed country in Europe and Asia has at least 75 percent or higher acceptance of evolution. These are also the countries ranked highest in science literacy by numerous studies. The United States is down at the bottom with Cyprus and Turkey, countries with heavy influence of religion like Greek Orthodoxy or Islam, yet the United States has a much higher GNP and spends much more per child than almost every country on this list except Norway. (Modified from Prothero 2013a).

evolve resistance to whatever we throw at them. If creationists ran the labs that produce these protective chemicals, do you think we would have a chance when the next deadly pest hits us?

A more concrete example happened in 1984, when a surgeon at Loma Linda University in California attempted to replace the defective heart of "Baby Fae" with the heart of a baboon. Not surprisingly, the poor baby died a few days later due to immune rejection. An Australian radio crew interviewed the surgeon, Dr. Leonard Bailey, and asked him why he didn't use a more closely related primate, such as a chimpanzee, and avoid the possibility of immune rejection, given the baboon's great evolutionary distance from humans. Bailey said, "Er, I find that difficult to answer. You see, I don't believe in evolution." If Bailey had performed the same experiment in any other medical institution except Loma Linda (which is run by the creationist Seventh-Day Adventist Church), his experiments would be labeled dangerous and unethical, and he would have been sued for malpractice and his medical license revoked. But under the cover of religion, his unscientific beliefs caused an innocent baby to die of immune rejection, when other alternatives might have been available. Bailey was never prosecuted for this unethical and shocking example of medical malpractice and kept doing surgery at Loma Linda until he retired.

7. Allowing ideologues of any type to suppress science through political means is deadly for a society as well.

The classic illustration of this is the infamous case of Trofim Lysenko, who became Stalin's favorite scientist and wielded almost absolute power over Soviet science from 1927 until 1964. By all accounts, he was a mediocre geneticist who held onto discredited notions of how Lamarckian inheritance might improve Soviet crop yields and stave off famine. Most of his results were inconclusive or outright fraudulent, yet he told Stalin that he could produce incredible crop yields. Consequently, he rose to power in the Soviet scientific establishment then used his clout with the brutal dictator to suppress the legitimate Mendelian geneticists, who *did* understand how inheritance worked. Most of them were killed outright, sent to concentration camps, or driven into exile, forever destroying the vitality and strength of Soviet genetics and biology. Soviet biology fell decades behind that of the rest of the world until the 1960s, when Lysenko was finally denounced, his work was discredited, and he fell from power.

The point here is that science cannot be subservient to ideology and be forced to compromise the truth in order to please the political leadership. Lysenko and Stalin did not want to believe in Mendelian genetics or Darwinian biology, and they murdered hundreds of legitimate scientists who had the temerity to disagree with them. Other regimes (such as the Nazis) have distorted science to support their ideas, but ultimately scientific reality must win.

Some might argue that we're not in the Soviet Union of Stalin, and that the United States has safeguards against such oppression of scientific ideas. But the Bush administration was well documented (see Mooney 2005, and Shulman 2007) as interfering with legitimate scientists, rewriting reports by federal scientists that disagreed with their right-wing ideology, encouraging fringe scientists to testify as legitimate equals with well-regarded scientists in order to cancel out politically inconvenient messages, and generally ignoring the conclusions of scientists who didn't agree with them. Already the stem-cell research program in the United States has been set back compared to that in other countries, as our best scientists go to countries with less political oppression. Likewise, the foot-dragging and denials of global warming by the Bush administration and the flunkies of the oil industry in Congress cost the world valuable time in addressing this serious crisis.

And now we have a Trump Administration in power that is run by science deniers from top to bottom, with a Congress controlled by climate deniers as well. Many are strident creationists, such as Vice President Mike Pence and Secretary of Education Betsy DeVos, as well as climate deniers like EPA head Mike Pruitt, Secretary of the Interior Ryan Zinke, and of course Donald Trump himself. Who knows what will happen to science education when a creationist like DeVos tries to cripple public education while funneling tax dollars to private religious schools that teach creationism, not science?

When the prophetess Cassandra told the Trojans what they didn't want to hear, they ignored her and were eventually destroyed. If science tells us that we have evolved from the animal kingdom, or that microbes are evolving resistances to all our medicines, or that our wasteful society is destroying our planet, we had better learn from it, rather than shooting the messenger—and letting our children pay the ultimate price for our folly.

Polling Problems—and a Ray of Hope

Having battled creationism for most of my professional career (more than 40 years now), I sometimes find myself despairing that nothing ever seems to change or get better. For decades now, the Gallup Poll has surveyed Americans about their belief in evolution and creation. Year in and year out, the numbers seem to remain constant: about 40–45 percent of Americans appear to be young-earth creationists (YEC). The exact phrasing of the question in the Gallup Poll is as follows:

> Which of the following statements comes closest to your views on the origin and development of human beings: human beings have evolved over millions of years from other forms of life and God guided this process, human beings have evolved over millions of years from other forms of life, but God had no part in this process, or God created human beings in their present form at one time within the last 10,000 years.

For decades about 44 percent of the respondents agree to the last answer (YEC), another 37 percent chose the first answer (theistic evolution, ID creationist), and only 12 percent favor the second answer (nontheistic evolution). Gallup wrote these questions decades ago, before there was much understanding of how the framing of a question can bias the answer, and for decades, they have kept the question the same, so comparisons remain consistent. But social scientists know that polls can be very misleading, especially in the way the question is framed to force certain responses. For example, the Gallup poll only gives us three possibilities and loads two of the answers with "God," which is an obvious bias right from the start. In addition, there is good evidence to suggest that human evolution is the real sticking point and that most people don't care one way or another if nonhuman creatures evolve or not. What if we asked people what they thought about specific scientific ideas, independent of emotional issues like "God" and "humans"?

As Josh Rosenau of the National Center for Science Education (NCSE) pointed out:

> In 2009, Pew stripped away the religious issues and explicit reference to the age of the earth by asking people if they agreed that "Humans and other living things have evolved over time due to natural processes" or alternatively "existed in their present form since the beginning of time." Six in ten opted for evolution.
>
> In 2005, when the Harris Poll asked people "Do you think human beings developed from earlier species or not," 38 percent agreed that humans did develop from early species, but in the same survey, 49 percent agreed with evolution when asked: "Do you believe all plants and animals have evolved from other species or not?" So explicitly mentioning human evolution led to 11 percent of people switching from pro-evolution to anti-evolution. In a 2009 survey, Harris asked a Gallup-like ques-

tion, in which only 29 percent agreed that "Human beings evolved from earlier spe-cies," but in a separate question from the same poll, 53 percent said that they "believe Charles Darwin's theory which states that plants, animals and human beings have evolved over time." Placing the issue in a scientific context, with no overt religious context, yields higher support for evolution.

The National Science Board's biennial report on Science and Engineering Indica-tors includes a survey on science literacy which, since the early 1980s, has asked if people agree that "Human beings, as we know them today, developed from earlier species of animals." About 46 percent of the American public consistently agree with that option, about the same number who back the middle option in Gallup's surveys.

Clearly, people respond to these subtle shifts in how the question is framed, tak-ing a harder stance toward human evolution than to the idea that animals and plants evolve, and stepping away from evolution if it is pitched in opposition to religion. Pollster George Bishop surveyed the diversity of survey responses in 2006 and con-cluded: "All of this goes to show how easily what Americans appear to believe about human origins can be readily manipulated by how the question is asked."

In 2009, Bishop ran a survey that clarifies how many people really think the earth is only 10,000 years old. In survey results published by *Reports of NCSE*, Bishop found that 18 percent agreed that "the earth is less than 10,000 years old." But he also found that 39 percent agreed "God created the universe, the earth, the sun, moon, stars, plants, animals, and the first two people within the past 10,000 years." Again, ques-tion wording and context clearly both matter a lot.

For more evidence that the number of true YECs is fairly small, consider another question from the survey run by the National Science Board since the early '80s. In that survey, about 80 percent consistently agree "The continents on which we live have been moving their locations for millions of years and will continue to move in the future." Ten percent say they don't know, leaving only about 10 percent reject-ing continental drift over millions of years. Though young-earth creationists often latch onto continental drift as a sudden process during Noah's flood (as a way to explain how animals could get from the Ark to separate continents), they certainly don't think the continents moved over millions of years. This question puts a cap of about 10 percent on the number of committed young-earth creationists, lower even than what Bishop found. More people in the NSB science literacy survey didn't know that the father's genes determine the sex of a baby, thought all radioactivity came from human activities, or disagreed that the earth goes around the sun.

This is a very different picture than the Gallup polls suggest. Most people don't regard plate tectonics and continental drift as controversial (YECs must deny its existence), don't have any problem with the evolution of nonhuman animals and plants, or an earth more than 10,000 years old. On average, this suggests that the true YECs are only about 10 percent of the American population (31 million people), another 25 percent prefer creationism but not necessarily a young earth. That's about 35 percent creationists total, not the 45 percent Gallup suggests. About 10 percent of Americans (another 31 million people) are nontheistic evolutionists, another 33 percent or so lean toward evolution, giving us about 35 percent evolutionists, not the 12 percent suggested by Gallup. The remaining third in the middle also seem to accept evolution, but believe God or gods were involved somehow. Thus, about

65 percent of Americans seem to accept evolution in some form, not the 55 percent that Gallup suggested. The wording of the poll makes all the difference.

Yet another set of polls seem to confirm that the number of YECs is much smaller than Gallup suggests and is also declining. A combined CBS/YouGov poll showed that between 2004 and 2013, the number of people accepting the statement "Human beings evolved from less advanced life forms over millions of years, and God did not directly guide this process" jumped from 13 percent to 21 percent. Meanwhile, the percentage of people agreeing with the statement "God created human beings in their present form within the last ten thousand years" dropped from 55 percent to 37 percent over the same interval (2004–2013). According to the analysis:

> The demographics of the respondents is fairly predictable. Fewer women (37 percent) accept some form of evolution than men (56 percent) and fewer women (13 percent) tend to identify themselves as non-religious than men (20 percent). Older respondents favored creationism, while respondents under the age of 30 favored evolution, whether guided by a deity or not. The largest number of strict evolutionists was among this youngest age group, which tells us that insisting on keeping science in science class is working. Unsurprisingly, only 5 percent of Republicans agreed that evolution happens without a deity guiding it. The additional 30 percent of Republicans who agreed evolution is a thing believe that their god directs it. Democrats (28 percent) are closely followed by political independents (26 percent) in their acceptance of non-divine evolution, while an additional 25 percent and 21 percent, respectively, think God drives the evolution train. This means that more than half of non-Republicans accept evolutionary science. Among Republicans, 55 percent believe the earth is less than 10,000 years old and a god created human beings in their present form. The respondents most strongly denying evolution were Muslims, with 64 percent believing young-earth creationism and 36 percent uncertain. None of the respondents identifying as Muslim would admit that they accepted evolution. Protestant (59 percent) and the various Orthodox churches (53 percent) tied for the next largest group of evolution deniers. The strongest supporters of evolution? Believe it or not, it isn't the religiously unaffiliated. All of the Buddhists polled accepted evolution, although 13 percent of them said a deity guided it. Agnostics (85 percent) accept evolution, 17 percent of whom say God guided it. The remaining 15 percent aren't sure. The atheist respondents throw a curve to the poll, though. Two percent of those identifying as atheist also claim to be young earth creationists. Since 48 atheists responded to the survey, that means one person in there somewhere is either very confused or clicked the wrong radio button.
>
> Other demographics spread pretty much as we might expect: the more educated the respondent, the less likely to believe in creationism. The coasts, made up mostly of blue states, are more accepting of evolution than the mostly-red Midwest and Southern states. People identifying as white were more likely than Hispanics to accept evolution, while only 6 percent of black people participating in the poll did. The percentage of respondents who favor teaching creationism in public schools (40 percent) followed the same trends among the different groupings of respondents. Younger people opposed teaching creationism in larger numbers (42 percent), as did Democrats (29 percent) and Independents (31 percent). The more educated respondents disapproved of creationism in public schools more strongly than the less educated.

In short, not only are the polls skewed by the way questions are written, but the trends are positive. YECs are nowhere near as numerous as Gallup suggested, their numbers are declining rapidly, and the YECs are older and dying off. In every other developed nation in the world—Canada, northern Europe, Japan, Australia, and others—creationism has no influence on public policy. This is striking contrast to the United States, where (despite the fact that YECs are a small minority according to these polls), creationists form the majority of the House and Senate science committees and are the majority of GOP presidential candidates in the past three elections.

The other encouraging sign is the change in the religious composition of the U.S. population. The United States is the last major developed nation in the world that has such a high degree of religiosity. If you travel in Canada, northern Europe, or the United Kingdom, you'll find that nearly everyone there is secular now, and that religion has just about vanished from the cultural landscape. Spectacular churches and cathedrals all over Scandinavia, Germany, the United Kingdom, and much of northern Europe have lost their flocks, and are now being repurposed as public meeting places or bars and restaurants or just sit empty and serve as tourist attractions—but no one worships there any more.

According to a 2013 Pew poll in the United States, the "religiously unaffiliated" are now about 20–30 percent of the general population, outnumbering nearly every other group (Jews, Muslims, Buddhists, Hindus, Mormons, and most Protestant denominations) by a big margin (each of the rest of these groups is 2 percent or less of the population). Only the Catholics and the Southern Baptists are still more numerous, and both of these are losing ground. And of those who said that they had "nothing in particular" in the way of a religious affiliation, 88 percent also said they were "not looking." Thus, the decline in religiosity across the board in the United States is not some sort of hippy-dippy movement to "New Age" religions from the old stale Protestant churches—but a movement away from *any* form of organized religion, especially fundamentalism.

Even more striking is how this breaks down demographically. The most remarkable of the trends is how much it is stratified by age. Young people are becoming increasingly secular and nonreligious, so much so that the youngest cohort (the "young millennials," born 1990–1994) are 34 percent nonreligious! About 30 percent of the "older millennials" (born 1980–1989) are also nonreligious, while the "Gen Xers" (born 1965–1980) report 21 percent "unaffiliated," so the percentage declines only very slightly as the cohorts age. All of these people together contribute to the overall 20–30 percent of "unaffiliated" in the poll and will increase through time. Clearly, organized religion is fading rapidly in this country, driven by a combination of young people who see no need of it and the dying off of the older generations that were raised in a strongly religious society. As sociologists like Phil Zuckerman have shown, such trends have already happened in most of the western industrialized nations (especially those in Scandinavia), as the benefits of a modern secular society and modern medicine and science become more central to their lives.

We can all speculate about why younger generations are alienated from organized religion, and certainly there are many reasons. But knowing the current political trends in this country, we might suggest that one factor of great importance is how "organized religion" in this country is largely dominated by the shrill and intolerant evangelicals and their hate-filled message against science, gays, women, and minorities. With the incredibly rapid shift in this country toward majority acceptance of gays (who are overwhelmingly supported by young people, among whom homophobia and religious intolerance is rare), it might seem that such an issue is driving people away from religious zealots in politics, and their causes. Sure enough, that is confirmed by recent polling. The Pew Study cited earlier shows almost mirror-image percentages: those who are "unaffiliated" are largely supportive of gay rights and abortion rights; those who are religious are just the opposite. Another study drives the point home in stark relief. The single biggest factor driving people away from churches is indeed the intolerance and hatred shown by the evangelicals, and how they have manifested this whenever they have secured political power. As the *Los Angeles Times* describes it, this is a striking change from only 30 years ago:

> During the 1980s, the public face of American religion turned sharply right. Political allegiances and religious observance became more closely aligned, and both religion and politics became more polarized. Abortion and homosexuality became more prominent issues on the national political agenda, and activists such as Jerry Falwell and Ralph Reed began looking to expand religious activism into electoral politics. Church attendance gradually became the primary dividing line between Republicans and Democrats in national elections.
>
> This political "God gap" is a recent development. Up until the 1970s, progressive Democrats were common in church pews and many conservative Republicans didn't attend church. But after 1980, both churchgoing progressives and secular conservatives became rarer and rarer. Some Americans brought their religion and their politics into alignment by adjusting their political views to their religious faith. But, surprisingly, more of them adjusted their religion to fit their politics.
>
> We were initially skeptical about that proposition, because it seemed implausible that people would make choices that might affect their eternal fate based on how they felt about George W. Bush. But the evidence convinced us that many Americans now are sorting themselves out on Sunday morning on the basis of their political views. For example, in our Faith Matters national survey of 3,000 Americans, we observed this sorting process in real time, when we interviewed the same people twice about one year apart.
>
> For many religious Americans, this alignment of religion and politics was divinely ordained, a long-sought retort to the immorality of the 1960s. Other Americans were not so sure.
>
> Throughout the 1990s and into the new century, the increasingly prominent association between religion and conservative politics provoked a backlash among moderates and progressives, many of whom had previously considered themselves religious. The fraction of Americans who agreed "strongly" that religious leaders should not try to influence government decisions nearly doubled from 22 percent in 1991 to 38 percent in 2008, and the fraction who insisted that religious leaders should not try to influence how people vote rose to 45 percent from 30 percent.
>
> This backlash was especially forceful among youth coming of age in the 1990s and just forming their views about religion. Some of that generation, to be sure, held deeply conservative moral and political views, and they felt very comfortable in the ranks of increasingly conservative churchgoers. But a majority of the Millennial generation was liberal on most social issues, and above all, on homosexuality. The fraction of twentysomethings who said that homosexual relations were "always" or "almost always" wrong plummeted from about 75 percent in 1990 to about 40 percent in 2008. (Ironically, in polling, Millennials are actually more uneasy about abortion than their parents.)

Hemant Mehta argues that the regressive social policies of fundamentalists aren't the only factors. He points to the fact that younger generations are more likely to learn from the Internet, and less likely to obey everything their parents tell them, especially when they have questions for which organized religion has no good answers. Certainly, the virtual community of Web-enabled young people can explore and learn about topics like secularism and evolution in a way that would have been impossible in many small religious American towns just a generation ago. Even if the social pressure of the conservative community censors or hushes up these topics in school and at the library, the Internet opens a window that cannot be shut by local authorities—and younger people are more likely to find their own answers this way than ever before.

For those of us who value science and science education in this country, this is good news. As I've argued in this chapter, the single biggest factor that causes us to fall behind nearly all the other westernized industrial nations (including Japan, South Korea, China, and Singapore, along with most of Europe) in science literacy is fundamentalism and creationism. When you break down the polling, it's always questions about evolution, the age of the earth, cosmology, and human evolution that nearly always cause Americans to flunk science literacy tests. These are all questions that reflect the creationist-evangelical influence on our culture. Thankfully, it is apparently declining. The United States probably won't become Denmark during my lifetime, but I'm optimistic that the never-ending battles with creationism in the United States will gradually end as all the old evangelicals die off and they are not replaced by a comparable cohort of the younger generation that was similarly brainwashed. One can only hope . . .

Choices for the Future

It is to be hoped that the ID movement, because of the very publicity that it has sought and achieved, will be seen by the majority of Americans for the giant step backward that it is. Our children are literally the future of our nation, which will increasingly need competent scientists and engineers to guide us through the coming technological revolutions—revolutions that are already under way all around us. There are examples in history of the collapse of great civilizations. There is no particular reason that the United State should be exempt from historical forces. The Visigoths are at the gate. Will we let them in?

—John Brockman, *Intelligent Thought: Science Versus the Intelligent Design Movement*

When all is said and done, we have just a few choices. We can let the creationists further damage our scientific literacy and technological and scientific advantages, or we can try to bring back rigorous science education in our schools and make it our priority. We can reject evolution, astronomy, geology, paleontology, and anthropology for some minority's religious viewpoint, or we can accept what science has taught us about the world and see ourselves in this humbling new light. As George Gaylord Simpson wrote in 1961, "One hundred years without Darwin [*and now over 150 years*] is enough!"

Instead of the narrow, claustrophobic extremist worldview of the creationists, we can accept the vastness of the universe and the immense length of geologic time and reach a humbler, less anthropocentric, less arrogant attitude about our place in nature. We can embrace the fact that we are part of the biosphere and need to shepherd and care for this planet before we destroy it. Many scientists and authors have written about how uplifting and liberating the scientific worldview can be for humankind, especially in comparison to the vengeful God of the Old Testament or the hateful practices of many religions that persecute and sometimes murder people in the name of their faith. Michael Shermer gives a good argument for why the evolutionary and scientific worldview is not only no threat to true religion or spirituality but actually helps us better understand our spirituality when complemented by our scientific understanding of the world. As Shermer (2006:159–161) writes,

> Does a scientific explanation for the world diminish its spiritual beauty? I think not. Science and spirituality are complementary, not conflicting; additive, not detractive. Anything that generates a sense of awe may be a source of spirituality. Science does this in spades. I am deeply moved, for example, when I observe through my Meade eight-inch reflecting telescope in my backyard the fuzzy little patch of light that is the Andromeda galaxy. It is not just because it is lovely, but because I also understand that the photons of light landing on my retina left Andromeda 2.9 million years ago, when our ancestors were tiny-brained hominids roaming the plains of Africa. . . .
>
> Herein lies the spiritual side of science—sciensuality, if you will pardon the awkward neologism but one that echoes the sensuality of discovery. If religion and spirituality are supposed to generate awe and humility in the face of the creator, what could be more awesome and humbling than the deep space discovered by Hubble and the cosmologists, or the deep time discovered by Darwin and the evolutionists?
>
> Darwin matters because evolution matters. Evolution matters because science matters. Science matters because it is the preeminent story of our age, an epic saga about who we are, where we came from, and where we are going.

In his famous television series *Cosmos* (1980), the late great Carl Sagan put it beautifully:

> The universe is all that is, or ever was, or ever will be. Our contemplations of the cosmos stir us. There's a tingling in the spine, a catch in the voice, a faint sensation as of a distant memory of falling from a great height. We know we are approaching the grandest of mysteries. . . . We've begun at last to wonder about our origins, star stuff contemplating the stars, organized collections of ten billion billion billion atoms contemplating the evolution of matter, tracing that long path by which it arrived at consciousness here on the planet Earth and perhaps through the cosmos. Our obligation to survive and flourish is owed not just to ourselves but also to that cosmos, ancient and vast, from which we spring.

Darwin (1859) said it best in the concluding paragraph of *On the Origin of Species*,

> There is grandeur in this view of life, with its several powers, having been originally breathed into a few forms or into one; and that, whilst this planet has gone cycling on according to the fixed law of gravity, from so simple a beginning endless forms most beautiful and most wonderful have been, and are being, evolved.

For Further Reading

Brown, B., and J. P. Alson. 2007. *Flock of Dodos: Behind Modern Creationism, Intelligent Design, and the Easter Bunny.* Cambridge, U.K.: Cambridge House.

Ehrlich, P. R., and A. H. Ehrlich. 1996. *Betrayal of Science and Reason: How Anti-environmental Rhetoric Threatens Our Future.* Washington, D.C.: Island Press.

Humes, E. 2007. *Monkey Girl: Evolution, Education, Religion, and the Battle for America's Soul.* New York: Ecco.

Kitcher, P. 2007. *Living with Darwin: Evolution, Design, and the Future of Faith.* Oxford: Oxford University Press.

Levine, G. 2006. *Darwin Loves You: Natural Selection and the Re-enchantment of the World.* Princeton, N.J.: Princeton University Press.

Lipps, J. H. 1999. Beyond reason: science in the mass media. In *Evolution! Facts and Fallacies*, ed. J. W. Schopf. San Diego, Calif.: Academic, 71–90.

Matzke, N. J. 2016. The evolution of antievolution policies after *Kitzmiller vs. Dover. Science* 351(6268):28–30.

Mooney, C. 2005. *The Republican War on Science.* New York: Basic.

Prothero, D.R. 2013. *Reality Check: How Science Deniers Threaten Our Future.* Bloomington, Ind.: Indiana University Press.

Pigliucci, M. 2002. *Denying Evolution: Creationism, Scientism, and the Nature of Science.* Sunderland Mass.: Sinauer.

Sagan, C. 1996. *The Demon-Haunted World: Science as a Candle in the Dark.* New York: Ballantine.

Shermer, M. 1997. *Why People Believe Weird Things: Pseudoscience, Superstition, and Other Confusions of Our Time.* New York: Freeman.

Shermer, M. 2006. *Why Darwin Matters: The Case Against Intelligent Design.* New York: Times Books.

Shulman, S. 2007. *Undermining Science: Suppression and Distortion in the Bush Administration.* Berkeley: University of California Press.

Scientific Literacy Ranking Web Links www.nationmaster.com/graph/edu_sci_lit-education-scientific -literacy www.livescience.com/humanbiology/060810_evo_rank.html seattletimes.nwsource .com/html/opinion/2002887594_sundaypnnl26.html.

Taibbi, M. 2008. *The Great Derangement: A Terrifying True Story of War, Politics, and Religion.* New York: Spiegel & Grau.

Zuckerman, P. 2010. *Society Without God: What the Least Religious Nations Can Tell Us about Contentment.* New York: NYU Press.

Zuckerman, P. 2011. *Faith No More: Why People Reject Religion.* Oxford: Oxford University Press.

BIBLIOGRAPHY

Adoutte, A., G. Balavoine, N. Lartillot, O. Lespinet, B. Prudhomme, and R. de Rosa. 2000. The new animal phylogeny: reliability and implications. *Proceedings of the National Academy of Sciences USA* 97:4453–4456.

Ahlberg, P. E., and N. H. Trewin. 1995. The postcranial skeleton of the Middle Devonian lungfish *Dipterus valenciennesi. Transactions of the Royal Society of Edinburgh, Earth Sciences* 85:159–175.

Alters, B., and S. Alters. 2001. *Defending Evolution.* Sudbury, Mass.: Jones and Bartlett.

Anderson, J. S., R. R. Reisz, D. Scott, N. B. Fröbisch, and S. S. Sumida. 2008. A stem batrachian from the Early Permian of Texas and the origin of frogs and salamanders. *Nature* 453:515–518.

Andrews, R. C. 1921. A remarkable case of external hind limbs in a humpback whale. *American Museum Novitates,* no. 9.

Archibald, J. D. 1996. Fossil evidence for the Late Cretaceous origin of "hoofed" mammals. *Science* 272:1150–1153.

Arthur, J. 1996. Creationism: bad science or immoral pseudoscience? *Skeptic* 4(4):88–93.

Arthur, W. 1997. *The Origin of Animal Body Plans: A Study in Evolutionary Developmental Biology.* New York: Cambridge University Press.

Ashton, J. F. 2000. *In Six Days: Why 50 Scientists Choose to Believe in Creation.* Frenchs Forest, Australia: New Holland.

Ayala, F. J., and A. Rzhetsky. 1998. Origins of the metazoan phyla: molecular clocks confirm paleontological estimates. *Proceedings of the National Academy of Sciences USA* 95:606–611.

Bar-Maor, J. A., K. M. Kesner, and J. K. Kaftori. 1980. Human tails. *Journal of Bone and Joint Surgery* 62-B(4):508–510.

Barnes, L. G. 1989. A new enaliarctine pinniped from the Astoria Formation, Oregon, and a classification of the Otariidae (Mammalia: Carnivora). *Natural History Museum of Los Angeles County Contributions in Science* 403:1–26.

Barnes, R. D. 1986. *Invertebrate Zoology.* 5th ed. Philadelphia, Pa.: Saunders.

Baskin, J. A. 1998a. Procyonidae. In *Evolution of Tertiary Mammals of North America,* ed. C. Janis, K. M. Scott, and L. Jacobs. New York: Cambridge University Press, 144–151.

Baskin, J. A. 1998b. Mustelidae. In *Evolution of Tertiary Mammals of North America,* ed. C. Janis, K. M. Scott, and L. Jacobs. New York: Cambridge University Press, 152–173.

Beard, K. C. 2004. *The Hunt for the Dawn Monkey: Unearthing the Origin of Monkeys, Apes, and Humans.* Berkeley: University of California Press.

Behe, M. 1996. *Darwin's Black Box.* New York: Free Press.

Benton, M. J., ed. 1988a. *The Phylogeny and Classification of the Tetrapods.* Vol. 1, *Amphibians, Reptiles, Birds.* Oxford, U.K.: Clarendon.

Benton, M. J., ed. 1988b. *The Phylogeny and Classification of the Tetrapods.* Vol. 2, *Mammals.* Oxford, U.K.: Clarendon.

Benton, M. J. 2000. *Vertebrate Palaeontology.* 2nd ed. New York: Chapman & Hall.

Benton, M. J. 2005. *Vertebrate Palaeontology.* 3rd ed. New York: Chapman & Hall.

Benton, M. J. 2014. *Vertebrate Palaeontology.* 4th ed. New York: Wiley-Blackwell.

Benton, M. J., and P. N. Pearson. 2001. Speciation in the fossil record. *Trends in Ecology and Evolution* 16:405–411.

Berra, T. 1990. *Evolution and the Myth of Creationism.* Stanford, Calif.: Stanford University Press.

Berta, A., and C. E. Ray. 1990. Skeletal morphology and locomotor capabilities of the archaic pinniped *Enaliarctos mealsi. Journal of Vertebrate Paleontology* 10:141–157.

Berta, A., C. E. Ray, and A. R. Wyss. 1989. Skeleton of the oldest known pinniped, *Enaliarctos. Science* 244:60–62.

Beus, S., and M. Morales, eds. 1990. *Grand Canyon Geology.* Oxford: Oxford University Press.

Boardman, R. S., A. H. Cheetham, and A. J. Rowell, eds. 1987. *Fossil Invertebrates.* Cambridge, Mass.: Blackwell.

Bridgwater, D., J. H. Allart, and J. W. Schopf. 1981. Microfossil-like objects from the Archean of Greenland: a cautionary note. *Nature* 289:51–53

Briggs, D. E. G., D. L. Bruton, and H. B. Whittington. 1979. Appendages of the fossil arthropod *Aglaspis spinifer* (Upper Cambrian, Wisconsin) and their significance. *Palaeontology* 22:167–180.

Briggs, D. E. G., and R. A. Fortey. 1989. The early radiation and relationships of the major arthropod groups. *Science* 256:241–243.

Briggs, D. E. G., and R. E. Fortey. 2005. Wonderful strife: systematics, stem groups, and the phylogenetic signal of the Cambrian radiation. *Paleobiology* 31:94–112.

Brockman, J., ed. 2006. *Intelligent Thought: Science Versus the Intelligent Design Movement.* New York: Vintage.

Brown, B., and J. P. Alson. 2007. *Flock of Dodos: Behind Modern Creationism, Intelligent Design, and the Easter Bunny.* Cambridge, U.K.: Cambridge House.

Brunet, M., et al. 2002. A new hominid from the upper Miocene of Chad, central Africa. *Nature* 418:145–151.

Brusca, R. C., and G. J. Brusca. 1990. *Invertebrates*. Sunderland, Mass.: Sinauer.

Burns, J. 1975. *Biograffiti: A Natural Selection*. Cambridge: Harvard University Press.

Cairns-Smith, A. G. 1985. *Seven Clues to the Origin of Life*. New York: Cambridge University Press.

Caldwell, M. W., and M. S. Y. Lee. 1997. A snake with legs from the marine Cretaceous of the Middle East. *Nature* 386:705–709.

Callaway, J. M., and E. M. Nicholls. 1996. *Ancient Marine Reptiles*. San Diego, Calif.: Academic.

Campbell, J. 1949. *The Hero with a Thousand Faces*. Princeton, N.J.: Princeton University Press.

Campbell, J. 1982. Autonomy in evolution. In *Perspectives on Evolution*, ed. R. Milkman. Sunderland, Mass.: Sinauer, 190–200.

Cao, Y., J. Adachi, T. Yano, and M. Hasegawa. 1994. Phylogenetic place of guinea pigs: no support of the rodent polyphyly hypothesis from maximum-likelihood analyses of multiple protein sequences. *Molecular Biology and Evolution* 11:593–604.

Carroll, R. L. 1988. *Vertebrate Paleontology and Evolution*. New York: Freeman.

Carroll, R. L. 1992. The primary radiation of terrestrial vertebrates. *Annual Review of Earth and Planetary Sciences* 20:45–84.

Carroll, R. L. 1996. Mesozoic marine reptile as models of long-term large-scale evolutionary phenomena. In *Ancient Marine Reptiles*, ed. J. M. Callaway and E. M. Nicholls. San Diego, Calif.: Academic, 467–487.

Carroll, R. L. 1997. *Patterns and Processes of Vertebrate Evolution*. New York: Cambridge University Press.

Carroll, S. 2005. *Endless Forms Most Beautiful: The New Science of Evo/Devo*. New York: Norton.

Chiappe, L. M. 1995. The first 85 million years of avian evolution. *Nature* 378:349–355.

Chiappe, L. M., and G. J. Dyke. 2002. The Mesozoic radiation of birds. *Annual Review of Ecology and Systematics* 33:91–124.

Chiappe, L. M., and Meng Qingjin. 2016. *Birds of Stone: Chinese Avian Fossils from the Age of Dinosaurs*. Baltimore, Md.: Johns Hopkins University Press.

Chiappe, L. M., and L. M. Witmer, eds. 2002. *Mesozoic Birds: Above the Heads of Dinosaurs*. Berkeley: University of California Press.

Cifelli, R. 1969. Radiation of the Cenozoic planktonic foraminifera. *Systematic Zoology* 18:154–168.

Clack, J. A. 2002. *Gaining Ground: The Origin and Early Evolution of Tetrapods*. Bloomington: Indiana University Press.

Clarkson, E. N. K. 1998. *Invertebrate Palaeontology and Evolution*. 4th ed. Oxford, U.K.: Blackwell Science.

Clyde, W. C., and D. C. Fisher. 1997. Comparing the fit of stratigraphic and morphologic data in phylogenetic analysis. *Paleobiology* 23:1–19.

Colbert, E. H., and M. Morales. 1991. *Evolution of the Vertebrates*. New York: Wiley.

Cone, J. 1991. *Fire Under the Sea: The Discovery of the Most Extraordinary Environment on Earth—Volcanic Hot Springs on the Ocean Floor*. New York: Morrow.

Conroy, G. C. 1990. *Primate Evolution*. New York: Norton.

Conway Morris, S. 1998. *The Crucible of Creation*. Oxford: Oxford University Press.

Conway Morris, S. 2000. The Cambrian "explosion": slow-fuse or megatonnage? *Proceedings of the National Academy of Sciences USA* 97:4426–4429.

Cook, G. 2013. Doubting "Darwin's doubt." *New Yorker*, July 2, 2013.

Costanza, G., S. Pino, F. Ciciriello, and E. Di Mauro. 2009. Generation of long RNA chains in water. *Journal of Biological Chemistry* 284:33206–33216.

Cuppy, W. 1941. *How to Become Extinct*. Chicago: University of Chicago Press.

Currie, P. J., E. B. Koppelhus, M. A. Shugar, and J. L. Wright, eds. 2004. *Feathered Dragons: Studies on the Transition from Dinosaurs to Birds*. Bloomington: Indiana University Press.

Daeschler, E. B., N. H. Shubin, and F. A. Jenkins Jr. 2006. A Devonian tetrapod-like fish and the evolution of the tetrapod body plan. *Nature* 440:757–773.

Dalrymple, G. B. 1991. *The Age of the Earth*. Stanford, Calif.: Stanford University Press.

Dalrymple, G. B. 2004. *Ancient Earth, Ancient Skies: The Age of Earth and Its Cosmic Surroundings*. Stanford, Calif.: Stanford University Press.

Darwin, C. R. 1859. *On the Origin of Species*. London: John Murray.

Darwin, C. R. 1871. *The Descent of Man and Selection in Relation to Sex*. London: John Murray.

Darwin, E. 1794. *Zoonomia*. Vol. 1, *The Laws of Organic Life*. London: John Murray.

Darwin, F., ed. 1888. *The Life and Letters of Charles Darwin: Including an Autobiographical Chapter*, Volume 2. London: D. Appleton.

Davis, P., and D. Kenyon. 2004. *Of Pandas and People: The Central Question of Biological Origins*. 2nd ed. Dallas, Tex.: Haughton.

Dawkins, R. 1986. *The Blind Watchmaker*. New York: Norton.

Dawkins, R. 1996. *Climbing Mount Improbable*. New York: Norton.

Dawkins, R. 2004. *The Ancestor's Tale: A Pilgrimage to the Dawn of Evolution*. Boston: Houghton Mifflin.

Dawkins, R. 2006. *The God Delusion*. Boston: Houghton Mifflin.

DeBraga, M., and R. L. Carroll. 1993. The origin of mosasaurs as a model of macroevolutionary patterns and processes. *Evolutionary Biology* 27:245–322.

Delson, E. C. 1985. *Ancestors: The Hard Evidence*. New York: Liss.

Desmond, A., and J. Moore. 1991. *Darwin: The Life of a Tormented Evolutionist*. New York: Warner.

Dial, K. P. 2003. Wing-assisted incline running and the evolution of flight. *Science* 299:402–405.

Diamond, J. 1992. *The Third Chimpanzee: The Evolution and Future of the Human Animal*. New York: HarperCollins.

Dingus, L., and T. Rowe. 1997. *The Mistaken Extinction*. New York: Freeman.

Dobzhansky, T. 1973. Nothing in biology makes sense except in the light of evolution. *American Biology Teacher* 35:125–129.

Dodson, P. 1996. *The Horned Dinosaurs*. Princeton, N.J.: Princeton University Press.

Domning, D. P. 1981. Sea cows and sea grasses. *Paleobiology* 7:417–420.

Domning, D. P. 1982. Evolution of manatees: a speculative history. *Journal of Paleontology* 56:599–619.

Domning, D. P. 2001. The earliest known fully quadrupedal sirenian. *Nature* 413:625–627.

Domning, D. P., C. E. Ray, and M. C. McKenna. 1986. Two new Oligocene desmostylians and a discussion of tethytherian systematics. *Smithsonian Contributions to Paleobiology* 59:1–56.

Dumeril, A. 1867. Métamorphoses des batraciens urodèles à branchies extérieures du Mexique dits axolotls, observées à la Menagérie des Reptiles du Muséum d'Histoire Naturelle. *Annales Scientifique Naturale Zoologique* 7:229–254.

Ecker, R. L. 1990. *Dictionary of Science and Creationism*. Buffalo, N.Y.: Prometheus.

Ehrlich, P. R., and A. H. Ehrlich. 1996. *Betrayal of Science and Reason: How Anti-environmental Rhetoric Threatens Our Future*. Washington, D.C.: Island Press.

Eldredge, N. 1977. Trilobites and evolutionary patterns. In *Patterns of Evolution as Illustrated in the Fossil Record*, ed. A. Hallam. New York: Elsevier, 305–332.

Eldredge, N. 1982. *The Monkey Business: A Scientist Looks at Creationism*. New York: Pocket Books.

Eldredge, N. 1985a. *Time Frames*. New York: Simon & Schuster.

Eldredge, N. 1985b. *Unfinished Synthesis*. New York: Oxford University Press.

Eldredge, N. 2000. *The Triumph of Evolution and the Failure of Creationism*. New York: Freeman.

Eldredge, N., and S. J. Gould. 1972. Punctuated equilibria: an alternative to phyletic gradualism. In *Models in Paleobiology*, ed. T. J. M. Schopf. San Francisco: Freeman Cooper, 82–115.

Eldredge, N., and S. M. Stanley, eds. 1984. *Living Fossils*. New York: Springer Verlag.

Engelmann, G. F., and E. O. Wiley. 1977. The place of ancestor-descendant relationships in phylogeny reconstruction. *Systematic Zoology* 26:1–11

Ernst, G. 1970. Zur Stammgeschichte und stratigraphischen Bedeutung der Echiniden-Gattung *Micraster* in der nordwest deutschen Oberkreide. *Mittelungen der Geologischer-Paläontologische Institut der Universität Hamburg* 39:11–135.

Erwin, D., and J. W. Valentine. 2013. *The Cambrian Explosion: The Construction of Biodiversity*. New York: Roberts.

Fastovsky, D. E., and D. B. Weishampel. 2005. *The Evolution and Extinction of the Dinosaurs*. 2nd ed. New York: Cambridge University Press.

Fisher, D. C. 1982. Phylogenetic and macroevolutionary patterns within the Xiphosurida. *Proceedings of the Third North American Paleontological Convention* 1:175–180.

Fisher, D. C. 1984. The Xiphosurida: archetypes of bradytely? In *Living Fossils*, ed. N. Eldredge and S. M. Stanley. New York: Springer Verlag, 196–213.

Fisher, D. C. 1994. Stratocladistics: morphological and temporal patterns and their relation to the phylogenetic process. In *Interpreting the Hierarchy of Nature*, ed. L. Grande and O. Rieppel. San Diego, Calif.: Academic, 133–171.

Fitch, W. M., and E. Margoliash. 1967. Construction of phylogenetic trees. *Science* 155:279–284.

Foote, M. 1996. On the probability of ancestors in the fossil record. *Paleobiology* 22:141–151.

Forey, P., and P. Janvier. 1984. Evolution of the earliest vertebrates. *American Scientist* 82:554–565.

Forrest, B., and P. R. Gross. 2004. *Creationism's Trojan Horse: The Wedge of Intelligent Design*. Oxford: Oxford University Press.

Forster, C. A., S. D. Sampson, L. M. Chiappe, and D. W. Krause. 1998. The theropod ancestry of birds: new evidence from the Late Cretaceous of Madagascar. *Science* 279:1915–1919.

Fortey, R. A., and R. P. S. Jefferies. 1982. Fossils and phylogeny—a compromise approach. *Systematics Association Special Volume* 21:197–234.

Fortey, R. A., and R. A. Owens. 1990. Trilobites. In *Evolutionary Trends*, ed. K. J. McNamara. Tucson: University of Arizona Press, 121–142.

Franz, M.-L. von. 1972. *Creation Myths*. Zurich: Spring.

Freud, S. 1917. Lecture 18: fixation to traumas—the unconscious. In *Lectures Introducing Psychoanalysis*. London: George Allen & Unwin.

Friedman, R. 1987. *Who Wrote the Bible?* New York: Harper & Row.

Froehlich, D. J. 2002. Quo vadis *Eohippus*? The systematics and taxonomy of the early Eocene equids (Perissodactyla). *Zoological Journal of the Linnaean Society of London* 134:141–256.

Fry, I. 2000. *The Emergence of Life on Earth: A Historical and Scientific Overview.* Piscataway, N.J.: Rutgers University Press.

Frye, R. M., ed. 1983. *Is God a Creationist? The Religious Case Against Creation-Science.* New York: Scribner.

Futuyma, D. 1983. *Science on Trial: The Case for Evolution.* New York: Pantheon.

Gagneux, P., C. Wills, U. Gerloff, D. Tautz, P.A. Morin, C. Boesch, B. Fruth, G. Hohmann, O.A. Ryder, and D.S. Woodruff. 1999. Mitochondrial sequences show diverse evolutionary histories of African hominoids." *Proceedings of the National Academy of Sciences USA* 93.

Gardner, M. 1952. *Fads and Fallacies in the Name of Science.* New York: Dover.

Gardner, M. 1981. *Science: Good, Bad, and Bogus.* Buffalo, N.Y.: Prometheus.

Gauthier, J. A. 1986. Saurischian monophyly and the origin of birds. *California Academy of Sciences Memoir* 8:1–56.

Gauthier, J. A., and L. F. Gall, eds. 2001. *New Perspectives on the Origin and Early Evolution of Birds.* New Haven, Conn.: Yale University Press.

Gauthier, J. A., A. G. Kluge, and T. Rowe. 1988. The early evolution of the Amniota. In *The Phylogeny and Classification of the Tetrapods.* Vol. 1, *Amphibians, Reptiles, Birds,* ed. M. J. Benton. Oxford, U.K.: Clarendon, 103–155.

Gee, H. 1997. *Before the Backbone: Views on the Origin of Vertebrates.* New York: Chapman & Hall.

Geisler, J. H., and M. D. Uhen. 2005. Phylogenetic relationships of extinct cetartiodactyls: results of simultaneous analyses of molecular, morphological, and stratigraphic data. *Journal of Mammalian Evolution* 12:145–160.

Gheerbrant, E., J. Sudre, and H. Cappetta. 1996. Palaeocene proboscidean from Morocco. *Nature* 383:68–70.

Gingerich, P. D. 1980. Evolutionary patterns in early Cenozoic mammals. *Annual Review of Earth and Planetary Sciences* 8:407–424

Gingerich, P. D. 1989. New earliest Wasatchian mammalian fauna from the Eocene of Northwestern Wyoming: composition and diversity in a rarely sampled high-floodplain assemblage. *University of Michigan Papers in Paleontology* 28:1–97.

Gingerich, P. D., M. Haq, I. S. Zalmout, I. H. Khan, and M. S. Malkani. 2001. Origin of whales from early artiodactyls; hands and feet of Eocene Protocetidae from Pakistan. *Science* 293:2239–2242.

Gingerich, P. D., B. H. Smith, and E. L. Simons. 1990. Hind limbs of Eocene *Basilosaurus:* evidence of feet in whales. *Science* 249:154–157.

Gingerich, P. D., N. A. Wells, D. E. Russell, and S. M. Ibrahim Shah. 1983. Origin of whales in epicontinental remnant seas: new evidence from the early Eocene of Pakistan. *Science* 220:403–406.

Gish, D. 1972. *Evolution? The Fossils Say NO!* San Diego, Calif.: Creation-Life.

Gish, D. 1973. Creation, evolution, and the historical evidence. *American Biology Teacher* 35(3):132–140.

Gish, D. 1978. *Evolution? The Fossils Say NO!* (Public School Edition) San Diego, Calif.: Creation-Life.

Gish, D. 1995. *Evolution, the Fossils Still Say NO!* San Diego, Calif.: Creation-Life.

Gittleman, J., ed. 1996. *Carnivore Biology, Behavior, and Evolution.* Ithaca, N.Y.: Cornell University Press.

Glaessner, M. F. 1984. *The Dawn of Animal Life.* New York: Cambridge University Press.

Godfrey, L., ed. 1983. *Scientists Confront Creationism.* New York: Norton.

Goll, R. M. 1976. Morphological intergradation between modern populations of *Lophospyris* and *Phormospyris* (Trissocyclidae, Radiolaria). *Micropaleontology* 22:379–418.

Gosse, P. H. 1907. *Father and Son.* New York: Norton.

Gould, S. J. 1972. Allometric fallacies and the evolution of *Gryphaea. Evolutionary Biology* 6:91–119.

Gould, S. J. 1977. *Ever Since Darwin.* New York: Norton.

Gould, S. J. 1977. *Ontogeny and Phylogeny.* Cambridge: Harvard University Press.

Gould, S. J. 1980a. Is a new and general theory of evolution emerging? *Paleobiology* 6:119–130.

Gould, S. J. 1980b. *The Panda's Thumb.* New York: Norton.

Gould, S. J. 1981. Evolution as fact and theory. *Discover* 2:34–37.

Gould, S. J. 1982. Darwinism and the expansion of evolutionary theory. *Science* 216:380–387.

Gould, S. J. 1984. *The Flamingo's Smile.* New York: Norton.

Gould, S. J. 1989. *Wonderful Life: The Burgess Shale and the Nature of History.* New York: Norton.

Gould, S. J. 1992. Punctuated equilibria in fact and theory. In *The Dynamics of Evolution,* ed. A. Somit and S. A. Peterson. Ithaca, N.Y.: Cornell University Press, 54–84.

Gould, S. J. 1993. "Cordelia's Dilemma." *Natural History* 102:10–18.

Gould, S. J., ed. 1995. Hooking Leviathan by its past. In *Dinosaur in a Haystack.* New York: Norton, 375–396.

Gould, S. J. 2002. *The Structure of Evolutionary Theory.* Cambridge: Harvard University Press.

Gould, S. J., and N. Eldredge. 1977. Punctuated equilibria: the tempo and mode of evolution reconsidered. *Paleobiology* 3:115–151.

Graur, D., W. A. Hide, and W. H. Li. 1991. Is the guinea pig a rodent? *Nature* 351:649–652.

Graves, R., and R. Patai. 1963. *Hebrew Myths: The Book of Genesis.* New York: McGraw-Hill.

Grotzinger, J. P., S. A. Bowring, B. Z. Saylor, and A. J. Kaufman. 1995. Biostratigraphic and geochronologic constraints on early animal evolution. *Science* 270:598–604.

Hallam, A. 1968. Morphology, palaeoecology, and evolution of the genus *Gryphaea* in the British Lias. *Philosophical Transactions of the Royal Society of London B* 254:91–128.

Hallam, A., ed. 1977. *Patterns of Evolution as Illustrated in the Fossil Record*. New York: Elsevier.

Hallam, A. 1982. Patterns of speciation in Jurassic *Gryphaea*. *Paleobiology* 8:354–366.

Hallam, A., and S. J. Gould. 1974. The evolution of British and American middle and upper Jurassic *Gryphaea*: a biometric study. *Proceedings of the Royal Society of London B* 189:511–542.

Haq, B. U., and A. Boersma, eds. 1978. *Introduction to Marine Micropaleontology*. New York: Elsevier.

Harris, D. J. 2003 Codon bias variation in C-mos between squamate families might distort phylogenetic inferences. *Molecular Phylogenetics and Evolution* 27:540–554.

Hazen, R. M. 2005. *Gen-e-sis: The Scientific Quest for Life's Origins*. Washington, D.C.: Joseph Henry.

Heidel, A. 1942. *The Babylonian Genesis*. Chicago: University of Chicago Press.

Heidel, A. 1946. *The Gilgamesh Epic and Old Testament Parallels*. Chicago: University of Chicago Press.

Heilmann, G. 1926. *The Origin of Birds*. London: Witherby.

Hennig, W. 1966. *Phylogenetic Systematics*. Urbana: University of Illinois Press.

Hillis, D. M., and C. Moritz, eds. 1990. *Molecular Systematics*. Sunderland, Mass.: Sinauer.

Hitchin, R., and M. J. Benton. 1997. Congruence between parsimony and stratigraphy: comparison of three indices. *Paleobiology* 23:20–32.

Honey, J., J. A. Harrison, D. R. Prothero, and M. S. Stevens. 1998. Camelidae. In *Evolution of Tertiary Mammals of North America*, ed. C. Janis, K. M. Scott, and L. Jacobs. New York: Cambridge University Press, 439–462.

Hooker, J. J. 1989. Character polarities in early perissodactyls and their significance for *Hyracotherium* and infraordinal relationships. In *The Evolution of Perissodactyls*, ed. D. R. Prothero and R. M. Schoch. New York: Oxford University Press, 79–101.

Hopson, J. A. 1994. Synapsid evolution and the radiation of non-eutherian mammals. In *Major Features of Vertebrate Evolution*, ed. D. R. Prothero and R. M. Schoch. Paleontological Society Short Course 7:190–219.

Hou, L.-H., Z. Zhou, L. D. Martin, and A. Feduccia. 1995. A beaked bird from the Jurassic of China. *Nature* 377:616618.

Huelsenbeck, J. P. 1994. Comparing the stratigraphic record to estimates of phylogeny. *Paleobiology* 20:470–483.

Huelsenbeck, J. P., and B. Rannata. 1997. Maximum likelihood estimation of phylogeny using stratigraphic data. *Paleobiology* 23:174–180.

Humes, E. 2007. *Monkey Girl: Evolution, Education, Religion, and the Battle for America's Soul*. New York: Ecco.

Hutchinson, J. E. 1959. Homage to Santa Rosalia or why are there so many kinds of animals? *American Naturalist* 93:145–159.

Huxley, A. 1939. *After Many a Summer Dies the Swan*. New York: Harper.

Huxley, T. H. 1863. *Zoological Evidences of Man's Place in Nature*. London: John Murray.

Huxley, T. H. 1870. Further evidence of the affinity between dinosaurian reptiles and birds. *Quarterly Journal of the Geological Society of London* 26:12–31.

Isaak, M. 2002. A Philosophical Premise of "Naturalism"? September 24. www.talkdesign.org/faqs/naturalism.html.

Isaak, M. 2006. *The Counter-Creationism Handbook*. Berkeley: University of California Press.

Janis, C., K. M. Scott, and L. L. Jacobs, eds. 1998. *Evolution of Tertiary Mammals of North America*. Vol. 1, *Terrestrial Carnivores, Ungulates and Ungulate-like Mammals*. New York: Cambridge University Press.

Janis, C., G. F. Gunnell, and M. D. Uhen, eds. 2008. *Evolution of Tertiary Mammals of North America*. Vol. 2, *Small Mammals, Xenarthrans, and Marine Mammals*. New York: Cambridge University Press.

Ji, Q., Z.-X. Luo, C-X. Yuan, J. R. Wible, J.-P. Zhang, and J. A. Georgi. 2002. The earliest known eutherian mammal. *Nature* 416:816–822.

Johanson, D., and B. Edgar. 1996. *From Lucy to Language*. New York: Simon & Schuster.

Johnson, P. E. 1991. *Darwin on Trial*. Washington, D.C.: Regnery Gateway.

Kardong, K. 1995. *Vertebrates*. Dubuque, Iowa: W. C. Brown, 94.

Kelley, P. H. 1983. Evolutionary patterns of eight Chesapeake Group molluscs; evidence for the model of punctuated equilibria. *Journal of Paleontology* 57:581–598.

Kemp, T. 1982. *Mammal-like Reptiles and the Origin of Mammals*. San Diego, Calif.: Academic.

Kemp, T. S. 2005. *The Origin and Evolution of Mammals*. Oxford: Oxford University Press.

Kennett, J. P., and M. S. Srinivasan. 1983. *Neogene Planktonic Foraminifera: A Phylogenetic Atlas*. Stroudsburg, Pa.: Hutchinson Ross.

Kermack, K. 1954. A biometrical study of *Micraster coranguinum* and *M. (Isomicraster) senonis*. *Philosophical Transactions of the Royal Society B* 237:375–428.

Kielan-Jaworowska, Z., R. L. Cifelli, and Z.-X. Luo. 2004. *Mammals from the Age of Dinosaurs: Origins, Evolution, and Structure*. New York: Columbia University Press.

Kier, P. M. 1965. Evolutionary trends in Paleozoic echinoids. *Journal of Paleontology* 39:436–465.

Kier, P. M. 1975. Evolutionary trends and their functional significance in post-Paleozoic echinoids. *Paleontological Society Memoir* 5:1–95.

Kier, P. M. 1982. Rapid evolution in echinoids. *Palaeontology* 25:1–10.

Kirschvink, J. L., R. L. Ripperdan, and D. A. Evans. 1997. Evidence for a large-scale reorganization of Early Cambrian continental masses by inertial interchange true polar wander. *Science* 277:541–545.

Kitcher, P. 1982. *Abusing Science: The Case Against Creationism.* Cambridge: MIT Press.

Kitcher, P. 2007. *Living with Darwin: Evolution, Design, and the Future of Faith.* Oxford: Oxford University Press.

Knoll, A. H. 2003. *Life on a Young Planet: The First Three Billion Years of Evolution on Earth.* Princeton, N.J.: Princeton University Press.

Knoll, A. H., and S. B. Carroll. 1999. Early animal evolution: emerging views from comparative biology and geology. *Science* 284:2129–2137.

Kukalova-Peck, J. 1978. Origin and evolution of insect wings and their relation to metamorphosis, as documented by the fossil record. *Journal of Morphology* 156:53–125.

Kun, A., M. Santos, and E. Szathmary, E. 2005. Real ribozymes suggest a relaxed error threshold. *Nature Genetics* 37:1008–1011.

Kunkle, R. P. 1958. Permian stratigraphy of the Paradox Basin. In *Guidebook to the Geology of the Paradox Basin: Ninth Annual Field Conference Guidebook,* ed. A. F. Sanborn. Salt Lake City, Utah: Intermountain Association of Petroleum Geologists, 163–168.

Lack, D. 1947. *Darwin's Finches.* New York: Cambridge University Press.

Larson, E. 1985. *Trial and Error: The American Controversy Over Creation and Evolution.* New York: Oxford University Press.

Larson, E., and L. Witham. 1997. Scientists are still keeping the faith. *Nature* 386:435–436.

Laurin, M., and R. R. Reisz. 1996. A re-evaluation of early amniote phylogeny. *Zoological Journal of the Linnean Society of London* 113:165–223.

Lazarus, D. B. 1983. Speciation in pelagic Protista and its study in the microfossil record: a review. *Paleobiology* 9:327–340.

Lazarus, D. B. 1986. Tempo and mode of morphologic evolution near the origin of the radiolarian lineage *Pterocanium prismatium. Paleobiology* 12:175–189.

Lazarus, D. B., H. Hilbrecht, C. Spencer-Cervato, and H. Thierstein. 1995. Sympatric speciation and phyletic change in *Globorotalia truncatulinoides. Paleobiology* 21:975–978.

Lazarus, D. B., and D. R. Prothero. 1984. The role of stratigraphic and morphologic data in phylogeny reconstruction. *Journal of Paleontology* 58:163–172.

Lazarus, D. B., R. P. Scherer, and D. R. Prothero. 1985. Evolution of the radiolarian species-complex *Pterocanium*: a preliminary survey. *Journal of Paleontology* 59:183–221.

Lehmann, J., M. Cibils, and A. Libchaber. 2009. Emergence of a code in the polymerization of amino acids along RNA templates. *PLoS ONE* 4: e5773.

Levine, G. 2006. *Darwin Loves You: Natural Selection and the Re-enchantment of the World.* Princeton, N.J.: Princeton University Press.

Levinton, J. 2001. *Genetics, Paleontology, and Macroevolution.* 2nd ed. New York: Cambridge University Press.

Lewin, R. 1987. *Bones of Contention: Controversies in the Search for Human Origins.* Chicago: University of Chicago Press.

Lewin, R. 1988. *In the Age of Mankind: A Smithsonian Book on Human Evolution.* Washington, D.C.: Smithsonian Institution Press.

Lewin, R. 1989. *Human Evolution: An Illustrated Introduction.* Cambridge, Mass.: Blackwell Scientific.

Lewin, R. 1998. *Principles of Human Evolution: A Core Textbook.* New York: Blackwell.

Lewis, D. L., M. DeCamillis, and R. L. Bennett. 2000. Distinct roles of the homeotic genes *Ubx* and *abd-A* in beetle embryonic abdominal appendage development. *Proceedings of the National Academy of Sciences USA* 97:4504–4509.

Lewontin, R. J., and J. L. Hubby. 1966. A molecular approach to the study of genic heterozygosity in natural populations. *Genetics* 54:595–605.

Li, C., X.-C. Wu, O. Rieppel, L.-T. Wang, and Li-Jun Zhao. 2008. An ancestral turtle from the Late Triassic of southwestern China. *Nature* 456:497–501.

Li, C. K., R. W. Wilson, and M. R. Dawson. 1987. The origin of rodents and lagomorphs. *Current Mammalogy* 1:97–108.

Lieberman, B. S. 2003. Taking the pulse of the Cambrian radiation. *Integrative and Comparative Biology* 43:229–237.

Lipps, J. H. 1999. Beyond reason: science in the mass media. In *Evolution! Facts and Fallacies,* ed. J. W. Schopf. San Diego, Calif.: Academic, 71–90.

Long, J., and H. Schouten. 2008. *Feathered Dinosaurs: The Origin of Birds.* New York: Oxford University Press.

Long, J. A. 2010. *The Rise of Fishes.* 2nd ed. Baltimore, Md.: Johns Hopkins University Press.

Long, M. 2001. Evolution of novel genes. *Current Opinions in Genetics and Development.* 11:673–680.

Long, M., E. Betran, K. Thornton, and W. Wang. 2003. The origin of new genes: glimpses from the young and old. *Nature Review of Genetics* 4:865–875.

Lovejoy, A. O. 1936. *The Great Chain of Being: A Study in the History of an Idea*. Cambridge: Harvard University Press.

Luckett, W. P., and J.-L. Hartenberger, eds. 1985. *Evolutionary Relationships Among Rodents*. New York: Plenum.

Luo, Z.-X., P. Chen, G. Li, and M. Chen. 2007. A new eutriconodont mammal and evolutionary development in early mammals. *Nature* 446:288–293.

Luo, Z.-X., Q. Ji, J. R. Wible, and C.-X. Yuan. 2003. An Early Cretaceous trinosphenic mammal and metatherian evolution. *Science* 302:1934–1940.

MacFadden, B. J. 1984. Systematics and phylogeny of *Hipparion, Neohipparion, Nannippus,* and *Cormohipparion* (Mammalia, Equidae) from the Miocene and Pliocene of the New World. *Bulletin of the American Museum of Natural History* 179:1–196.

MacFadden, B. J. 1992. *Fossil Horses*. Cambridge: Cambridge University Press.

Maddox, B. 2007. Blinded by science. *Discover*, February, 28–29.

Mader, B. J. 1989. The Brontotheriiae: a systematic revision and preliminary phylogeny of North American genera. In *The Evolution of Perissodactyls*, ed. D. R. Prothero and R. M. Schoch. Oxford University Press: New York, 458–484.

Madsen, O., M. Scally, C. J. Douady, D. J. Kao, R. W. DeBry, R. M. Adkins, H. Amrine-Madsen, M. J. Stanhope, W. W. de Jong, and M. S. Springer. 2001. Parallel adaptive radiations in two major clades of placental mammals. *Nature* 409:610–614.

Mahboubi, M., R. Ameur, J.-Y. Crochet, and J.-J. Jaeger. 1984. Earliest known proboscidean from early Eocene of North Africa. *Nature* 308:543–544.

Maisey, J. 1994. Gnathostomes. In *Major Features of Vertebrate Evolution*, ed. D. R. Prothero and R. M. Schoch. Paleontological Society Short Course 7:38–56.

Maisey, J. G. 1996. *Discovering Fossil Fishes*. New York: Holt.

Malmgren, B. A., and W. A. Berggren. 1987. Evolutionary change in some late Neogene planktonic foraminifera lineages and their relationships to paleoceanographic change. *Paleoceanography* 2:445–456.

Malmgren, B. A., W. A. Berggren, and G. P. Lohmann. 1983. Evidence for punctuated gradualism in the Late Neogene *Globorotalia tumida* lineage of planktonic foraminifera. *Paleobiology* 9:377–389.

Malmgren, B. A., and J. P. Kennett. 1981. Phyletic gradualism in a Late Cenozoic planktonic foraminiferal lineage, DSDP Site 284, southwest Pacific. *Paleobiology* 7:230–240.

Margulis, L. 1981. *Symbiosis in Cell Evolution*. San Francisco: Freeman.

Margulis, L. 1982. Early animal evolution: emerging view from comparative biology and geology. *Science* 284:2129–2137.

Margulis, L. 2000. *Symbiotic Planet: A New Look at Evolution*. New York: Basic.

Marks, J. 2002. *What It Means to be 98 Percent Chimpanzee*. Berkeley: University of California Press.

Marsh, O. C. 1892. Recent polydactyle horses. *American Journal of Science* 43:339–355.

Marshall, C. R. 2013. When prior beliefs trump scholarship. *Science* 341:1344.

Martill, D. M., H. Tischling, and H. R. Longrich. 2015. A four-legged snake from the Early Cretaceous of Gondwana. *Science* 349:416–419.

Martin, R. 2004. *Missing Links: Evolutionary Concepts and Transitions Through Time*. Sudbury, Mass.: Jones and Bartlett.

Matthew, W. D. 1926. The evolution of the horse: a record and its interpretation. *Quarterly Review of Biology* 1:139–185.

Mayr, E. 1942. *Systematics and the Origin of Species*. New York: Columbia University Press.

Mayr, E. 1966. *Systematic Zoology*. New York: McGraw-Hill.

McGowan, C. 1983. *The Successful Dragons: A Natural History of Extinct Reptiles*. Toronto: Stevens.

McGowan, C. 1984. *In the Beginning: A Scientist Shows Why the Creationists Are Wrong*. Buffalo, N.Y.: Prometheus.

McKenna, M. C. 1975. Toward a phylogenetic classification of the Mammalia. In *Phylogeny of the Primates: A Multidisciplinary Approach*, ed. W. P. Luckett and F. S. Szalay. New York: Plenum, 21–46.

McKenna, M. C., and S. K. Bell. 1997. *Classification of Mammals*. New York: Columbia University Press.

McKenna, M. C., M. Chow, S. Ting, and Z. Luo. 1989. *Radinskya yupingae*, a perissodactyl-like mammal from the late Paleocene of China. In *The Evolution of Perissodactyls*, ed. D. R. Prothero and R. M. Schoch. New York: Oxford University Press, 24–36.

McKenna, M. C., and E. M. Manning. 1977. Affinities and biogeographic significance of the Mongolian Paleocene genus *Phenacolophus*. *Géobios, Memoire Spécial* 1:61–85.

McLoughlin, J. C. 1980. *Synapsida: A New Look Into the Origin of Mammals*. New York: Viking.

McMenamin, M. A. S. 1998. *The Garden of Ediacara*. New York: Columbia University Press.

McMenamin, M. A. S., and D. L. S. McMenamin. 1990. *The Emergence of Animals, the Cambrian Breakthrough*. New York: Columbia University Press.

McNamara, K. J., ed. 1990. *Evolutionary Trends*. Tucson: University of Arizona Press.

Meng J., Y.-M. Hu, and C.-K. Li. 2003. The osteology of *Rhombomylus* (Mammalia:Glires): implications for the phylogeny and evolution of Glires. *Bulletin of the American Museum of Natural History* 275:1–247.

Meng, J., and A. R. Wyss. 2005. Glires (Lagomorpha, Rodentia). In *The Rise of Placental Mammals*, ed. K. D. Rose and J. D. Archibald. Baltimore, Md.: Johns Hopkins University Press, 145–158.

Miller, K. 1999. *Finding Darwin's God: A Scientist's Search for Common Ground Between God and Evolution*. New York: HarperCollins.

Miller, K. 2004. The flagellum unspun: the collapse of "irreducible complexity." In *Debating Design: From Darwin to DNA*, ed. M. Ruse and W. Dembski. New York: Cambridge University Press, 81–97.

Miller, S. L. 1953. A production of amino acids under possible primitive earth conditions. *Science* 117:528–529.

Mindell, D. P. 2006. *The Evolving World: Evolution in Everyday Life*. Cambridge: Harvard University Press.

Mitchell, E. D., Jr., and R. H. Tedford. 1973. The Enaliarctinae: a new group of extinct aquatic Carnivora, and a consideration of the origin of the Otariidae. *Bulletin of the American Museum of Natural History* 151:205–284.

Miyazaki, J. M., and M. F. Mickevich. 1982. Evolution of *Chesapecten* (Mollusca: Bivalvia, Miocene-Pliocene) and the Biogenetic Law. *Evolutionary Biology* 15:369–409.

Mojzsis, S. J., G. Arrhenius, K. D. McKeegan, T. M. Harrison, A. P. Nutman, and C. R. L. Friend. 1996. Evidence for life on Earth by 3800 million years ago. *Nature* 384:55–59.

Mooi, R. 1990. Paedomorphosis, Aristotle's lantern, and the origin of sand dollars (Echinodermata: Clypeasteroidea). *Paleobiology* 16:25–48.

Mooney, C. 2005. *The Republican War on Science*. New York: Basic.

Moore, R. 1983. The impossible voyage of Noah's ark. *Creation/Evolution* 11:1–40.

Morris, H. 1970. *Biblical Cosmology and Modern Science*. Nutley, N.J.: Craig.

Morris, H. 1972. *The Remarkable Birth of Planet Earth*. San Diego, Calif.: Creation-Life.

Morris, H. 1974. *Scientific Creationism*. San Diego, Calif.: Creation-Life.

Morris, J. 1986. Article 151. The Paluxy River mystery. *Acts/Facts/Impacts* 15:1–4.

Moy-Thomas, J., and R. S. Miles. 1971. *Palaeozoic Fishes*. Philadelphia: Saunders.

Murphy, W. J., E. Eizirik, W. E. Johnson, Y. P. Zhang, O. A. Ryder, and H. P. O'Brien. 2001a. Molecular phylogenetics and the origins of placental mammals. *Nature* 409:614–618.

Murphy, W. J., E. Eizirik, S. J. O'Brien, O. Madsen, M. Scally, C. J. Douady, E. C. Teeling, O. A. Ryder, M. J. Stanhope, W. W. de Jong, and M. S. Springer. 2001b. Resolution of the early placental mammal radiation using Bayesian phylogenetics. *Science* 294:2348–2351.

Naef, A. 1926. Über die Urformen der Anthropomorphen und die Stammesgeschichte des Menschenschädels. *Naturwissenschaften* 14:445–452.

Nagy, B., M. H. Engel, and J. H. Zumberger. 1981. Amino acids and hydrocarbons approximately 3,800-Myr old in the Isua rocks, southwestern Greenland. *Nature* 289:353–356.

Naish, D., and P. Barrett. 2016. *Dinosaurs: How They Lived and Evolved*. Washington, D.C.: Smithsonian Books.

Narbonne, G. M. 1998. The Ediacara biota: a terminal Neoproterozoic experiment in the evolution of life. *GSA Today* 8(2):1–6.

Neville, G. T. Fossils in evolutionary perspective. *Science Progress* 48:1–3.

Newell, N. D. 1959. The nature of the fossil record. *Proceedings of the American Philosophical Society* 103:264–265.

Nichols, D. 1959. Changes in the chalk heart-urchin *Micraster* interpreted in relation to living forms. *Philosophical Transactions of the Royals Society of London B* 242:347–437.

Nielsen, C. 2001, *Animal Evolution: Interrelationships of the Living Phyla*. 2nd ed. New York: Oxford University Press.

Norell, M. 2005. *Unearthing the Dragon: The Great Feathered Dinosaur Discoveries*. New York: Pi.

Norman, D. 1985. *The Illustrated Encyclopedia of Dinosaurs*. New York: Crescent.

Norman, J. R., and P. H. Greenwood. 1975. *A History of Fishes*. London: Ernest Benn.

Novacek, M. J. 1992. Mammalian phylogeny: shaking the tree. *Nature* 356:121–125.

Novacek, M. J. 1994. The radiation of placental mammals. In *Major Features of Vertebrate Evolution*, ed. D. R. Prothero and R. M. Schoch. Paleontological Society Short Course 7:220–237.

Novacek, M. J., and A. R. Wyss. 1986. Higher-level relationships of Recent eutherian orders: morphological evidence. *Cladistics* 2:257–287.

Novacek, M. J., A. R. Wyss, and M. C. McKenna. 1988. The major groups of eutherian mammals. In *The Phylogeny and Classification of the Tetrapods*, vol. 2, ed. M. J. Benton. Oxford, U.K.: Clarendon, 31–73.

Numbers, R. 1992. *The Creationists: The Evolution of Scientific Creationism*. New York: Knopf.

Nutman, A. P., V. C. Bennett, C. R. L. Friend, M. J. van Kranendonk, and A. R. Chivas. 2016. Rapid emergence of life shown by 3700-million-year-old microbial structures. *Nature* 537:535–538.

Olasky, M., and J. Perry. 2005. *Monkey Business: The True Story of the Scopes Trial*. New York: B&H.

Orwell, G. 1946. In front of your nose. *Tribune*, March 22, 1946.

Osborn, H. F. 1929. The titanotheres of ancient Wyoming, Dakota, and Nebraska. *U S. Geological Survey Monographs* 55:1–93.

Ostrom, J. H. 1974. *Archaeopteryx* and the origin of flight. *Quarterly Review of Biology* 49:27–47.

Ostrom, J. H. 1976. *Archaeopteryx* and the origin of birds. *Biological Journal of the Linnean Society* 8:91–182.

Owen, R. 1861. *Palaeontology*. 2nd ed. Edinburgh, U.K.: Adam and Charles Black.

Padian, K., and L. M. Chiappe. 1998. The origin of birds and their flight. *Scientific American*, 278:28–37.

Patterson, C. 1981. Significance of fossils in determining evolutionary relationships. *Annual Review of Ecology and Systematics* 12:195–223.

Patterson, C., ed. 1987. *Molecules or Morphology in Evolution: Conflict or Compromise?* New York: Cambridge University Press.

Patthy, L. 2003. Modular assembly of genes and the evolution of new functions. *Genetica* 118:217–231

Paul, C. R. C. 1992. How complete does the fossil record have to be? *Revista Española de Paleontologia* 7:127–133.

Pearson, J. C., D. Lemons, and W. McGinnis. 2005. Modulating *Hox* gene functions during animal body patterning. *Nature Reviews Genetics* 6:893–904.

Pearson, P. N. 1993. A lineage phylogeny for the Paleogene planktonic foraminifera. *Micropaleontology* 39:193–232.

Pearson, P. N. 1998. The glorious fossil record. *Nature*, November. 19. www.nature.com/nature/debates/fossil/fossil_1.html.

Pearson, P. N., N. J. Shackleton, and M. A. Hall. 1997. Stable isotopic evidence for the sympatric divergence of *Globigerinoides trilobus* and *Orbulina universa* (planktonic foraminifera). *Journal of the Geological Society of London* 154:295–302.

Pelikan, J. 2005. *Whose Bible Is It? A History of the Scriptures Through the Ages*. New York: Viking.

Pennock, R. 1999. *Tower of Babel: The Evidence Against the New Creationism*. Cambridge: MIT Press.

Perakh, M. 2004. *Unintelligent Design*. Buffalo, N.Y.: Prometheus.

Peters, D. 1991. *From the Beginning: The Story of Human Evolution*. New York: Morrow.

Peterson, K., M. A. McPeek, and D. A. D. Evans. 2005. Tempo and mode of early animal evolution: inferences from rocks, Hox and molecular clocks. *Paleobiology* 31:36–55.

Petto, A. 2005. The art of debate. *Reports of the National Center for Science Education* 24(6):43.

Philip, G. M. 1962. The evolution of *Gryphaea*. *Geological Magazine* 99:327–343.

Philip, G. M. 1967. Additional observations on the evolution of *Gryphaea*. *Geological Journal* 5:329–338.

Pickrill, J. 2014. *Flying Dinosaurs: How Reptiles Became Birds*. New York: Columbia University Press.

Pierce, K. 1981. Putting Darwin back on the dock. *Time*, March 16, 80–82.

Pigliucci, M. 2002. *Denying Evolution: Creationism, Scientism, and the Nature of Science*. Sunderland, Mass.: Sinauer.

Pino, S., F. Ciciriello, G. Costanzo, and E. Di Mauro, E. 2008. Nonenzymatic RNA ligation in water. *Journal of Biological Chemistry* 283:36494–36503.

Popper, K. 1935. *The Logic of Scientific Discovery*. London: Routledge Classics.

Popper, K. 1963. *Conjectures and Refutations: The Growth of Scientific Knowledge*. London: Routledge Classics.

Pough, F. H., C. M. Janis, and J. B. Heiser. 2002. *Vertebrate Life*. 6th ed. Upper Saddle River, N.J.: Prentice Hall.

Prothero, D. R. 1990. *Interpreting the Stratigraphic Record*. New York: Freeman.

Prothero, D. R. 1992. Punctuated equilibria at twenty: a paleontological perspective. *Skeptic* 1(3):38–47.

Prothero, D. R. 1993. Ungulate phylogeny: morphological vs. molecular evidence. In *Mammal Phylogeny*. Vol. 2, *Placentals*, ed. F. S. Szalay, M. J. Novacek, and M. C. McKenna. New York: Springer-Verlag, 173–181.

Prothero, D. R. 1994a. *The Eocene-Oligocene Transition: Paradise Lost*. New York: Columbia University Press.

Prothero, D. R. 1994b. Mammalian evolution. In *Major Features of Vertebrate Evolution*, ed. D. R. Prothero and R. M. Schoch. Paleontological Society Short Course 7:238–270.

Prothero, D. R. 1996. Camelidae. In *The Terrestrial Eocene-Oligocene Transition in North America*, ed. D. R. Prothero and R. J. Emry. New York: Cambridge University Press, 591–633.

Prothero, D. R. 1999. Does climatic change drive mammalian evolution? *GSA Today* 9(9):1–5.

Prothero, D. R. 2005. *The Evolution of North American Rhinoceroses*. New York: Cambridge University Press.

Prothero, D. R. 2006. *After the Dinosaurs: The Age of Mammals*. Bloomington: Indiana University Press.

Prothero, D. R. 2013a. *Bringing Fossils to Life: An Introduction to Paleobiology*. 3rd ed. New York: Columbia University Press.

Prothero, D. R. 2013b. Stephen Meyer's fumbling bumbling Cambrian follies: a review of *Darwin's Doubt* by Stephen Meyer. *Skeptic* 18(4):50–53.

Prothero, D. R. 2016. *The Princeton Field Guide to Prehistoric Mammals*. Princeton, N.J.: Princeton University Press.

Prothero, D. R., and S. Foss, eds. 2007. *The Evolution of Artiodactyls*. Baltimore, Md.: Johns Hopkins University Press.

Prothero, D. R., and T. H. Heaton. 1996. Faunal stability during the early Oligocene climatic crash. *Palaeogeography, Palaeoclimatology, Palaeoecology* 127:239–256.

Prothero, D. R., and D. B. Lazarus. 1980. Planktonic microfossils and the recognition of ancestors. *Systematic Zoology* 29:119–129.

Prothero, D. R., and R. M. Schoch, eds. 1989. *The Evolution of Perissodactyls*. New York: Oxford University Press.

Prothero, D. R., and R. M. Schoch. 2002. *Horns, Tusks, and Flippers: The Evolution of Hoofed Mammals*. Baltimore, Md.: Johns Hopkins University Press.

Prothero, D. R., and F. Schwab. 2013. *Sedimentary Geology*. 3rd ed. New York: Freeman.

Prothero, D. R., and N. Shubin, 1989. The evolution of Oligocene horses. In *The Evolution of Perissodactyls*, ed. D. R. Prothero and R. M. Schoch. New York: Oxford University Press, 142–175.

Prothero, D. R., E. Manning, and M. Fischer. 1988. The phylogeny of the ungulates. In *The Phylogeny and Classification of the Tetrapods*, vol. 2, ed. M. J. Benton. Oxford, U.K.: Clarendon, 201–234.

Prothero, D. R., E. Manning, and C. B. Hanson. 1986. The phylogeny of the Rhinocerotoidea (Mammalia, Perissodactyla). *Zoological Journal of the Linnean Society of London* 87:341–366.

Prum, R. O., and A. H. Brush. 2003. Which came first, the feather or the bird? *Scientific American* 288:84–93.

Rachootin, S. P., and K. S. Thomson.1981 Epigenetics, paleontology and evolution. *Evolution Today* 2:181–193.

Radinsky, L. B. 1966. The families of the Rhinocerotoidea. *Journal of Mammalogy* 47:631–639.

Radinsky, L. B. 1969. The early evolution of the Perissodactyla. *Evolution* 23:308–328.

Raff, R. A. 1998. *The Shape of Life: Genes, Development, and the Evolution of Animal Form*. Chicago: University of Chicago Press.

Raup, D. M., and J. J. Sepkoski Jr. 1984. Periodicity of extinctions in the geologic past. *Proceedings of the National Academy of Sciences USA* 81:805–801.

Raup, D. M., and J. J. Sepkoski Jr. 1986. Periodicity of extinction of families and genera. *Science* 231:833–836.

Retallack, G. J. 1983. Late Eocene and Oligocene paleosols from Badlands National Park, South Dakota. *Geological Society of America Special Paper* 193.

Ridley, M. 1996. *Evolution*. 2nd ed. Cambridge, Mass.: Blackwell.

Rieppel, O. 1988. A review of the origin of snakes. *Evolutionary Biology* 25:37–130.

Rieppel, O., et al. 2003. The anatomy and relationships of *Haasiophis terrasanctus*, a fossil snake with well-developed hind limbs from the mid-Cretaceous of the Middle East. *Journal of Paleontology* 77:536–558.

Rodda, P. U., and W. L. Fisher. 1964. Evolutionary features of *Athleta* (Eocene, Gastropoda) from the Gulf Coastal Plain. *Evolution* 18:235–244.

Romanes, G. J. 1910. *Darwin and After Darwin*. Chicago: Open Court.

Romer, A. S. 1959. *The Vertebrate Story*. Chicago: University of Chicago Press.

Ronshaugen, M., N. McGinnis, and W. McGinnis. 2002. Hox protein mutation and macroevolution of the insect body plan. *Nature* 415:914–917.

Rose, K. D. 1987. Climbing adaptations of the early Eocene mammal *Chriacus* and the origin of the Artiodactyla. *Science* 236:314–316.

Rose, K. D., and J. D. Archibald, eds. 2005. *The Rise of Placental Mammals*. Baltimore, Md.: Johns Hopkins University Press.

Ross, C., and R. Rezak. 1959. The rocks and fossils of Glacier National Park: the story of their origin and history. *U.S. Geological Survey Professional Paper* 294:401–439.

Rowe, A. W. 1899. An analysis of the genus *Micraster* as determined by rigid zonal collecting from the zone of *Rhynchonella cuvieri* to that of *Micraster coranguinum*. *Quarterly Journal of the Geological Society of London* 55:494–547.

Runnegar, B. 1992. Evolution of the earliest animals. In *Major Events in the History of Life*, ed. J. W. Schopf. New York: Jones and Bartlett, 65–94.

Runnegar, B., and J. W. Schopf, eds. 1988. *Molecular Evolution in the Fossil Record*. Lancaster, Pa.: Paleontological Society Short Course Notes 1.

Ruse, M. 1982. *Darwinism Defended*. New York: Addison-Wesley.

Ruse, M. 1988. *But Is It Science? The Philosophical Questions in the Creation/Evolution Controversy*. Buffalo, N.Y.: Prometheus.

Ruse, M. 2003. *Darwin and Design: Does Evolution Have a Purpose?* Cambridge: Harvard University Press.

Ruse, M. 2005. *The Evolution-Creation Struggle*. Cambridge: Harvard University Press.

Sagan, C. 1996. *The Demon-Haunted World: Science as a Candle in the Dark*. New York: Ballantine.

Sarfati, J. 1999. *Refuting Evolution*. Green Forest, Ark.: Master Books.

Sarfati, J. 2002. *Refuting Evolution 2*. Green Forest, Ark.: Master Books.

Sarna, N. 1966. *Understanding Genesis: The Heritage of Biblical Israel*. New York: Schocken.

Savage, R. J. G., and M. R. Long. 1986. *Mammal Evolution: An Illustrated Guide*. New York: Facts-on-File.

Schaeffer, B., M. K. Hecht, and N. Eldredge. 1972. Phylogeny and paleontology. *Evolutionary Biology* 6:31–46.

Schaeffer, B., and D. E. Rosen. 1961. Major adaptive levels in the evolution of actinopterygian feeding mechanisms. *American Zoologist* 1:187–204.

Scheele, W. E. 1955. *The First Mammals*. New York: World Press.

Schidlowski, M., P. W. U. Appel, R. Eichmann, and C. E. Junge. 1979. Carbon isotope geochemistry of the 3.7×10^9 yr old Isua sediments, West Greenland; implications for the Archaean carbon and oxygen cycles. *Geochimica Cosmochimica Acta* 43:189–200.

Schoch, R., and H. D. Sues. 2015. A Middle Triassic stem-turtle and the evolution of the turtle body plan. *Nature* 523:584–587.

Schoch, R. M. 1986. *Phylogeny Reconstruction in Paleontology*. New York: Van Nostrand Reinhold.

Schopf, J. W., ed. 1983. *Earth's Earliest Biosphere: Its Origin and Development*. Princeton, N.J.: Princeton University Press.

Schopf, J. W. 1999. *Cradle of Life: The Discovery of the Earth's Earliest Fossils*. Princeton, N.J.: Princeton University Press.

Schopf, J. W. 2002. *Life's Origin: The Beginnings of Biological Evolution.* Berkeley: University of California Press.

Schopf, J. W., and C. Klein, eds. 1992. *The Proterozoic Biosphere, a Multidisciplinary Study.* Cambridge: Cambridge University Press.

Schultze, H.-P., and L. Trueb, eds. 1991. *Origins of the Higher Groups of Tetrapods: Controversy and Consensus.* Ithaca, N.Y.: Cornell University Press.

Schwartz, J. 1999. *Sudden Origins: Fossils, Genes, and the Emergence of Species.* New York: Wiley.

Scott, E. C. 2005. *Evolution vs. Creationism: An Introduction.* Berkeley: University of California Press.

Scott, W. B. 1913. *The History of the Land Mammals in the Western Hemisphere.* New York: Macmillan.

Seilacher, A. 1989. Vendozoa: organismic construction in the Proterozoic biosphere. *Lethaia* 22:229–239.

Seilacher, A. 1992. Vendobionta and Psammocorallia. *Journal of the Geological Society of London* 149:607–613.

Sepkoski, J. J., Jr. 1989. Periodicity in extinction and the problem of catastrophism in the history of life. *Journal of the Geological Society of London* 145:7–19.

Sepkoski, J. J., Jr. 1993. Foundation: life in the oceans. In *The Book of Life,* ed. S. J. Gould. New York: Norton, 37–64.

Sereno, P. C., and C. Rao. 1992: Early evolution of avian flight and perching: new evidence from the Lower Cretaceous of China. *Nature* 255:845–848.

Shanks, N. 2004. *God, the Devil, and Darwin: A Critique of Intelligent Design Theory.* Oxford: Oxford University Press.

Shapiro, R. 1986. *Origins, A Skeptic's Guide to the Creation of Life on Earth.* New York: Summit.

Sheldon, P. R. 1987. Parallel gradualistic evolution of Ordovician trilobites. *Nature* 330:561–563.

Shermer, M. 1997. *Why People Believe Weird Things: Pseudoscience, Superstition, and Other Confusions of Our Time.* New York: Freeman.

Shermer, M. 2005. *Science Friction: Where the Known Meets the Unknown.* New York: Times Books.

Shermer, M. 2006. *Why Darwin Matters: The Case Against Intelligent Design.* New York: Times Books.

Shipman, P. 1988. *Taking Wing: Archaeopteryx and the Evolution of Bird Flight.* New York: Simon & Schuster.

Shu, D.-G., H.-L. Luo, S. Conway Morris, X.-L. Zhang, S.-X. Hu, L. Chen, J. Han, M. Zhu, Y. Li, and L.-Z. Chen. 1999. Lower Cambrian vertebrates from China. *Nature* 402:42–46.

Shubin, N. H., and P. Alberch. 1986. A morphogenetic approach to the origin and basic organization of the tetrapod limb. *Evolutionary Biology* 20:318–390.

Shubin, N. H., E. B. Daeschler, and F. A. Jenkins Jr. 2006. The pectoral fin of *Tiktaalik roseae* and the origins of the tetrapod limb. *Nature* 440:764–771.

Shulman, S. 2007. *Undermining Science: Suppression and Distortion in the Bush Administration.* Berkeley: University of California Press.

Sibley, C. G., and J. E. Ahlquist. 1984. The phylogeny of hominoid primates, as indicated by DNA-DNA hybridization. *Journal of Molecular Evolution* 20:2–15.

Siegler, H. R. 1978. A creationist's taxonomy. *Creation Research Society Quarterly* 15:36–38.

Simmons, N. 2005. Chiroptera. In *The Rise of Placental Mammals,* ed. K. D. Rose and J. D. Archibald. Baltimore, Md.: Johns Hopkins University Press, 159–174.

Simmons, N., and J. H. Geisler. 1998. Phylogenetic relationships of *Icaronycteris, Archaeonycteris, Hassianycteris,* and *Palaeochiropteryx* to extant bat lineages, with comments on the evolution of echolocation and foraging strategies in Microchiroptera. *Bulletin of the American Museum of Natural History* 235:1–182.

Simpson, G. G. 1944. *Tempo and Mode in Evolution.* New York: Columbia University Press.

Simpson, G. G. 1961. *Principles of Animal Taxonomy.* New York: Columbia University Press.

Simpson, G. G., and W. S. Beck. 1965. *Life: An Introduction to Biology.* New York: Harcourt, Brace, & World.

Smith, A. B. 1984. *Echinoid Palaeobiology.* London: George Allen and Unwin.

Smith, A. B. 1994. *Systematics and the Fossil Record: Documenting Evolutionary Patterns.* London: Blackwell.

Smith, A. B., and K. J. Peterson. 2002. Dating the time of origin of major clades: molecular clocks and the fossil record. *Annual Reviews of Earth and Planetary Sciences* 30:65–88.

Smith, C. M., and C. Sullivan. 2007. *The Top Ten Myths About Evolution.* Buffalo, N.Y.: Prometheus.

Smith, H. 1952. *Man and His Gods.* New York: Little, Brown.

Smith, J. L. B. 1956. *Old Fourlegs: The Story of the Coelacanth.* London: Longman Green.

Smith, J. M. 1958. *The Theory of Evolution.* New York: Penguin.

Smithson, T. R., R. L. Carroll, A. L. Panchen, and S. M. Andrews. 1994. *Westlothiana lizziae* from the Visean of East Kirkton, West Lothian, Scotland, and the amniote stem. *Transactions of the Royal Society of Edinburgh* 84:383–412.

Solounias, N. 1999. The remarkable anatomy of the giraffe's neck. *Journal of Zoology* 247:257–268.

Solounias, N. 2007. Giraffidae. In *The Evolution of Artiodactyls,* ed. D. R. Prothero and S. Foss. Baltimore, Md.: Johns Hopkins University Press, 257–277.

Springer, M. S., and J. A. W. Kirsch. 1993. A molecular perspective on the phylogeny of placental mammals based on the mitochondrial 12S rDNA sequence, with special reference to the problem of Paenungulata. *Journal of Mammalian Evolution* 1:149–168.

Springer, M. S., W. J. Murphy, E. Eizirik, and S. J. O'Brien. 2005. Molecular evidence for major placental clades. In *The Rise of Placental Mammals: Origins and Relationships of Major Clades,* ed. K. D. Rose and J. D. Archibald. Baltimore, Md.: Johns Hopkins University Press, 37–49.

Springer, M. S., M. J. Stanhope, O. Madsen, and W. W. de Jong. 2004. Molecules consolidate the placental mammal tree. *Trends in Ecology and Evolution* 19:430–438.

Standen, E. M., T. Y. Du, and H. C. E. Larsson. 2014. Developmental plasticity and the origin of tetrapods. *Nature* 513:54–58.

Stanhope, M. J., W. J. Bailey, J. Czelusnaik, M. Goodman, J.-S. Si, J. Nickerson, J. G. Sgouros, G. A. M. Singer, and T. K. Kleinschmidt. 1993. A molecular view of primate supraordinal relationships from the analysis of both nucleotide and amino acid sequences. In *Primates and Their Relatives in Phylogenetic Perspective*, ed. R. D. E. MacPhee. New York: Plenum, 251–292.

Stanhope, M. J., M. R. Smith, V. G. Waddell, C. A. Porter, M. S. Shivig, and M. Goodman. 1996. Mammalian evolution and the interphotoreceptor retinoid binding protein (IRBP) gene: convincing evidence for several supraordinal clades. *Journal of Molecular Evolution* 43:83–92.

Stanley, S. M. 1979. *Macroevolution: Patterns and Process*. New York: Freeman.

Stanley, S. M. 1981. *The New Evolutionary Timetable*. New York: Basic.

Stanley, S. M. 1990. Delayed recovery and the spacing of major extinctions. *Paleobiology* 16:401–414.

Steele, E. 1979. *Somatic Selection and Adaptive Evolution: On the Inheritance of Acquired Characters*. Chicago: University of Chicago Press.

Steele, E., R. Lindley, and R. Blanden. 1998. *Lamarck's Signature: How Retrogenes Are Changing Darwin's Natural Selection Paradigm*. Reading, Mass.: Perseus.

Stokes, R. B. 1977. The echinoids *Micraster* and *Epiaster* from the Turonian and Senonian of southern England. *Palaeontology* 20:805–821.

Stringer, C., and C. Gamble. 1993. *In Search of Neanderthals: Solving the Puzzle of Human Origins*. London: Thames and Hudson.

Sumida, S., and K. L. M. Martin, eds. 1997. *Amniote Origins: Completing the Transition to Land*. San Diego, Calif.: Academic.

Swisher, C. C., III, G. H. Curtis, and R. Lewin. 2000. *Java Man*. New York: Scribner.

Szalay, F. S., M. J. Novacek, and M. C. McKenna, eds. 1993. *Mammal Phylogeny*. New York: Springer-Verlag.

Tattersall, I. 1993. *The Human Odyssey: Four Million Years of Human Evolution*. Upper Saddle River, N.J.: Prentice-Hall.

Tattersall, I. 2015. *The Strange Case of the Rickety Cossack and Other Cautionary Tales from Human Evolution*. New York: St. Martin's.

Tattersall, I., and N. Eldredge. 1977. Fact, theory and fantasy in human paleontology. *American Scientist* 65:204–211.

Tattersall, I., and J. Schwartz. 2000. *Extinct Humans*. New York: Westview.

Tchernov, E., O. Rieppel, and H. Zaher. 2000. A fossil snake with limbs. *Science* 287:2010–2012.

Thewissen, J. G. M., ed. 1998. *The Emergence of Whales: Evolutionary Patterns in the Origin of Cetacea*. New York: Plenum.

Thewissen, J. G. M., S. T. Hussain, and M. Arif. 1994. Fossil evidence for the origin of aquatic locomotion in archaeocete whales. *Science* 263:210–212.

Thewissen, J. G. M., E. M. Williams, L. J. Roe, and S. T. Hussain. 2001. Skeletons of terrestrial cetaceans and the relationship of whales to artiodactyls. *Nature* 413:277–281.

Thomson, K. S. 1991. *Living Fossil*. New York: Norton.

Trueman, A. E. 1922. The use of *Gryphaea* in the correlation of the Lower Lias. *Geological Magazine* 59:256–268.

Tudge, C. 2000. *The Variety of Life: A Survey and a Celebration of All the Creatures That Have Ever Lived*. Oxford: Oxford University Press.

Turner, A., and M. Anton. 2004. *National Geographic Prehistoric Mammals*. Washington, D.C.: National Geographic Society.

Valentine, J. W. 2004. *On the Origin of Phyla*. Chicago: University of Chicago Press.

Valley, J. W., W. H. Peck, E. M. King, and S. A. Wilde. 2002. A cool early earth. *Geology* 30:351–354.

Van Valen, L. 1966. Deltatheridia, a new order of mammals. *Bulletin of the American Museum of Natural History* 132:1–126.

Wächtershäuser, G. 2006. From volcanic origins of chemoautotrophic life to Bacteria, Archaea, and Eukarya. *Philosophical Transactions of the Royal Society of London B* 361:1787–1806.

Wächtershäuser, G. 2008. Origin of life: life as we don't know it. *Science* 289:1307–1308.

Ward, L. W., and B. W. Blackwelder. 1975. *Chesapecten*, a new genus of Pectinidae (Mollusca: Bivalvia) from the Miocene and Pliocene of eastern North America. *U.S. Geological Survey Professional Paper* 861.

Ward, R. R. 1965. *In the Beginning*. Grand Rapids, Mich.: Baker.

Weinberg, S. 2000. *A Fish Caught in Time: The Search for the Coelacanth*. New York: HarperCollins.

Weiner, J. 1994. *The Beak of the Finch: A Story of Evolution in Our Own Time*. New York: Knopf.

Weiner, J. 2005. Evolution in action. *Natural History* 115(9):47–51.

Weishampel, D. B., P. Dodson, and H. Osmolska, eds. 1990. *The Dinosauria*. 1st ed. Berkeley: University of California Press.

Weishampel, D. B., P. Dodson, and H. Osmolska, eds. 2004. *The Dinosauria*. 2nd ed. Berkeley: University of California Press.

Wells, J. 2000. *Icons of Evolution: Science or Myth? Why Much of What We Teach About Evolution Is Wrong.* Washington, D.C.: Regnery.

Wesson, R. 1991. *Beyond Natural Selection.* Cambridge: MIT Press.

Whitcomb, J. C., Jr., and H. M. Morris. 1961. *The Genesis Flood.* Nutley, N.J.: Presbyterian and Reformed Publishing.

Wilde, S. A., J. W. Valley, W. H. Peck, and C. M. Graham. 2001. Evidence from detrital zircons for the existence of continental crust and oceans on the earth 4.4 Gyr ago. *Nature* 400:175–181.

Wiley, E. O. 1981. *Phylogenetics: The Theory and Practice of Phylogenetic Systematics.* New York: Wiley Interscience.

Wills, C. 1989. *The Wisdom of the Genes: New Pathways in Evolution.* New York: Basic.

Wills, C., and J. Bada. 2000. *The Spark of Life: Darwin and the Primeval Soup.* New York: Perseus.

Winchester, S. 2002. *The Map That Changed the World: William Smith and the Birth of Modern Geology.* New York: Harper.

Woese, C. R., and G. E. Fox. 1977. Phylogenetic structure of the prokaryotic domain: the primary kingdoms. *Proceedings of the National Academy of Sciences USA* 274:5088–5090.

Wray, G. A., J. S. Levinton, and L. H. Shapiro. 1996. Molecular evidence for deep Precambrian divergences among metazoan phyla. *Science* 74:568–573.

Wyss, A. R. 1987. The walrus auditory region and the monophyly of pinnipeds. *American Museum Novitates* no. 2871.

Wyss, A. R. 1988. Evidence from flipper structure for a single origin of pinnipeds. *Nature* 334:427–428.

Xu, X., C. A. Forster, J. M. Clark, and J. Mo. 2006. A basal ceratopsian with transitional features from the Late Jurassic of northwestern China. *Proceedings of the Royal Society of London B* 273:2135–2140.

Young, M., and T. Edis, eds. 2005. *Why Intelligent Design Fails: A Scientific Critique of the New Creationism.* Piscataway, N.J.: Rutgers University Press.

Zimmer, C. 1998. *At the Water's Edge: Macroevolution and the Transformation of Life.* New York: Free Press.

Zuckerkandl, E., and L. Pauling. 1965. Evolutionary divergence and convergence in proteins. In *Evolving Genes and Proteins.* San Diego, Calif.: Academic.

INDEX